The Handbook for Quality Management

About the Authors

Thomas Pyzdek is a Six Sigma consultant with more than 40 years of experience in the field. His clients include Ford, McDonald's, Intuit, Boeing, Seagate, Avon Products, and many other companies. Mr. Pyzdek is a recipient of the American Society for Quality Edwards Medal for outstanding contributions to the practice of quality management and the E.L. Grant Medal for outstanding leadership in the development and presentation of meritorious educational programs in quality. He has also received a Lean Six Sigma Leadership award from the American Quality Institute.

Paul Keller is president and chief operating officer with Quality America, Inc. He has developed and implemented successful Six Sigma and quality improvement programs in service and manufacturing environments. He is the author of several books, including *The Six Sigma Handbook*, Third Edition (coauthor), and *Six Sigma Demystified*.

The Handbook for Quality Management

A Complete Guide to Operational Excellence

Thomas Pyzdek

Paul Keller

Second Edition

McGraw Hill

New York Chicago San Francisco
Lisbon London Madrid Mexico City
Milan New Delhi San Juan
Seoul Singapore Sydney Toronto

The McGraw-Hill Companies

Cataloging-in-Publication Data is on file with the Library of Congress.

The Handbook for Quality Management

1 2 3 4 5 6 7 8 9 0 DOC/DOC 1 9 8 7 6 5 4 3 2

ISBN 9781265829230
MHID 1265829233

Sponsoring Editor
Judy Bass

Acquisitions Coordinator
Bridget Thoreson

Editorial Supervisor
David E. Fogarty

Project Manager
Vastavikta Sharma, Cenveo Publisher Services

Copy Editor
Kate Bresnahan

Proofreader
Surendra Nath Shivam, Cenveo Publisher Services

Indexer
ARC Films Inc.

Production Supervisor
Pamela A. Pelton

Composition
Cenveo Publisher Services

Art Director, Cover
Jeff Weeks

Contents

Preface

Thank you for your interest in McGraw Hill's *The Handbook for Quality Management*.

The original version of the text, first released in 1996 by Quality Publishing, was written exclusively by Tom Pyzdek. I had the pleasure of editing a revision released in 2000, which included Six Sigma and Lean method chapters (written by myself), as well as Bill Dettmer's Constraint Management material, which is repeated in this edition. The early editions sold several thousand copies by the end of 2000, establishing the *Handbook* as an essential desktop reference for the quality professional.

The earlier versions relied heavily on the American Society for Quality (ASQ) body of knowledge for quality managers, even to the extent that the chapter headings and sub-headings matched those in the body of knowledge. Although this may have helped those seeking to check off items they learned, it tended to disrupt the flow of the topics. A main objective of *this* edition was the reorganization of the material into more naturally flowing discussions of the concepts and methods essential to quality management and operational excellence. For those who want to use this as a reference for the ASQ CMQ/OE exam, the information is still in the book, with sample questions at the back, and answers available on the affiliated website: www.mhprofessional.com/HQM2

The essential body of knowledge for achieving operational excellence is heavily influenced by the works of Deming and Juran, most of which date from the period of 1950 through the mid 1980s. These authors spent their careers advocating a scientific approach to quality, displacing the widely held notion that quality assurance inspections prevalent in the post-war era were sufficient or even credible approaches to achieving quality.

Over the last 40 years, the quality management discipline has undergone steady evolution from internally focused command-and-control to more proactive, customer-focused functions. The market certainly encouraged that, as economies shifted from dominance of product-based manufacturers to more heavily depend on service-based solution providers. It seems reasonable that service economies will naturally tend toward customer-focus, since much of the service involves direct customer contact. Feedback can be bitterly honest, yet also quickly addressed (compared with poor manufacturing quality). Aspects of quality management are becoming integral to business operations; quality ratings and awards are a competition, and success is marketed as a sign of commitment to the customer; innovation is a constant refrain in business journals and even advertisements; customer surveys are endemic; data is rampant, so differentiating between real change and random variation becomes a core competency; and so on. The cost of poor quality is realized in real time as loss of market share or profitability.

This latest edition expands on the historical notions of Juran's quality trilogy to describe business transformation through innovative customer-driven strategy, meaningful process control using statistics, and management-sponsored, focused improvements in core products and services. Deming's teachings on management responsibilities and systems are integrated throughout.

The manager in today's world must implement cost-reducing quality initiatives that increase market share in spite of competitive forces. This text seeks to demystify the science of quality management for effective use and benefit across the organization.

We hope you enjoy it.

Paul Keller

PART I

Business-Integrated Quality Systems

Modern organizations trace their roots to the Industrial Revolution, which provided the impetus for movement from a tradition of craftsmen to that of mechanized industries. Rapid advances in mobile power sources, such as the steam engine, improved transportation, gas lighting, advances in metallurgical and chemical processing, and so on led to both supply of material, methods, and infrastructure and a demand for business innovation to meet the needs of a growing market. As businesses grew, smaller (often family-run) businesses were replaced by larger corporations, who could raise the capital necessary to grow rapidly.

In industrialized countries, organizations changed completely, giving rise to the bureaucratic form of organization. This organizational form is characterized by the division of activities and responsibilities into departments managed by full-time management professionals who had no other source of livelihood other than the organization.

CHAPTER 1

Organizational Structures

Organizations exist because they serve a useful purpose. The transaction-cost theory of a firm (Coase, 1937) postulates that there are costs associated with market transactions, and organizations prosper only when they provide a cost advantage. Examples of these costs include the cost of discovering market prices, negotiation and contracting costs, sales taxes and other taxes on exchanges between firms, cost of regulation of transactions between firms, and so on.

Transaction-cost theory offers a framework for understanding limits on the size of a firm. As firms grow, it becomes more costly to organize additional transactions within the firm, called "decreasing returns to management." When the cost of organizing an additional transaction equals the cost of carrying out the transaction in the open market, growth of the firm will cease. Of course, these costs are also affected by technology: facsimile machines (in their day), satellites, computers, and more recently the Internet each altered the cost of organization, impacting the optimal size of the firm accordingly. Such inventions simultaneously impact the cost of using external markets, so the relative impact of the technology on market costs and organization costs determines the overall impact on the organization. Clearly, the ability to efficiently carry out market transactions, with minimal bureaucratic overhead, impacts an organization's usefulness to the market, and its prosperity and eventual life span.

General Theory of Organization Structure

Organizations consist of systems of relationships that direct and allocate resources; therefore the purpose of organization structure is to develop relationships that perform these functions well. There are several possible ways in which these relationships can be viewed. The most common is the reporting relationship view. Here the organization is viewed as an entity consisting of people who have the authority to direct other people, their "reports." In this view the organization appears as a stratified triangle, with the positions higher in a given strata of the triangle having the authority to direct the lower positions. In modern organizations, the authority to set policy and plan strategic direction is vested in the highest level of the

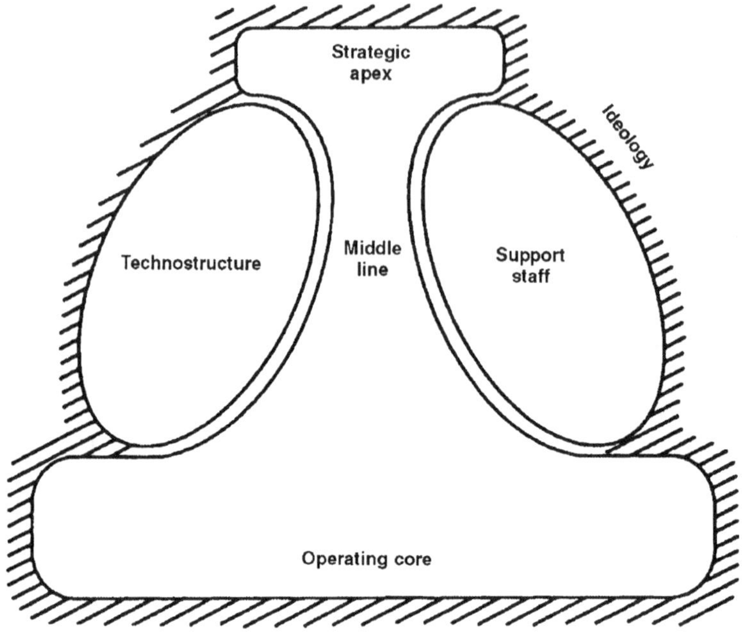

FIGURE 1.1 The six basic parts of the organization (Mintzberg and Quinn, 1991).

structure: the strategic apex. The middle line consists of management personnel who deploy the policy and plan to the operating core (at the bottom of the structure). Technological expertise and support are provided by groups of professionals not directly involved in operations. The entire organization is held together by a common set of beliefs and shared values known as the organization's ideology. Figure 1.1 illustrates these ideas.

The Functional/Hierarchical Structure

The traditional organization that results from the above view of the organization is the functional/hierarchical structure. This is a command and control structure with ancient military origins. In this type of organization, work is divided according to function, for example, marketing, engineering, finance, manufacturing, etc. A stratum within the organization is given responsibility for a particular function. Work is delegated from top to bottom within the stratum to personnel who specialize in the function. An example of the traditional functional hierarchical organization chart is shown in Fig. 1.2.

A key component of the hierarchal structure is its command and control elements, facilitated by the theories of scientific management developed by Frederick Taylor. Taylor believed that management could never effectively control the workplace unless it controlled the work itself, that is, the specific

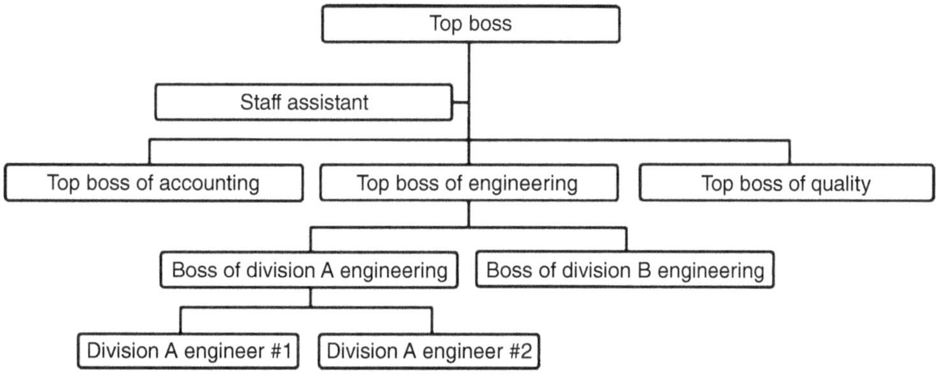

FIGURE 1.2 Functional/hierarchical organization chart.

tasks performed by the workers to get the job done. Management could improve the efficiency of work, to the benefit of both management and workers, by applying the methods of science in (1) selecting the individuals best suited to a particular job and (2) identifying the optimal way in which the jobs could be performed. Henry Ford further advanced this de-skilling of the workforce through production mechanization.

In spite of resistance from craftsmen and machinists, who understood the value of their knowledge and skill in terms of monetary rewards and job security, the reduction of work to a series of simple tasks done with relatively small investment in training is one of the major results of scientific management. The ramifications of these efforts includes better management oversight, reduced investment in worker training, and easier replacement of those who did unsatisfactory work (with employee incentives to improve performance). Unfortunately, the de-skilled work is usually far more boring, leading to a variety of problems such as high levels of stress and employee turnover.

The legacy of de-skilling is that the workforce is less able to change as new conditions arise. Whereas a machinist could work for any number of companies in many industries, machine loaders had limited mobility outside their current employer, thus increasing worker demands for job security. In the modern era, lack of generalized employee skills can be a major impediment to a quick reaction to rapidly changing market conditions. When rapid change creates new tasks, the workers' previous experience does not help them adapt to the new circumstance; they must be constantly "retrained."

Organizationally, the introduction of scientific management perpetuated the growth of the bureaucratic form, and increasingly led to larger and larger organizational support structures. On the technical side, organizational units were formed to codify the detailed knowledge of necessary work practices, including manufacturing engineering, industrial engineering, quality control, human resources, and cost accounting. This de-skilling of the workforce creates an increasingly large number of transactions to manage, which leads in turn to larger bureaucracies and decreasing returns to management, an issue described earlier by Coase.

The traditional organization structure has come under pressure in recent years. One problem with the structure is that it tends to produce a "silo mentality" among those who work in a particular stratum: they tend to see the company from the perspective of an "accountant" or an "engineer" rather than from a companywide perspective. This produces a tendency to optimize their function without regard for the effect on the rest of the organization—a tendency that produces markedly suboptimal results when viewed from a holistic perspective. Cooperation is discouraged in such an organization. In these structures, employees tend to think of their superiors as their "customers." The focus becomes pleasing one's boss rather than pleasing the external customer. Finally, the top-down arrangement often results in resource allocation that does not optimally meet the needs of external customers, who are generally served by processes that cut across several different functions.

Given these problems, one might wonder why such organizations still dominate the business scene. There are several reasons, chief among them the comfort level employees have with this model: this has been the dominant model for decades, so there is an organizational resistance to change. Furthermore, such organizations maximize the development and utilization of specialized skills. They produce a cost-effective division of labor within the subprocess (but not necessarily across the system). In many organizations, particularly larger ones, the functional/hierarchical structure provides economies of scale for specialized activities. Finally, these organizations provide clear career paths for specialists. A case in point is the quality function, where one can enter into the specialty out of high school and potentially advance to progressively higher positions throughout one's career.

Matrix Organizations

In a matrix organization the functional hierarchy remains intact but a horizontal cross-functional team structure is superimposed on the functional hierarchy. The matrix form is depicted in Fig. 1.3.

The matrix form was used extensively in the 1970s as a general method of organizing work. Most businesses concluded that organizing routine work in this way was impractical. Still, because of this experience, the matrix structure is well understood. Also, the matrix did prove to be useful as a method of conducting large, cross-functional projects. To an extent, the matrix form overcomes the "silo" mentality of the functional hierarchy by creating cross-functional teams.

When used for projects, the matrix approach creates structures that are focused (on the project) and can exist temporarily. In fact, most large,

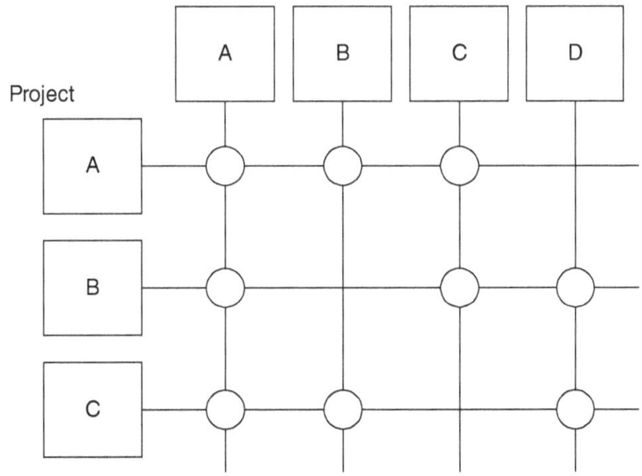

Figure 1.3 Matrix organization structure.

multifunctional quality improvement projects are organized using the matrix form. This approach to project management organization is discussed in greater detail in Chap. 15.

Cross-Functional Organization Structure

As discussed earlier, a major problem with the functional/hierarchical structure is the proliferation of focused, departmental perspectives. This invariably results in neglect of company-wide issues. Cross-functional structures provide a way of breaking down this mind-set. Figure 1.4 shows the basic layout of a cross-functional organization structure. Note that the appearance

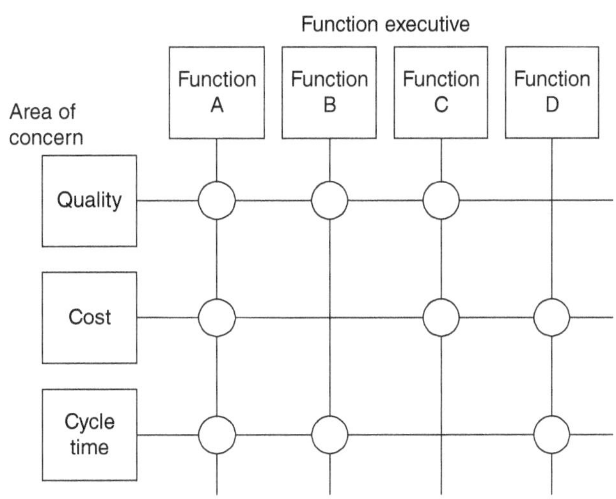

Figure 1.4 Cross-functional organization structure.

is similar to that of the matrix structure. However, there are a number of important differences between matrix and cross-functional structures:

- *Scope.* Cross-functional organizations deal with company-wide issues, while matrix organizations focus on specific tasks, goals, or projects.
- *Duration.* Matrix organizations are temporary, while cross-functional organiza-tions are often permanent.
- *Focus.* Cross-functional organizations often deal with external groups such as customers, society at large, or regulators. Matrix organizations are typically focused on internal concerns.
- *Membership.* Membership in cross-functional organizations typically consists of high-level functional executives. Membership in matrix organizations usually consists of personnel with technical skills needed to complete a specific task.

Compared with traditional organizations, cross-functional organizations offer better coordination and integration of work, faster response times, simplified cost controls, greater use of creativity, and higher job satisfaction. It should be noted that cross-functional organizations are an addition to, rather than a replacement for, traditional organizations.

Process- or Product-Based (Horizontal) Organization Structures

Process-based and product-based "horizontal organizations" present an entirely different focus than traditional organizations. The basis of this organizational structure is the goal of the work being organized, that is, the product or service being created. This differs markedly from the traditional structure, which is based on reporting relationships. An example of a customer process–focused organization structure is shown in Fig. 1.5, which is a "patient-focused" labor and delivery process in a hospital.

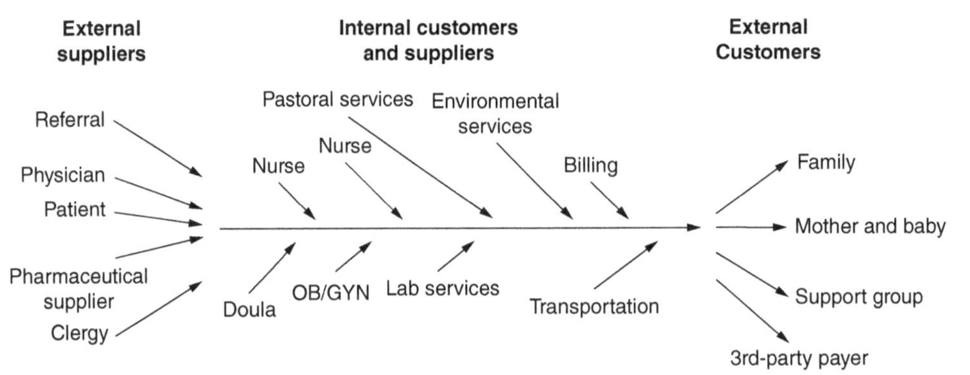

FIGURE 1.5 Patient-focused care-organization structure.

The knowledgeable quality manager will immediately recognize the similarity of Fig. 1.5 to the cause-and-effect diagram. This is a useful analogy. The "effects" being sought must be clearly defined before the design of this type of organization can proceed. The "causes" are built into the organization such that the desired effects can be consistently and economically produced. Note that the design can accommodate multiple customers, suppliers, and internal subprocesses; in this example the mother and baby are the primary customers. The scope is neither internal nor external: it embraces the entire process.

Also noteworthy is the complete absence of reporting relationships. The foundation of this type of organization is work flow, not authority. In effect, everyone "reports" to the customer. This blurring of lines of authority is a characteristic of this type of organization, which can be a source of discomfort for those accustomed to the clear chain of command inherent in traditional organizations. Clearly this involves a significant cultural change. Another cultural change is the obliteration of the professional reference group. In functional organizations, professionals (e.g., accountants, nurses, doctors, engineers) report to and work with others in the same profession and are often more loyal to their profession than to their employer. This is changed dramatically in horizontal organizations. The transition from a traditional management approach to a horizontal structure must deal explicitly with the cultural aspects of the change.

Horizontal organizations maximize core competencies, rather than suboptimizing quasi-independent functions. For example, in the patient-focused-care example several support activities are involved in the delivery of care (lab services, transportation, etc.). In a traditional organization there would be a tendency for the laboratory manager to optimize the laboratory, the transportation manager to optimize transportation, etc. However, in the horizontal organization the optimization is focused on delivery of care. This may well result in a perceived "suboptimal" performance of support activities, if each are (inappropriately) viewed in isolation.

Experience has shown that horizontal organizations have achieved dramatically improved efficiencies, compared to traditional hierarchal organizations. One reason is in the intelligent reintegration of work to correct the disintegrated work practices advocated by Taylor's scientific management theories. This segregation of work was done partly in response to conditions that no longer exist: a better-educated workforce combined with modern technology makes it possible to design integrated processes that combine related tasks and bring the needed resources under local control. In addition to improved efficiencies, the new approach to work creates other welcome results, notably: improved employee morale, increased customer satisfaction, and greater supplier loyalty and cooperation.

Table 1.1 summarizes the changing pattern of the marketplace. In some ways the changing business environment involves a return to the

Was	Is
National markets	International markets
National competition	International competition
Control the business environment	Adapt to the environment rapidly
Homogeneous product	Customized product
De-skilled jobs	Complex jobs
Product-specific capital	Flexible systems
Maintain status quo	Continuous improvement
Management by control	Management by planning

TABLE 1.1 The Changing Business Environment

craftsman era of the past: more complex jobs with the resulting need for workers with a broader repertoire of skills. Other tendencies are continuations of past trends: international markets are the next logical step after moving from local markets to national markets. In other ways the new world of business is simply different: modern flexible systems diverge in fundamental ways from previous systems.

It follows that yesterday's organizations, which evolved in response to the realities of the past, might not be suited to the changing reality. In fact, there is strong evidence to suggest that organizations that do not adapt will simply disappear. Over 40 percent of the 1979 list of the *Fortune* 500 had disappeared by 1990 (Peters, 1990). The organizations that have managed to progress have not stood still.

Forms of Organization

In addition to describing organizations in terms of their structures, Mintzberg (1994) also describes them in terms of forms. Mintzberg proposes a framework of five basic forms of organization:

1. *The Machine Organization*. Classic bureaucracy, highly formalized, specialized, and centralized, and dependent largely on the standardization of work processes for coordination. Common in stable and mature industries with mostly rationalized, repetitive operating work (as in airlines, automobile companies, retail banks).
2. *The Entrepreneurial Organization*. Nonelaborated, flexible structure, closely and personally controlled by the chief executive, who coordinates by direct supervision. Common in start-up and turnaround situations as well as in small business.
3. *The Professional Organization*. Organized to carry out the expert work in relatively stable settings, hence emphasizing the standardization

of skills and the pigeonholing of services to be carried out by rather autonomous and influential specialists, with the administrators serving for support more than exercising control; common in hospitals, universities, and other skilled and craft services.

4. *The Adhocracy Organization*. Organized to carry out expert work in highly dynamic settings, where the experts must work cooperatively in project teams, coordinating the activities by mutual adjustment, in flexible, usually matrix forms of structure; found in "high technology" industries such as aerospace and in project work such as filmmaking, as well as in organizations that have to truncate their more machinelike mature operations in order to concentrate on product development.
5. *The Diversified Organization.* Any organization split into semi-autonomous divisions to serve a diversity of markets, with the "headquarters" relying on financial control systems to standardize the outputs of the divisions, which tend to take on the machine form.

CHAPTER 2

The Quality Function

As discussed in Chap. 1, organizations are traditionally structured according to functional specializations, for instance, marketing, engineering, purchasing, manufacturing. Conceptually, each function performs an activity essential in delivering value to the customer. In the past, these activities were performed sequentially. As shown in Fig. 2.1, Shewhart, Deming, and Juran all depict these activities as forming a circle or a spiral, where each cycle incorporates information and knowledge acquired during the previous cycle.

Juran Trilogy

Juran and Gryna (1988, p. 2.6) define the quality function as "the entire collection of activities through which we achieve fitness for use, no matter where these activities are performed." Quality is thus influenced by, if not the responsibility of, many different departments. In most cases, the quality department serves a secondary, supporting role. While the quality department is a specialized function, quality activities are dispersed throughout the organization. The term "quality function" applies to those activities, departmental and companywide, that collectively result in product or service quality. An analogy can be made with the finance department. Even though many specialized finance and accounting functions are managed by the finance department, every employee in the organization is expected to practice responsible management of his or her budgets and expenditures.

Juran and Gryna (1988) grouped quality activities into three categories, sometimes referred to as the Juran trilogy: planning, control, and improvement. *Quality planning* is the activity of developing the products and processes required to meet customers' needs. It involves a number of universal steps (Juran and DeFeo, 2010):

- Define the customers.
- Determine the customer needs.
- Develop product and service features to meet customer needs.
- Develop processes to deliver the product and service features.
- Transfer the resulting plans to operational personnel.

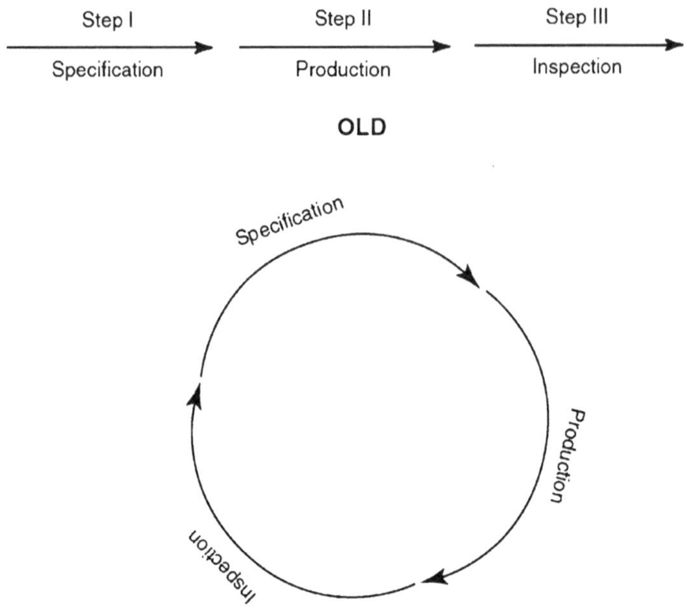

(a)

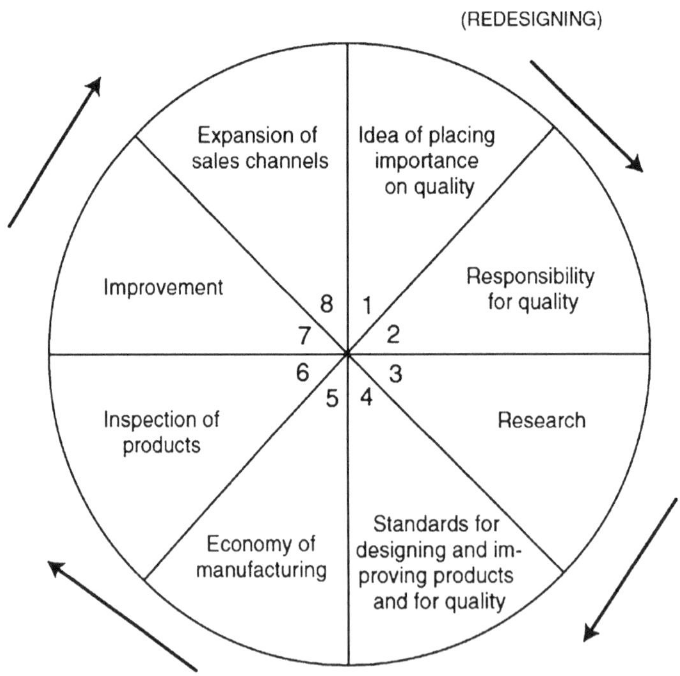

(b1)

FIGURE 2.1 (*a*) Representation of quality activities in the organization (Shewhart, 1939). (*b*) Deming's wheel of quality control (1986). (*c*) Juran's spiral of progress in quality (Juran and Gryna, 1988).

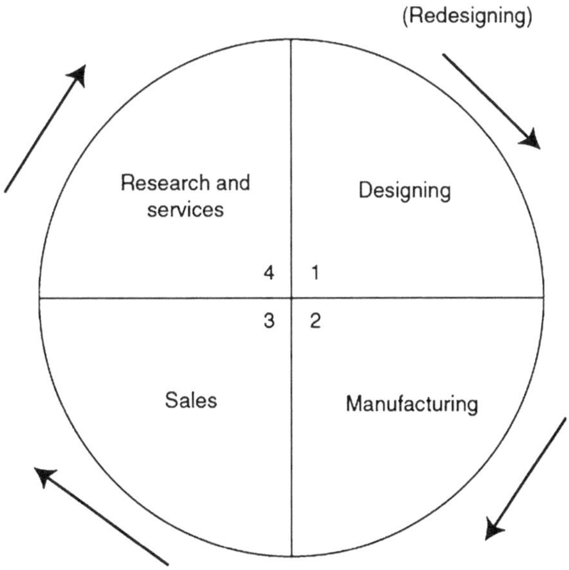

The idea of placing importance on quality
The responsibility for quality

(b2)

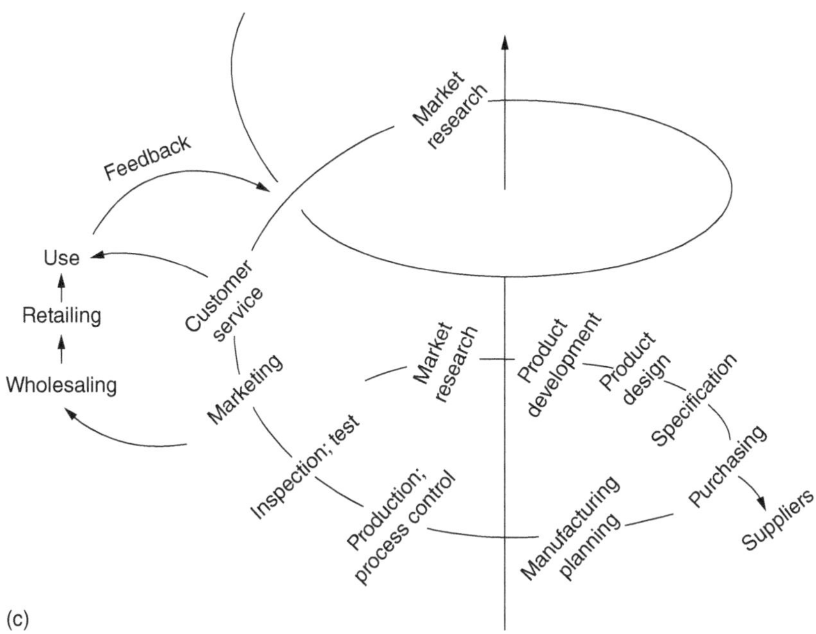

(c)

Figure 2.1 *(Continued)*

Quality control is the process used by operational personnel to ensure that their processes meet the product and service requirements (defined during the planning stage). It is based on the feedback loop and consists of the following steps:

- Evaluate actual operating performance.
- Compare actual performance with goals.
- Act on the difference.

Quality improvement aims to attain levels of performance that are unprecedented—levels that are significantly better than any past level. The methodologies recommended for quality improvement efforts utilize Six Sigma project teams, as described in Part IV. Notably, whereas earlier version of Juran's *Quality Handbook* did not specifically advocate cross-functional project-based teams for quality improvement efforts, the most recent sixth edition (2010) clearly prescribes their use.

The mission of the quality function is company-wide quality management. Quality management is the process of identifying and administering the activities necessary to achieve the organization's quality objectives. These activities will fall into one of the three categories in Juran's trilogy.

Since the quality function transcends any specialized quality department, extending to all of the activities throughout the company that affect quality, the primary role in managing the quality function is exercised by senior leadership. Only senior leadership can effectively manage the necessary cross-functional activities.

As the importance of quality has increased, the quality function has gained prominence within the organizational hierarchy. Figure 2.2 presents

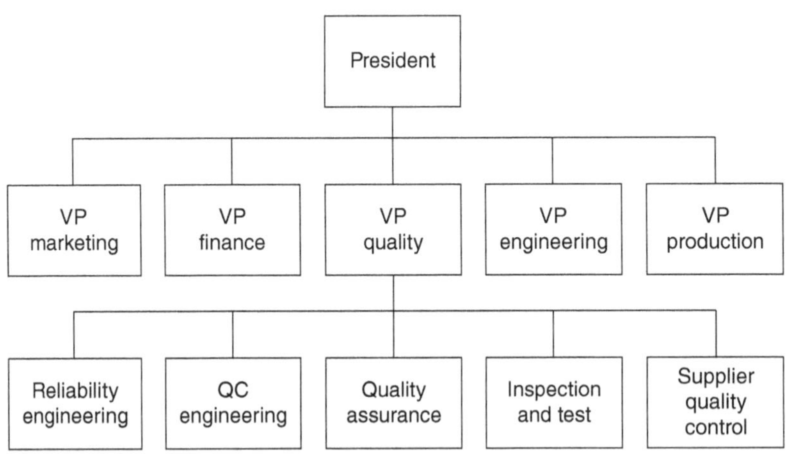

FIGURE 2.2 Quality within a traditional organization chart.

a prototypical modern organization chart for a hypothetical large manufacturing organization.

In this traditional structure, the quality specialists have no more than a secondary responsibility for most of the important tasks that impact quality. Table 2.1 lists the major work elements normally performed by these specialized departments.

Because the traditional, functionally specialized hierarchy creates a "silo mentality," each functional area tends to focus on its own function, often to the detriment of cross-functional concerns like quality. This is not a failing of the workforce, but a predictable result of the system in which these people work. The situation will not be corrected by exhortations to think or act differently. It can only be changed by modifying the system itself.

Several alternative organizational approaches to deal with the problems created by the traditional structure have already been discussed. The cross-functional organization is, as of this writing, the most widespread alternative structure. Quality "councils" or "steering committees" are cross-functional teams that set quality policy and, to a great extent, determine the role of the quality specialists in achieving the policy goals. The steering committee makes decisions regarding the totality of company resources (including those assigned to other functional areas) to be devoted to quality planning, improvement, and control.

Quality concerns must be balanced with other organizational concerns, such as market share, profitability, and development of new products and

Reliability Engineering	Establish reliability goals; Reliability apportionment; Stress analysis; Identification of critical parts; Failure Modes & Effects Analysis (FMEA); Reliability prediction; Design review; Supplier selection; Control of reliability during manufacturing; Reliability testing; Failure reporting and corrective action system
Quality Engineering	Process capability analysis; Quality planning; Establishing quality standards; Test equipment and gage design; Quality troubleshooting; Analysis of rejected or returned material; Special studies (measurement error, etc.)
Quality Assurance	Write quality procedures; Maintain quality manual; Perform quality audits; Quality information systems; Quality certification; Training; Quality cost systems
Inspection & Test	In-process inspection and test; Final product inspection and test; Receiving inspection; Maintenance of inspection records; Gauge calibration
Vendor Quality	Preaward vendor surveys; Vendor quality information systems; Vendor surveillance; Source inspection

Table 2.1 Quality Work Elements

services. Customer concerns must be balanced with the concerns of investors and employees. The senior leadership, consisting of top management and the board of directors, must weigh all of these concerns and arrive at a resource allocation plan that meets the needs of all stakeholders in the organization. The unifying principle for all stakeholders is the organization's purpose.

There are two basic ways to become (or remain) competitive: achieve superior perceived quality by developing a set of product specifications and service standards that more closely meet customer needs than competitors; and achieve superior conformance quality by being more effective than your competitors in conforming to the appropriate product specifications and service standards. These are not mutually exclusive; excellent companies do both simultaneously.

Research findings indicate that achieving superior *perceived quality* (that is, as perceived by customers), provides three options to a business—all of which are favorable to its competitiveness (Buzzell and Gale, 1987):

- You can charge a higher price for your superior quality and thus increase profitability.
- You can charge a higher price and invest the premium in R&D, thus ensuring higher perceived quality and greater market share in the future.
- You can charge the same price as your competitor for your superior product, building market share. Increased market share, in turn, means volume growth and rising capacity utilization (or capacity expansion), allowing you to lower costs (or increase profit).

Research also suggests additional benefits to companies that provide superior perceived quality, including higher customer loyalty; more repeat purchases; and lower marketing costs. Achieving superior *conformance quality* provides two key benefits:

- Lower cost of quality than competitors, which translates to lower overall cost.
- Since conformance quality is a factor in achieving perceived quality, it leads to the perceived quality benefits listed above.

Customer "satisfaction" does not simply happen; it is an effect. Quality is one important cause of the customer satisfaction effect, along with price, convenience, service, and a host of other variables. Quality and customer satisfaction are not synonyms; the former causes the latter. Generally businesses do not seek customer satisfaction as an end in itself. The presumption is that increased customer satisfaction will lead to higher revenues and higher profits, at least in the long term. This presumption has been validated by numerous studies, including the Profit Impact of

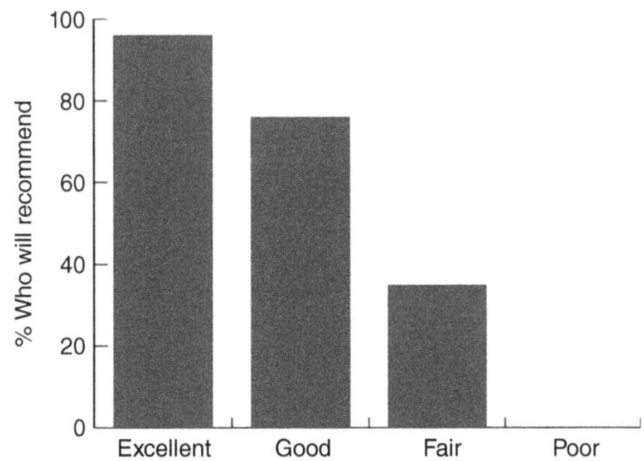

FIGURE 2.3 Customer satisfaction and sales.

Market Strategy (PIMS) studies (Buzzell and Gale, 1987). Since 1972 the PIMS Program, working with a database of 450 companies and 3000 business units, has developed a set of principles for business strategy based on the actual experiences of businesses. The principles drawn from this database provide a foundation for situation-specific analysis that managers perform to arrive at good decisions. The PIMS research indicates that quality is the major driver behind customer satisfaction, which in turn impacts a wide variety of other measures of organizational success. Figure 2.3, based on actual customer data, illustrates one important relationship: the percentage of customers who recommend the purchase of the firm's products or services to others.

Based on data such as these, and the relationships between such data and other measures of business success, the PIMS authors concluded: "The Customer is KING!" To best serve customers, the successful quality program will apply specific principles, techniques, and tools to better understand and serve their firm's royalty—the customer.

Related Business Functions

There are many related business functions within the organization that involve the quality mission in a significant capacity but which are not properly considered "quality functions."

Safety

A safety problem arises when a product, through use or foreseeable misuse, poses a hazard to the user or others. Clearly, the optimal approach to address safety issues is through prevention. Product and process-design review activities should include safety as a primary focus. Safety is quite simply a conformance requirement. The quality professional's primary role is the creation of systems for the prevention and detection of safety problems caused by nonconformance to established requirements, and development of systems for controlling the traceability of products that may have latent safety problems that might be discovered at a future date, or that may develop these problems as the result of unanticipated product usage.

There are a myriad of government agencies that are primarily concerned with safety, including the Consumer Product Safety Commission (CPSC), which deals with the safety of consumer products; the Defense Nuclear Facilities Safety Board Office of Environment, Safety, and Health; the Center for Food Safety and Applied Nutrition; the Mine Safety and Health Administration; the Occupational Safety and Health Administration (OSHA); the Office of System Safety; and the Defense Nuclear Facilities Safety Board, among others.

Regulatory Issues

For many years the fastest growing "industry" in the United States has been federal regulation of business. The U.S. Small Business Administration estimated that compliance with federal regulations alone consumed $1.75 trillion dollars in 2008 (Crain and Crain, 2010), or approximately 13 percent of 2008 GDP. Each year over 150,000 pages of new regulations are issued by government agencies. The quality manager will almost certainly be faced with regulatory compliance issues in his or her job. In some industries, compliance may be the major component of the quality manager's job.

Product Liability

The subject of quality and the law is also known as product liability. While the quality manager isn't expected to be an expert in the subject, the quality activities bear directly on an organization's product liability exposure and deserve the quality manager's attention. To understand product liability, one must first grasp the vocabulary of the subject. Table 2.2 presents the basic terminology (Thorpe and Middendorf, pp. 20–21).

There are three legal theories involved in product liability: breach of warranty, strict liability in tort, and negligence. Two branches of law deal with these areas, contract law and tort law.

A *contract* is a binding agreement for whose breach the law provides a remedy. Key concepts of contract law relating to product liability are those of *breach of warranty* and *privity of contract*.

Breach of warranty can occur from either an express warranty or an implied warranty. An express warranty is a part of the basis for a sale: the buyer agreed to the purchase on the reasonable assumption that the product would perform in the manner described by the seller. The seller's

Assumption of risk	The legal theory that a person who is aware of a danger and its extent and knowingly exposes himself to it assumes all risks and cannot recover damages, even though he is injured through no fault of his own.
Contributory negligence	Negligence of the plaintiff that contributes to his injury and at common law ordinarily bars him from recovery from the defendant although the defendant may have been more negligent than the plaintiff.
Deposition	The testimony of a witness taken out of court before a person authorized to administer oaths.
Discovery	Procedures for ascertaining facts prior to the time of trial in order to eliminate the element of surprise in litigation.
Duty of care	The legal duty of every person to exercise due care for the safety of others and to avoid injury to others whenever possible.
Express warranty	A statement by a manufacturer or seller, either in writing or orally, that his product is suitable for a specific use and will perform in a specific way.
Foreseeability	The legal theory that a person may be held liable for actions that result in injury or damage only when he was able to foresee dangers and risks that could reasonably be anticipated.
Great care	The high degree of care that a very prudent and cautious person would undertake for the safety of others. Airlines, railroads, and buses typically must exercise great care.
Implied warranty	An automatic warranty, implied by law, that a manufacturer's or dealer's product is suitable for either ordinary or specific purposes and is reasonably safe for use.
Liability	An obligation to rectify or recompense for any injury or damage for which the liable person has been held responsible or for failure of a product to meet a warranty.
Negligence	Failure to exercise a reasonable amount of care or to carry out a legal duty that results in injury or property damage to another.
Obvious peril	The legal theory that a manufacturer is not required to warn prospective users of products whose use involves an obvious peril, especially those that are well-known to the general public and that generally cannot be designed out of the product.
Prima facie	Such evidence as by itself would establish the claim or defense of the party if the evidence were believed.
Privity	A direct contractual relationship between a seller and a buyer. If A manufactures a product that is sold to dealer B, who sells it to consumer C, privity exists between A and B and between B and C, but not between A and C.
Proximate cause	The act that is the natural and reasonably foreseeable cause of the harm or event that occurs and injures the plaintiff.
Reasonable care	The degree of care exercised by a prudent person in observance of his legal duties toward others.
Res ipsa loquitur	The permissible inference that the defendant was negligent in that "the thing speaks for itself" when the circumstances are such that ordinarily the plaintiff could not have been injured had the defendant not been at fault.

Table 2.2 Fundamental Legal Terminology

Standard of reasonable prudence	The legal theory that a person who owes a legal duty must exercise the same care that a reasonably prudent person would observe under similar circumstances.
Strict liability in tort	The legal theory that a manufacturer of a product is liable for injuries due to product defects, without the necessity of showing negligence of the manufacturer.
Subrogation	The right of a party secondarily liable to stand in the place of the creditor after he has made payment to the creditor and to enforce the creditor's right against the party primarily liable in order to obtain indemnity from him.
Tort	A wrongful act or failure to exercise due care, from which a civil legal action may result.

Table 2.2 Fundamental Legal Terminology (*Continued*)

statement need not be written for the warranty to be an express warranty; his mere statement of fact is sufficient. An implied warranty is a warranty not stated by the seller, but implied by law. Certain warranties result from the simple fact that a sale has been made. One of the most important of the attributes guaranteed by an implied warranty is that of fitness for normal use. The warranty is that the product is reasonably safe.

Privity of contract means that a direct relationship exists between two parties, typically buyer and seller. At one time manufacturers were not held liable for products purchased from vendors or sold to a consumer through a chain of wholesalers, dealers, etc. Manufacturers were treated as third-party assignees and said to be not in privity with the end user. This concept began to deteriorate in 1905 when courts began to permit lawsuits against sellers of unwholesome food, whether or not they were negligent, and against original manufacturers, whether or not they were in privity with the consumers. The first recognition of strict liability for an express warranty without regard to privity was enunciated by a Washington court in 1932 in a case involving a Ford Motor Company express warranty that their windshields were "shatterproof." When the windshield shattered and injured a consumer, the court allowed the suit against Ford, ruling that even without privity the manufacturer was responsible for the misrepresentation, even if the misrepresentation was done innocently.

Under the rule of *strict liability* an innocent consumer who knows nothing about disclaimers and the requirement of giving notice to a manufacturer with whom he did not deal cannot be prevented from suing. The rule avoids the technical limits of privity, which can create a chain of lawsuits back to the party that originally put the defective product into the stream of commerce. The seller (whether a salesman or manufacturer) is liable even though he has been careful in handling the product and even if the consumer did not deal directly with him.

The first case to apply this modern rule was *Greenman vs. Yuba Power Products, Inc.*, in California in 1963. A party, Mr. Greenman, was injured when a work piece flew from a combination power tool purchased for him by his wife two years prior to the injury. He sued the manufacturer and produced witnesses to prove that the machine was designed with inadequate set screws.

The manufacturer, who had advertised the power tool as having "rugged construction" and "positive locks that hold through rough or precision work" claimed that it should not have to pay money damages because the plaintiff had not given it notice of breach of warranty within a reasonable time as required. Furthermore, a long line of California cases had held that a plaintiff could not sue someone not in privity with him unless the defective product was food.

The court replied that this was not a warranty case but a *strict liability case*. The decision stated that any "manufacturer is strictly liable … when an article he placed on the market, knowing that it is to be used without inspection for defects, proves to have a defect that causes injury to a human being."

The concept of strict liability was a turning point for both the consumer movement and quality control. The use of effective, modern quality control methods became a matter of paramount importance. The concept is also called *strict liability in tort*, which is virtually synonymous with the common usage of the term "product liability." A *tort* is a wrongful act or failure to exercise due care resulting in an injury, from which civil legal action may result. Tort law seeks to provide compensation to people who suffer loss because of the dangerous or unreasonable actions of others.

A related concept is that of *negligence*. Negligence occurs when one person fails to fulfill a duty owed to another or fails to act with due care. There are two elements necessary to establish negligence: a standard of care recognized by law, and a breach of the duty or requisite care. Also, the breach of duty must be the proximate cause of the harm or injury. The accepted standard of care is that of the "reasonable person." The court must measure the action of the parties involved relative to the actions expected from an imaginary reasonable person. To muddy the waters further, the court must weigh the risk or danger of the situation against the concept of "reasonable risk." Clearly, these concepts are far from cut and dried.

The case cited above, and many other developments since, have resulted in a feature that is unique to product liability law: namely, *the conduct of the manufacturer is irrelevant*.

The plaintiff in a product liability suit need not prove that the manufacturer failed to exercise due care; he need show only that the product was the proximate cause of harm, and that it was either defective or unreasonably dangerous. This is what is meant by "strict liability." In a sense, it is the product that is on trial and not the manufacturer. There are several areas in which engineering and management are vulnerable, including design; manufacturing and materials; packaging, installation, and application; and warnings and labels.

Designs that create hidden dangers to the user, designs that fail to comply with accepted standards, designs that exclude necessary safety features or devices, or designs that don't properly allow for possible unsafe *misuse or abuse* that is reasonably foreseeable to the designer are all suspect. Quality control includes design review as one of its major elements, and all designs should be carefully evaluated for these shortcomings. As always, the concept of reasonableness applies in all its ambiguity.

The application of quality control principles to manufacturing, materials, packaging, and shipping is probably the best protection possible against future litigation. Defect prevention is the primary objective of quality control and the defect that isn't made will never result in loss or injury. Bear in mind, however, a defect in quality control is usually defined as a non-conformance to requirements. There is no such definition in the law. Legal definitions of a defect are based on the concept of reasonableness and the need to consider the use of the product.

Environmental Issues Relating to the Quality Function

The primary connection between environmental issues and the quality function is the ISO 14000 standard, which covers six areas:

1. Environmental management systems
2. Environmental auditing
3. Environmental performance evaluation
4. Environmental labeling
5. Life-cycle assessment
6. Environmental aspects in product standards

The 14000 series standard mirrors the ISO 9001 quality standard in requiring a policy statement, top-down management commitment, document control, training, corrective action, management review, and continual improvement. Plans call for integrating ISO 9000 and ISO 14000 into one management standard that will also include health and safety. It is possible that eventually a single audit will cover both ISO 9000 and ISO 14000. ISO 14001—the environmental management system (EMS) specification—is intended to be the only standard establishing requirements against which companies will be audited for certification. The standard does not set requirements for results, only for the continuous improvement of a company's EMS. ISO 14001 is not a requirement; it is voluntary. ISO 14001 is a systems-based standard that gives companies a blueprint for managing their impact on the environment.

The requirements fall into five main areas:

- Senior management shall articulate the company's environmental policy. The policy will include commitments toward pollution prevention and continuous improvement of the EMS. The policy will be available to the public.
- Consistent with the environmental policy, you shall establish and maintain procedures to identify significant environmental aspects and their associated impacts. Procedures should include legal and other requirements. Objectives and targets will also be documented, including continual improvement and pollution prevention.
- Each employee's role and position must be clearly defined, and all employees must be aware of the impact of their work on the environment. Employees shall be adequately trained.
- The EMS should be set up to facilitate internal communication. To that end, all relevant documentation should be easily available and usable, in either print or electronic form.
- Organizations must continually monitor and document their environmental effects and periodically review them to ensure continual improvement and the effectiveness of the EMS. Management is responsible for an internal review of the EMS on a regular basis.

CHAPTER 3

Approaches to Quality

Traditional quality programs for the most of the twentieth century were focused on the command and control aspects inherent to the functional hierarchy organizational structure. The efforts of Taylor to both standardize and simplify work at least made this possible, if not enforced its legitimacy. In the aftermath of WWII, American business had extended control backward to the sources of supply and forward to the distribution and merchandising. To the extent that organizations succeeded in this endeavor, they reduced uncertainty in their environments and gained control over critical elements of their business.

Internal quality practices were largely associated with off-line inspection by trained quality inspectors, assigned to a Quality Control department. Operational personnel were responsible for their assigned functions, such as production; inspectors were responsible for ensuring conformance of the product to the customer requirement, usually just before the product was shipped to the customer. Although Shewhart had developed the statistical control chart in the 1920s, its use in industry was dwarfed by inspection sampling plans that better fit this organizational model. These sampling plans had become established as MIL-STD 105, a requirement of military suppliers, which had made them the de facto standard throughout the war years for all suppliers.

Unlike that of most of the industrialized world, the American infrastructure was undamaged by the war. While the rest of the world rebuilt, shortages were endemic, and American suppliers ramped up production to fill the void, resulting in a period of prosperity and profitability that further enforced the perception of well-designed, or at least adequate, systems. In reality, quality levels were poor, as is often the case during shortages (Juran, 1995).

By the late 1970s, however, market influences emerged to challenge the status quo, including (Juran, 1995):

- The growth of consumerism
- The growth of litigation over quality
- The growth of government regulation of quality
- The Japanese quality revolution

Deming's Approach

Deming is probably best known for his theory of management as embodied in his 14 points. According to Deming, "The 14 points all have one aim: to make it possible for people to work with joy." The 14 points are:

1. Create constancy of purpose for the improvement of product and service, with the aim to become competitive, stay in business, and provide jobs.
2. Adopt the new philosophy of cooperation (win-win) in which everybody wins. Put it into practice and teach it to employees, customers, and suppliers.
3. Cease dependence on mass inspection to achieve quality. Improve the process and build quality into the product in the first place.
4. End the practice of awarding business on the basis of price tag alone. Instead, minimize total cost in the long run. Move toward a single supplier for any one item, on a long-term relationship of loyalty and trust.
5. Improve constantly and forever the system of production, service, planning, or any activity. This will improve quality and productivity and thus constantly decrease costs.
6. Institute training for skills.
7. Adopt and institute leadership for the management of people, recognizing their different abilities, capabilities, and aspirations. The aim of leadership should be to help people, machines, and gadgets do a better job. Leadership of management is in need of overhaul, as well as leadership of production workers.
8. Eliminate fear and build trust so that everyone can work effectively.
9. Break down barriers between departments. Abolish competition and build a win-win system of cooperation within the organization. People in research, design, sales, and production must work as a team to foresee problems of production and use that might be encountered with the product or service.
10. Eliminate slogans, exhortations, and targets asking for zero defects or new levels of productivity. Such exhortations only create adversarial relationships, as the bulk of the causes of low quality and low productivity belong to the system and thus lie beyond the power of the workforce.
11. Eliminate numerical goals, numerical quotas, and management by objectives. Substitute leadership.

12. Remove barriers that rob people of joy in their work. This will mean abolishing the annual rating or merit system that ranks people and creates competition and conflict.
13. Institute a vigorous program of education and self-improvement.
14. Put everybody in the company to work to accomplish the transformation. The transformation is everybody's job.

These principles clearly define responsibilities for management, many of which were contradicted by the traditional functional hierarchy structure, as well as its command and control tendencies.

Deming also described a system of "profound knowledge." Deming's system of profound knowledge consists of four parts: appreciation for a system, knowledge about variation, theory of knowledge, and psychology.

A system is a network of interdependent components that work together to accomplish the aim of the system. The system of profound knowledge is itself a system. The parts are interrelated and cannot be completely understood when separated from one another. Systems must be managed. The greater the interdependence of the various system components, the greater the need for management. In addition, systems should be *globally* optimized; global optimization cannot be achieved by optimizing each component independent of the rest of the system.

Systems can be thought of as networks of intentional cause-and-effect relationships. However, most systems also produce *unintended* effects. Identifying the causes of the effects produced by systems requires understanding of variation—part 2 of Deming's system of profound knowledge. Without knowledge of variation people are unable to learn from experience. There are two basic mistakes made when dealing with variation: (1) reacting to an outcome as if it were produced by a special cause, when it actually came from a common cause, and (2) reacting to an outcome as if it were produced by a common cause, when it actually came from a special cause. The terms *special cause* and *common cause* are operationally defined by the statistical control chart, discussed in detail in Chap. 9.

Deming's theory of profound knowledge is based on the premise that management is prediction. Deming, following the teachings of the philosopher C. I. Lewis (1929), believed that prediction is not possible without theory. Deming points out that knowledge is acquired as one makes a rational prediction based on theory, then revises the theory based on comparison of prediction with observation. Knowledge is reflected in the new theory. Without theory, there is nothing to revise, that is, there can be no new knowledge, no learning. The process of learning is operationalized by Deming's Plan-Do-Study-Act cycle (a modification of Shewhart's Plan-Do-Check-Act cycle). It is important to note that information is not knowledge. Mere "facts" in and of themselves are not knowledge. Knowing what the Dow Jones Industrial Average is right now, or what it has been for the last 100 years, is not enough to tell us what it will be tomorrow.

Psychology is the science that deals with mental processes and behavior. In Deming's system of profound knowledge, psychology is important because it provides a theoretical framework for understanding the differences between people, and provides guidance in the proper ways to motivate them.

Total Quality Control In Japan

Japan is well-known for replacing its old reputation for terrible quality with a new reputation for excellence. The system they employed to accomplish this impressive feat is a uniquely Japanese version of a system that originated in America known as total quality control (TQC). TQC is a system of specialized quality control activities initially developed by Feigenbaum (1951, 1983). The Japanese took Feigenbaum's American version of quality control (which was very much a continuation of the scientific management approach) and made it their own. The Japanese rendition of TQC is described by Ishikawa (1985) as a "thought revolution in management," drawing heavily on contributions of American quality experts, especially Walter A. Shewhart, W. Edwards Deming, and Joseph M. Juran. However, there are some elements of the Japanese system that are purely Japanese in character. The thought revolution involves a transformation in six categories (Ishikawa, 1985):

1. Quality first—not short-term profit first.

   Management that stresses "quality first" can gain customer confidence step-by-step, resulting in a gradual increase in company sales with longer-term improvement to profitability and management stability. A company following the principle of "profit first," may obtain a quicker profit but will be unable to sustain competitiveness for longer periods of time.
2. Consumer orientation—not producer orientation. Ishikawa stressed thinking in terms of another party's position: listen to their opinions and act in a way respectful to their views.
3. The next process is your customer—breaking down the barrier of sectionalism.

   Especially within highly structured (functionally based) organizations, this approach is needed to drive company-wide quality and overcome silo-based mentality (i.e., each department looking after its own best interests). The company as a whole must look at its processes for delivering customer value, rather than a department or section separately.
4. Using facts and data to make presentations—utilization of statistical methods.

The importance of facts must be clearly recognized. Facts, in contrast to opinions, can be translated into data. If data are accurate, they can be analyzed using statistical methods and engineering or management models. This, in turn, forms the basis of decisions. Such decisions will, in the long run, prove to be better than decisions made without this process, that is, the decisions will produce results that more closely match management's goals.

5. Respect for humanity as a management philosophy—full participatory management. When management decides to make companywide quality its goal, it must standardize all processes and procedures and then boldly delegate authority to subordinates. The fundamental principle of successful management is to allow subordinates to make full use of their ability. The term *humanity* implies autonomy and spontaneity. People are different from animals or machines. They have their own minds, and they are always thinking. Management based on humanity is a system of management that lets the unlimited potential of human beings blossom.
6. Cross-functional management.

   From the perspective of companywide goals, the main functions are quality assurance, cost control, quantity control, and personnel control. The company must establish cross-functional committees to address these section-spanning issues. The committee chair should be a senior managing director. Committee members are selected from among those who hold the rank of director or above.

At the time, cross-functional management was a uniquely Japanese feature of quality management, requiring a fundamental modification of the bureaucratic (functional hierarchy) model of traditional organizations. It has long been known that this form of organization, sometimes called the "chimney stack model," results in isolation of the various functions from one another. This in turn results in parochialism and other behavior that, while optimal for a given function, is detrimental to the system as a whole. Most business texts address this problem superficially at best. The Japanese developed a formal approach for dealing with it, which is often referred to as *cross-functional management* (GOAL/QPC, 1990).

Another feature of the Japanese approach to management was the concentration on the "core business." The core business is the essence of what the company does; for example, Toyota might identify their core as the production of personal transportation vehicles. The company strives to provide a sense of family and belonging for full-time core employees. Lifelong service to the employer is expected, and the employer demonstrates similar loyalty to the employee, for example by providing lifelong job security.

How is this possible? After all, the normal business cycle rises and falls periodically and a company that is adequately staffed for peak production is overstaffed when production downturns occur. The Japanese manage to provide security in good times and bad in a number of ways. One is by massive outsourcing of non-core business activities to suppliers. It is not unusual to find as much as 80 percent of the value of the finished product in purchased materials. Another is by making use of a large buffer of part-time employees. In some Japanese companies as many as 50 percent of their employees during peak production periods are part-time workers.

Prevailing wisdom in America would suggest that this strategy would result in a great increase in uncertainty and an overall decline in quality. in Japan, however, suppliers are very tightly controlled using a system known as Keiretsu. Suppliers within the Keiretsu often have board members from the parent company, and from other members of the Keiretsu. Unlike American businesses, which tend to deal with their suppliers at arm's length, parent companies play a very active role in the affairs of their suppliers. While in the USA a firm might regularly appraise the quality of the product delivered by a supplier, in Japan the parent would also carefully evaluate how the product was made. If appropriate, the parent would suggest better, more economical ways to produce the product. Parent companies provide larger, longer-term contracts than their American counterparts, and they often demand steady price decreases. The supplier might be expected to become "dedicated" to its parent, providing product only to the parent and not to competitors. Continuous improvement in quality, cost, and delivery is expected of all Japanese suppliers, just as it is expected from the employees.

KAIZEN™ (a trademark of the KAIZEN Institute, Ltd.) is a philosophy of continuous improvement, a belief that all aspects of life should be constantly improved. In Japan, where the concept originated, KAIZEN applies to all aspects of life, not just the workplace. In America the term is usually applied to work processes.

The KAIZEN approach focuses on ongoing incremental improvement that involves all stakeholders. Over time these small improvements produce changes every bit as dramatic as the "big project" approach. KAIZEN does not concern itself with changing fundamental systems, but seeks to optimize existing systems.

All employees in an organization have responsibilities for two aspects of quality: process improvement and process control. Control involves taking action on deviations to maintain a given process state. In the absence of signals indicating that the process has gone astray, control is achieved by adhering to established standard operating procedures (SOPs). In contrast, improvement requires experimentally modifying the

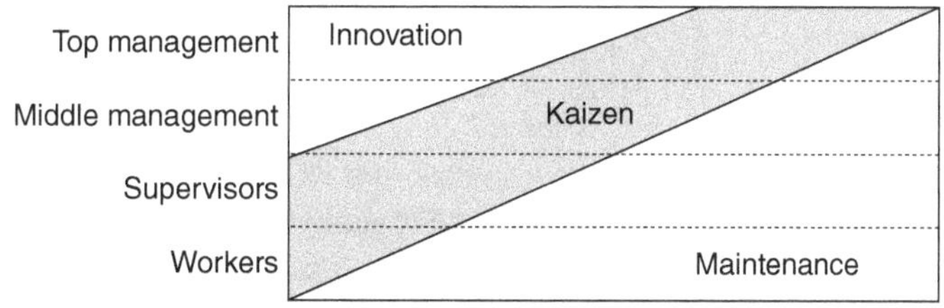

FIGURE 3.1 Responsibility for KAIZEN and KAIZEN's role in process improvement (Imai, 1986).

process to produce better results through innovation and KAIZEN. When an improvement has been identified, the SOPs are changed to reflect the new way of doing things. Imai (1986) illustrates the job responsibilities as shown in Figure 3.1.

The figure illustrates both the shared responsibility and the limited role of KAIZEN in excluding radical innovations (sometimes referred to as reengineering). More detailed responsibilities for KAIZEN are provided in Table 3.1.

Another rather considerable contribution from post-war Japan is the set of lean practices documented by Taiichi Ohno of Toyota. The lean methods are sometimes referred to as the Toyota Production System (due to their origins), and include principles and methodologies for improving cycle times and quality through the elimination of waste (also known by its Japanese name of *muda*). Lean distinguishes between activities that create value, and those that don't, with the objective to improve cycle times and efficiencies, reduce waste of resources, and increase value to the customer.

Taiichi Ohno of Toyota defined the following five types of waste (Womack and Jones (1996) added the sixth):

1. Errors requiring rework. (Rework refers to any activity required to fix or repair the result of another process step. In service processes, management intervention to resolve a customer complaint may be considered rework.)
2. Work with no immediate customer, either internal or external, resulting in work in progress or finished goods inventory.
3. Unnecessary process steps.
4. Unnecessary movement of personnel or materials.
5. Waiting by employees as unfinished work in an upstream process is completed.
6. Design of product or processes that do not meet the customer's needs.

Top management	• Be determined to introduce KAIZEN as a corporate strategy • Provide support and direction for KAIZEN by allocating resources • Establish policy for KAIZEN and cross-functional goals • Realize KAIZEN goals through policy deployment and audits • Build systems, procedures, and structures conducive to KAIZEN
Middle management and staff	• Deploy and implement KAIZEN goals as directed by top management through policy deployment and cross-functional management • Use KAIZEN in functional capabilities • Establish, maintain, and upgrade standards • Make employees KAIZEN-conscious through intensive training programs • Help employees develop skills and tools for problem solving
Supervisors	• Use KAIZEN in functional roles • Formulate plans for KAIZEN and provide guidance to workers • Improve communication with workers and sustain high morale • Support small group activities (such as quality circles) and the individual suggestion system • Introduce discipline in the workshop • Provide KAIZEN suggestions
Workers	• Engage in KAIZEN through the suggestion system and small group activities • Practice discipline in the workshop • Engage in continuous self-development to become better problem solvers • Enhance skills and job performance expertise with cross-education

From Imai (1986).

Table 3.1 Hierarchy of Kaizen Involvement

Value is the opposite of waste, and may be identified by considering:

1. Is this something the customer is willing to pay for?
2. Does the step change form, fit, or function of the product? Stated differently, does it convert input to output?

If the answer to both questions is "No," then it's likely the activity does not create value in the customer's eyes, even if it is necessary to ensure quality in the current process. Inspection and review activities, such as monitoring of sales calls or management sign-offs on exceptions, are examples of a non–value added waste. They do nothing to change the product (or service), and are only necessary to address the poor quality associated with the underlying process. Unfortunately, if their removal would degrade the quality of the product or service, then they are Type I waste, sometimes called Business Value Added, that is necessary given the current state of the business processes. In many cases, it is beneficial to change the process to remove the waste.

Lean thinking has been shown to reap dramatic benefits in organizations. Organizations are able to sustain production levels with half the manpower, improving quality and reducing cycle times from 50 to 90 percent (Womack and Jones, 1996).

Several of the lean methods are fairly well known in and of themselves. Just in Time (JIT), for example, has been a buzzword within American manufacturing since the 1980s. Other well-known methods include Kanban (Japanese for cards) and 5S. Unfortunately, practitioners often find they are unable to experience significant advances in any of these areas individually if they do not embrace the complete principles of Lean. Furthermore, many of the methods must be undertaken in conjunction with, or after appreciating results from, rigorous quality improvement. It would perilous to implement JIT if the underlying processes were not in statistical control: without statistical control, the process is not stable or predictable, so cannot be balanced to achieve JIT performance.

Although many of these techniques were initially applied to manufacturing applications, they are particularly well suited (and have broad usage) to address issues in transactional processes within service industries. Furthermore, they have origins and a strong track record in small job-shop type environments at Toyota and its suppliers, where production was often very low volume and far from mass production levels.

ISO 9000 Series

The best-known system of quality standards is the ISO 9000 series, published by ISO (the International Organization for Standardization). The standards were originally published in 1987 and subsequently updated in 1994, 2000, and 2008. The standard was initially based on the U.S. Department of Defense Mil-Q-9858, released in 1959.

The use of ISO 9000 is extremely widespread, with over 1.1 million organizations certified to the standard (as of 2010); 86 percent of the registrations are in Europe and the Far East. It's likely that the proliferation of registrations in the Far East results from the recognition in the supply chain that use of a common standard eliminates the need for multiple quality systems audits by their various customers. ISO 9000 registration is achieved by third-party registrar audits; that is, audits are not performed by customers but by specially trained, independent, third-party auditors. In the past, many firms had to deal with auditors from many different customers. Furthermore, the customers usually had their own specific requirements. This resulted in the development of parallel systems to meet the varied requirements of their key customers, which was costly, confusing, and ineffective. While some industry groups, notably in the aerospace and automotive industries, made progress in coordinating their requirements, ISO 9000 has greatly improved upon and extended the scope of acceptable quality systems standards. It now serves as a base for other industry-specific standards, including TL-9000 (for the telecommunications industry); AS9000 (for aerospace); and ISO/TS 16949:2009 (for automotive, replacing QS 9000 in the United States).

While ISO 9000 applies to any organization and to all product categories, it does not specify *how* the requirements are to be implemented. Also, the series specify requirements at a very high level; ISO 9000 does not replace product, safety, or regulatory requirements or standards. The concept that underlies the ISO 9000 standards is that consistently high quality is best achieved by a combination of technical product specifications and management system standards. ISO 9000 standards provide only the management systems standards.

It should be noted that ISO 9000 is designed as a *minimal* quality standard. According to A. Blanton Godfrey, who served 10 years on the technical committee TC176, which developed ISO 9000, the requirements of ISO 9000 represent a 1970s understanding of quality (Paton, 1995). Godfrey states

> *... in one way we created a very good minimal standard for companies who were doing nothing. ISO 9000 gave them a worldwide accepted definition of a quality system. On the other hand, we did a lot of harm because some people thought it was a world-class system. And those companies that stopped when they got their certificate had a rude awakening when they found out that that didn't mean they were competitive.*

These sentiments are echoed by nearly every quality expert. Stapp (2001) summarized the following issues with the pre-2000 standards:

1. *Not enough emphasis on preventive action.* This area has been the victim of "requirements creep" since the inception of ISO 9001. The 1987 edition contained little in the way of solid requirements for a preventive action system. This was addressed by adding the requirement to the 1994 edition, but organizations could comply with only a weak system of preventive action.
2. *No emphasis on continual improvement.* W. Edwards Deming and other gurus of the quality profession found this to be a fundamental weakness in an ISO-compliant quality management system. The lack of attention to conti-nual improvement allowed organizations to comply with the letter of the requirements without really understanding the purpose of implementing a quality management system in the first place.

3. *Fragmented approach to the quality management system.* The famous "20 Quality Elements" defined in the first two editions of ISO 9001 offered an easier (almost a checklist) approach to developing a quality management system. Unfortu-nately, few of the elements can exist by themselves, but instead form parts of a unified system. The fragmented elements do not describe how an organization actually operates.
4. *Not enough focus on human resources.* Other than the requirements for training, the first two releases of ISO 9001 dealt very little with human resources and the needs of an organization's employees. An organization could actually have dangerous working conditions or treat its employees unfairly and be registered in good standing to ISO 9001.
5. *Not enough emphasis on customer communication.* Other than in the area of resolving differences during the contract review phase, the first two editions of ISO 9001 dealt very little with the importance of customer communication. Organizations that had developed the most successful quality management systems understood the importance of keeping customer communication and satisfaction at the forefront of everything they did. These organizations especially felt that the first two editions lacked the necessary focus.
6. *An ISO 9000 quality management system does not improve product or service quality.* Expecting ISO 9000 to directly create improved quality is somewhat like expecting the accounting system to directly create profitability. In any case, disappointment has been widely expressed about organizations with ISO-compliant quality management systems producing inferior goods or services.
7. *Not enough emphasis on management using hard data in decision making.* Under the first two editions of ISO 9001, top management's involvement could be limited to periodic management reviews. In some organizations, top management tried to relinquish control over the quality management system by delegating this key management duty to the quality manager. This hands-off approach to quality system management was often manifested in the lack of consistent data collection and analysis.
8. *Not well integrated with ISO 14001.* Organizations that were involved with audits of their environmental management system encountered difficulties in integrating the system design and auditing between ISO 14001 and ISO 9001.
9. *Too prescriptive.* The wide variety of goods and services provided by the world's various organizations has resulted in widely disparate needs and expectations for quality management systems. As organizations attempted to accommodate the ISO 9000 family of standards

to their unique organizations, they encountered difficulties in achieving a "fit." Organizations that developed quality manage-ment systems in compliance with ISO 9001 tended to show similarities to each other, even though the organizations were markedly different.

10. *Does not easily fit service businesses.* The wording of the first two editions of the ISO 9000 family of standards fit manufacturing organizations well but required some imagination to apply to service, education, medical, and other types of organizations.

The ISO 9000:2000 revisions sought to address these concerns. Most notably, a systems approach became evident; emphasis was placed on process control and continuous improvement; and the mandate for management responsibility for the quality system compliance was strengthened. The ISO 9000:2000 revision (which was not substantially revised in the 2008 release), lists eight Quality Management Principles forming the basis for the revisions (Stapp, 2001):

1. *Customer focus.* Attention to your assorted customers' needs, including a continual attempt to meet their requirements and exceed their expectations, should be seen as central to your organization's objectives.
2. *Leadership.* Without leadership, your organization will not have an environment that fosters a constant purpose. Strong leadership creates an environment in which those in the organization can actively participate in the achievement of the organization's objectives.
3. *Involvement of people.* Leadership involves more than giving orders, but includes the involvement of people throughout the organization in achieving the organization's goals, using their talents to further the organization's purpose.
4. *Process approach.* While the previous revisions of ISO 9001 used the concept of 20 quality elements, the 2000 revision builds on the concept that anything an organization does, including the quality management system, should be viewed as a logical process. This process includes inputs and resources, and a desired result that occurs through proper management of the processes involved.
5. *System approach to management.* Only when the organization can identify and manage the systems of interrelated processes can objectives be met. By meeting those objectives, the organization becomes more efficient and effective.
6. *Continual improvement.* While the earlier revisions of ISO 9001 dealt with problem solving through corrective and preventive action, the 2000 revision expands on that concept. Central to any organization's objectives should be a commitment to continually

improve in all its activities: greater efficiency, lower rejection rates, increased customer satisfaction, etc.

7. *Factual approach to decision making.* None of these principles can be achieved if the organization does not include methods of gathering information about its systems of interrelated processes. That information becomes the source for ensuring ongoing customer satisfaction and implementing continual improve-ment efforts, both of which result from a properly run organization.
8. *Mutually beneficial supplier relationships.* The organization cannot succeed if it allows hostile or uncooperative relationships with its suppliers. Since suppliers comprise an integral part of the systems an organization must manage, creating a cooperative relationship with suppliers must not be minimized.

Whereas the previous standards provided a near-checklist of 20 quality elements, these elements are now dispersed through five main clauses, representing a more systems-focused approach (Stapp, 2001):

Clause 4: Quality management system. Includes general requirements, documentation requirements.

Clause 5: Management responsibility. Includes management commitment, customer focus, quality policy, planning, administration of the quality management system, and management review.

Clause 6: Resource management. Includes provision of resources, human resources, facilities, work environment.

Clause 7: Product realization. Includes planning of realization processes customer-related processes, design and development, purchasing, production and service operations, and control of monitoring and measuring devices.

Clause 8: Measurement, analysis, and improvement. Includes customer satisfaction measurement, measurement of process and product, control of nonconformance, analysis of data, continual improvement, and corrective/preventive action.

In some ways, the new standard makes it easier for organizations to attain higher levels of performance by adopting other models, such as the Baldrige criteria, which can now more easily coexist in the organization.

Malcolm Baldrige National Quality Award

Public Law 100-107, the Malcolm Baldrige National Quality Improvement Act of 1987, signed by President Reagan on August 20, 1987, established an annual U.S. National Quality Award. The purposes of the award are to promote awareness of quality excellence, to recognize quality achievements of U.S. companies, and to publicize successful quality strategies. The Secretary of Commerce and the National Institute of Standards and Technology (NIST, formerly the National Bureau of Standards) were given responsibilities to develop and administer the award with cooperation and financial support from the private sector.

Awards may be given each year in each of three categories: manufacturing companies or subsidiaries; service companies or subsidiaries; and small businesses. Fewer than two awards may be given in a category if the high standards of the award program are not met.

Seven areas are examined on a weighted scale, as indicated in Table 3.2. The weights assigned to each category provide an indication of the relative

Leadership	**120**
Senior Leadership	70
Governance and Societal Responsibilities	50
Strategic Planning	**85**
Strategy Development	40
Strategy Implementation	45
Customer Focus	**85**
Voice of the Customer	45
Customer Engagement	40
Measurement, Analysis, and Knowledge Management	**90**
Measurement, Analysis, and Improvement of Organizational Performance	45
Management of Information, Knowledge, and Information Technology	45
Workforce Focus	**85**
Workforce Environment	40
Workforce Engagement	45
Operations Focus	**85**
Work Systems	45
Work Processes	40
Results	**450**
Product and Process Outcomes	120
Customer-Focused Outcomes	90
Workforce-Focused Outcomes	80
Leadership and Governance Outcomes	80
Financial and Market Outcomes	80
TOTAL POINTS	**1,000**

Table 3.2 Baldrige Scoring Weights

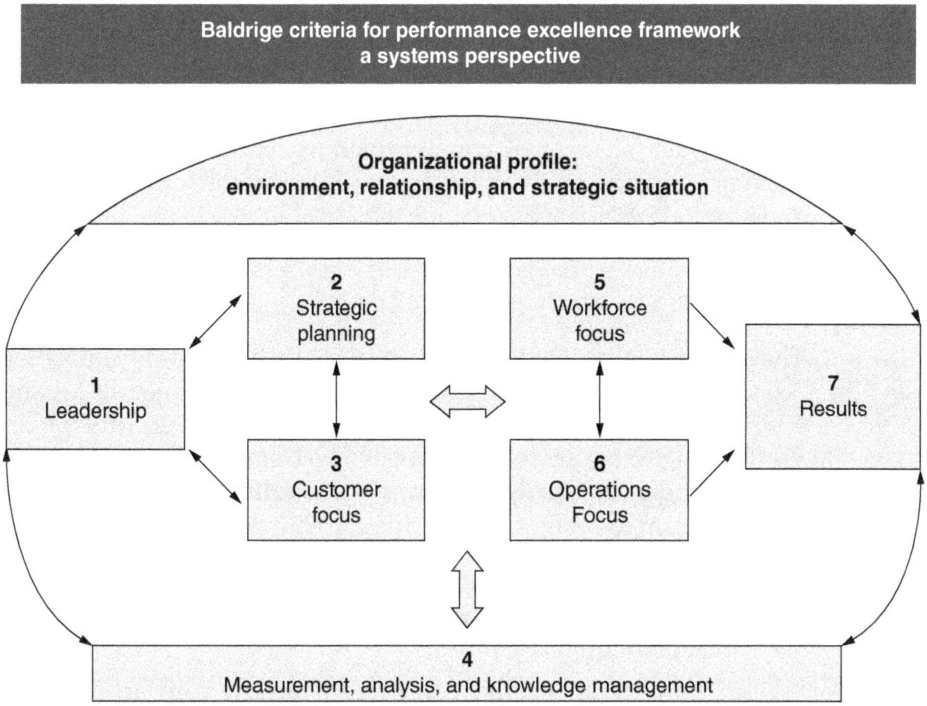

Figure 3.2 The Baldrige model of management.

importance of each item in the Baldrige model. Applicants must address a set of examination items within each of these categories. Heavy emphasis is placed on business excellence and quality achievement as demonstrated through quantitative data furnished by applicants. These criteria are integrated into a model of management, as shown in Figure 3.2.

Each written application is evaluated by members of the board of examiners. High-scoring applicants are selected for site visits by a panel of judges who recommend award recipients to the Secretary of Commerce from among the applicant sites visited. Applicants receive a written feedback summary of strengths and areas for improvement in their quality management. The American Society for Quality (ASQ) assists in the administration of the examination process.

The board of examiners is comprised of quality experts selected from industry, professional and trade organizations, universities, government agencies, education and health care organizations, and from the ranks of the retired. Those selected meet the highest standards of qualification and peer recognition. Examiners must take part in a preparation program based upon the criteria, the scoring system, and the examination process. Each fall applications are solicited from quality experts to serve as examiners for the following year.

The focus of the Baldrige Award is enhancing competitiveness. The award criteria reflect two key competitiveness thrusts: (1) delivery of ever-improving value to customers; and (2) improvement of overall operational performance. The award's central purpose is educational—to encourage sharing of competitiveness learning and to "drive" this learning, creating an evolving body of knowledge, nationally (Reimann and Hertz, 1993).

Deming Prize

The potential impact of the methods of Dr. W. Edwards Deming on Japan's economic success was recognized early on by the Union of Japanese Scientists and Engineers (JUSE). In 1951, JUSE passed a resolution to institute the Deming Prize in recognition of Dr. Deming's contributions to the cause of industrial quality control. The Deming Prize Committee issues awards for three categories (Sheridan, 1993):

- The Deming Prize for Individual Person is awarded to an individual who shows significant achievement in the theory or application of quality control.
- The Deming Application Prize is awarded to an enterprise that achieves the most distinctive improvement of performance through the application of statistical quality control. This award is further broken down into the Deming Application Prize for Small Enterprises and the Deming Application Prize for Division.
- The Deming Application Prize for Overseas Companies is awarded to an overseas company that displays the meritorious implementation of TQM.

Applicants must submit a detailed description of their TQM process. The report is reviewed by the application prize subcommittee consisting of professors and quality experts employed by nonprofit organizations and government entities (not corporations). If the subcommittee believes that the application warrants further review, a series of on-site audits are scheduled. The audits are performed by members of the subcommittee. The areas examined are:

- Policy
- Organizational design
- Education/training
- Information
- Analysis
- Standardization

- Control
- Quality assurance
- Effectiveness
- Future plans

Upon completing the audit, the subcommittee reviews their findings with the Deming Prize committee. The committee's goal is to determine whether the applicant applied the principles advanced by Dr. Deming to maximum benefit in their organization. Score sheets are used only in the preliminary stage. At the committee evaluation stage a discussion takes place regarding how well the Deming principles were used to support the organization.

The use of statistical quality control methods is evaluated in detail. The committee's focus is on the proper use of statistical principles as an aid to management decision making. The use of data and proper statistical analysis in decision making is evaluated at all levels of the organization.

Each year the Deming Prize committee and the application prize subcommittees select those organizations and individuals that display meritorious application of Deming's principles. In November of each year winners are announced in an elaborate ceremony. There are no limits on the number of Deming Prizes that can be awarded in a given year. Thus, there is no "competition" in the Baldrige sense. Instead, organizations are evaluated against the standard set by Deming's principles.

European Quality Award

The European Foundation for Quality Management (EFQM) was founded in 1988 "to enhance the position of European organizations and the effectiveness and efficiency of organizations generally by reinforcing the importance of quality in all aspects of the organization's activities and stimulating and assisting the development of quality improvement." The European Quality Prizes and European Quality Award were created by EFQM in 1991 to recognize and reward organizations that demonstrated a superior commitment to quality. Each year several European Quality Prizes are awarded to organizations that show their implementation of TQM has contributed significantly to their success over a period of years. The best of all the prize winners receives the coveted European Quality Award, recognizing them as "the most successful exponent of Total Quality Management in Europe." Figure 3.3 illustrates the EFQM model.

Applications are assessed and scored on a scale of 0 to 1000 points. According to the EFQM model the enablers on the left produce the results shown on the right. Leadership drives Policy and Strategy, People Management, and the management of Resources, which feed into Processes. Customer Satisfaction, People Satisfaction (employees), and Impact on

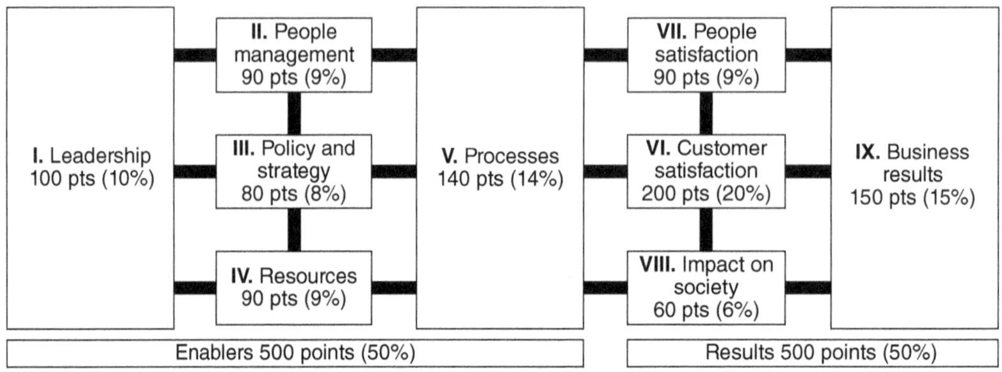

FIGURE 3.3 EFQM application assessment and scoring.

Society are achieved through Leadership. Excellent Business Results are the result of excellent performance in the preceding areas.

A summary of the criteria are provided below (Wendel, 1996).

Enabler Criteria (How *results are being achieved*)

- *Leadership*. How the executive team and all other managers inspire, drive, and reflect total quality (TQ) as the organization's fundamental process for continuous improvement.
- *Policy and strategy*. How the organization's policy and strategy reflect the concept of TQ and how the principles of TQ are used in formulation, deployment, review, and improvement of policy and strategy.
- *People management*. How the organization releases the full potential of its people to continuously improve its business.
- *Resources*. How the organization's resources are effectively deployed in support of policy and strategy.
- *Processes*. How processes are identified, reviewed, and, if necessary, revised to ensure continuous improvement.

Results Criteria (What *the organization has achieved and is achieving*)

- *Customer satisfaction*. What the organization is achieving in relation to the satisfaction of its external customers.
- *People satisfaction*. What the organization is achieving in relation to the satisfac-tion of its people.
- *Impact on society* What the organization is achieving in satisfying the expectations of the community at large. This includes perceptions of the organization's approach to quality of life and the environment.

- *Business results* What the organization is achieving in relation to its planned business objectives and in satisfying the needs and expectations of everyone with a financial interest or stake in the organization.

Total Quality Management (TQM)

The common thread in the evolution of quality management is that attention to quality has moved progressively further up in the organizational hierarchy. Quality was first considered the responsibility of the line worker, then the inspector, then the supervisor, the engineer, the middle manager and, in TQM, upper management.

In many ways, TQM is difficult to encapsulate, primarily because it was never clearly defined industry-wide. For some, it provided a framework for continuous improvement and an abundance of tools; for others, a philosophy of value to society; to others, more of the same experiences of the American post-war quality movement, repackaged under a different name.

Many organizations that implemented TQM were disappointed with the results (*The Economist*, 1992). A survey of 500 American manufacturing and service companies found that only a third felt their total-quality programs were having a "significant impact" on their competitiveness. A similar study in Britain revealed that only one-fifth of the 100 firms surveyed believed that their quality programs had produced any tangible benefits.

In contrast, a General Accounting Office (GAO) survey of 20 of the highest scoring applicants for the 1988 and 1989 Malcolm Baldrige National Quality Award found (Mendelowitz, 1991):

> *Companies that adopted quality management practices experienced an overall improvement in corporate performance. In nearly all cases, companies that used total quality management practices achieved better employee relations, higher productivity, greater customer satisfaction, increased market share, and improved profitability.*

What accounts for the differences in the results? There are a number of factors that seem to be related to success and failure:

Failure is likely if the techniques of TQM are implemented without a commitment to the underlying philosophy. TQM is a customer-focused, process-driven activity, yet many of the firms that experienced failures were not focused on customers and devoted too much resource to studying internal processes, or teaching quality tools to employees who rarely had the opportunity to use them.

TQM is a companywide activity. Those firms that approached TQM by beefing up their quality departments sent the wrong message. Successful TQM disperses the responsibility for quality to those outside of the quality department. It's likely a well-defined functional hierarchal organization structure severely hampered attempts to disperse quality throughout the organization.

TQM takes time. The GAO reports "Many different kinds of companies benefited from putting specific total quality management practices in place. However, none of these companies reaped those benefits immediately. Allowing sufficient time for results to be achieved was as important as initiating a quality management program."

Some TQM advocates suggested the need for a "total quality leader" (TQL) (Kendrick, 1992), who encourages the CEO to be an instrument for change and function as an extension of the CEO as a change agent (see Chap. 12). In this view, the CEO and the TQL must be closely allied, to build credibility for the TQL within the organization.

Six Sigma

Motorola, under the direction of Bob Galvin, developed the principles now known as Six Sigma in the 1980s. In 1981, Motorola set out to improve the quality of their products and services tenfold. This effort led to their acceptance of the 1988 Malcolm Baldrige National Quality Award, and the inception of the Six Sigma Quality movement. In the early years of the program, between 1983 and 1987, Motorola estimated they spent $70 million on quality-related employee education. (www.quality.nist.gov/winners/motorola.htm) Although this certainly represents a steep commitment, their benefits have soundly outweighed these costs (http:/mu.Motorola.com/Six Sigma/SixSigma.html):

- Productivity increased an average of 12.3 percent per year
- Cost of Quality reduced by more than 84 percent
- 99.7 percent of in-process defects eliminated
- $11 Billion in manufacturing costs saved
- Average annual compounded growth rate of 17 percent in earnings, revenues, and stock prices realized

Larry Bossidy, CEO of Allied Signal, began their Six Sigma program in 1994. In 1998, they achieved cost savings of $500 million directly attributable to their Six Sigma program; in 1999, the cost savings grew to $600 million. The total benefits greatly exceed these savings, as explained in their 1999 Annual Report:

> *... cost savings are only part of the story. Delighting customers and accelerating growth completes the picture. When we are more efficient and improve work flow throughout every function in the company, we provide tremendous added value to our customers—through higher quality solutions that are more competitively priced, delivered on time and invoiced correctly. That makes us a more desirable business partner.*

Allied Signal, which merged with Honeywell in 1999, emphasized cycle time reduction. In one example, two of their plants operating at full capacity could not satisfy customer demand. The Six Sigma methodology was employed to increase the production rate by 30 percent, with little to no additional costs. Mike Bonsignore, Honeywell CEO, pointed to the successful merger and integration of Allied Signal and Honeywell (viewed as a best practice benchmark by merger experts) as yet another example of a process improved by the application of Six Sigma tools.

General Electric, under Jack Welch's leadership, began their Six Sigma journey in the fall of 1995 after learning of Allied Signal's successes from Larry Bossidy, a former GE Vice Chairman and friend of Welch's. Their successes are perhaps the best documented:

- GE reported capacity improvements of 12–18 percent, a rise in operating margin to 16.7 percent, and $750 million in savings. (General Electric 1998 Annual Report to Shareholders)
- GE Plastics Singapore team, starting in July 1996, reduced color variation in plastic products. The team raised quality from 2 sigma to 4.9 sigma over 4 months, saving $400,000 a year for one plant. (Slater, 1999)
- In 1996, their first year of Six Sigma deployment, GE Plastics achieved benefits of $20 million. This is quite impressive given that the first year training costs substantially exceed subsequent year costs.
- A Six Sigma team at GE Capital Mortgage Insurance used Six Sigma methodology to cut defects 96 percent. Claim payments were reduced by $8 million, while borrowers were offered alternatives to foreclosure. (General Electric 1997 Annual Report to Shareholders) Overall, GE Capital reported a 160 percent increase in new transactions.
- GE Aircraft Engines in Canada reduced custom charges, and cut delays at the border by 50 percent, using Six Sigma tools to reduce defects in the paperwork needed when parts are imported into Canada. (General Electric 1997 Annual Report to Shareholders)
- GE Medical Systems developed a new ultrasound technology that allows medical personnel to more clearly diagnose risk factors contributing to stroke. This technology became available two years earlier than otherwise possible, due to GE's Design for Six Sigma deployment.

This list of companies benefiting from Six Sigma deployment is far from complete. Many other companies have implemented Six Sigma techniques, including industry leaders IBM, Bombadier, Asea Brown Boveri, DuPont, Compaq, and Texas Instruments. As with GE, Motorola, and Allied Signal, these companies have benefited from Six Sigma deployment across their operations, in both service and manufacturing applications.

Other examples of service-based deployments include GMAC Mortgage, Citibank, JP Morgan, and Cendant Mortgage.

It should be clear that Six Sigma doesn't cost—it pays. A typical deployment will emphasize Six Sigma training projects that save at least as much money for the company as the cost of the training. Larger organizations will spend several years building the program and training additional team members. With the proper deployment they can expect to reap rewards as they go, so as program maturity is neared the bottom line impacts grow.

A properly deployed Six Sigma program addresses the major issues encountered in TQM (Keller, 2011a):

- *Focus.* TQM often sought widespread adoption of quality techniques across the organization. Six Sigma deployment revolves around projects concentrating on one or more key areas: cost, schedule, and quality. Projects are directly linked to the strategic goals of the organization and approved for deployment by high-ranking sponsors, as documented in a project charter (a contract between the sponsor and the project team). The scope of a project is typically set for completion in a three- to four-month time frame, delivering a minimal annualized return of $100,000. Improvement is achieved one project at a time.
- *Organizational support and infrastructure.* TQM sought to diversify quality into the organization by training the masses, in the expectation they would use quality methods to make local process improvements. Middle management could easily thwart these efforts, usually on the sound premise that they interrupted operations. The Six Sigma deployment provides an infrastructure for success. As noted above, the deployment is led by the executive team, who use Six Sigma projects to further their strategic goals and objectives. Projects are actively championed by mid and upper level leaders in their functional areas to meet the challenges laid down by their divisional leaders (in terms of the strategic goals). Teams are led by Black Belts trained as full-time project leaders in the area of statistical analysis and problem solving, while process personnel are engaged as process experts (and trained in as Green Belts in the basic methods).
- *Methodology.* A standard methodology has been developed for Six Sigma projects: DMAIC, an acronym for Define, Measure, Analyze, Improve, and Control. When new products or services are designed, we can alternatively use the DMADV approach (replacing Improve with Design and Control with Verify), although the techniques are essentially the same. The importance of the methodology is in its structured approach, fundamentally based on

Shewhart's Plan-Do-Study-Act (PDSA). This discipline ensures that Six Sigma projects are clearly defined and implemented; that organizational buy-in is built among the key stakeholder groups; that data-driven decision making is used to analyze and improve the process; and that results are standardized into the daily operations, preventing only partial or short-lived project success. The objectives of each stage of DMAIC are summarized in Table 3.3.

- *Training.* A final key difference is the level and extent of training throughout the organization. A properly structured Deployment

DMAIC Stage	Objective
Define	Project Definition: Define project's scope, goals, and objectives; its team members and sponsors; its schedule and deliverables.
	Top-level Process Definition: Define the stakeholders, inputs and outputs, and broad functions.
	Team Formation: Assemble highly capable team from the key stakeholder groups; Create common understanding of issues and benefits for project.
Measure	Process Definition: Define the process at a detailed level, including decision points and functions.
	Metric Definition: Define metric to reliably establish process estimates.
	Process Baseline: Use the defined metrics to establish the current state of the process, which should verify the assumptions of the Define stage. Determine whether the process is in statistical control.
	Measurement Systems Analysis: Quantify errors associated with the metric.
Analyze	Value Stream Analysis: Determine value-producing activities.
	Analyze sources of process variation.
	Determine process drivers.
Improve/Design	Propose one or more solutions to sponsor; Quantify benefits of each; Reach consensus on solution.
	Investigate and address failure modes for new process/design; Define new operating/design conditions.
	Implement and verify new process/design.
Control/Verify	Standardize new procedures/product design elements.
	Continually verify project deliverables.
	Document lessons learned.

From Keller (2011a).

Table 3.3 DMAIC/DMADV Stage Objectives

> starts at the top, with training of key management as Six Sigma Champions. At the executive-level, they steer the program to achieve strategic objectives. At operational levels, they allocate resources to project teams, providing the authority, resources, and the far-reaching appreciation of business needs necessary for project success. Once Champions have been trained, and project selection criteria has been established, Black Belts are trained just in time in the application of DMAIC, including change management skills, problem solving, statistical and lean principles, and methods. Green Belts are selected from critical process areas, and trained to serve as process experts on specific process improvement projects.

The ultimate goal is data-driven decision making at all levels of the organization, focused on benefits to their three stakeholder groups: customers, shareholders, and employees.

CHAPTER 4

Customer-Focused Organizations

The importance of the quality function within the organization has been evolving along with that of the customer. Figure 4.1 illustrates the evolution of the quality function's role since the mid-1970s.

Edosomwan (1993) defines a customer- and market-driven enterprise as one that is committed to providing excellent quality and competitive products and services to satisfy the needs and wants of a well-defined market segment. This approach is in contrast to that of the traditional organization, as shown in Table 4.1.

Customer-driven organizations share certain traits.

Flattened hierarchies. When customers are the focus, a larger percentage of the resources are directly or indirectly involved with customers (see Figure 4.2), reducing the number of bureaucratic layers in the organization structure. Employees will be empowered to make decisions that immediately address customer issues, reducing the need for structured oversight. The traditional functional hierarchy, with departments focused on singular functions, is best replaced with horizontal process or product-based structures that can quickly respond to customer need.

Adaptable processes. Customers' demands are at times unpredictable, requiring adaptability and potential risk. Customer-driven organizations create adaptable systems that remove bureaucratic impediments such as formal approval mechanisms or excessive dependence on written procedures. Employees are encouraged to act on their own best judgments. If the organization's employees are unionized, the changing roles will require union partnering in the transformation process. Union representatives should be involved in all phases of the transformation, including planning and strategy development.

Effective communication. During the transformation the primary task of the leadership team is the clear, consistent, and unambiguous marketing of their vision to the organization.

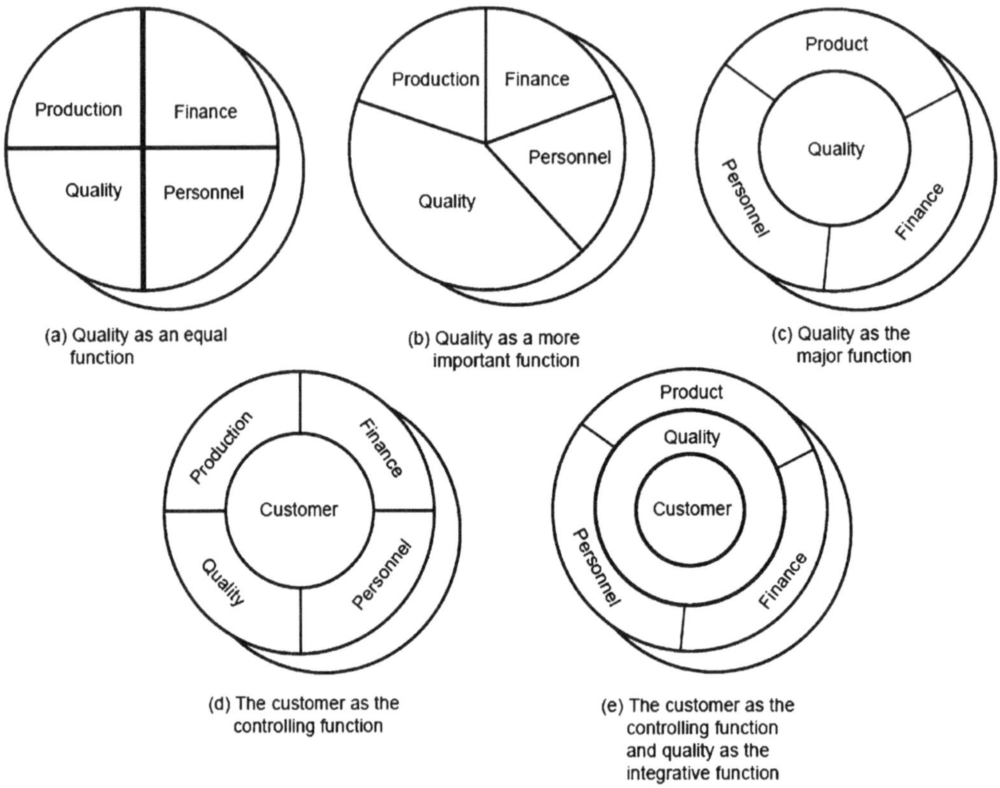

FIGURE 4.1 Evolving views of quality's role in the company (Kotler, 1991 by permission).

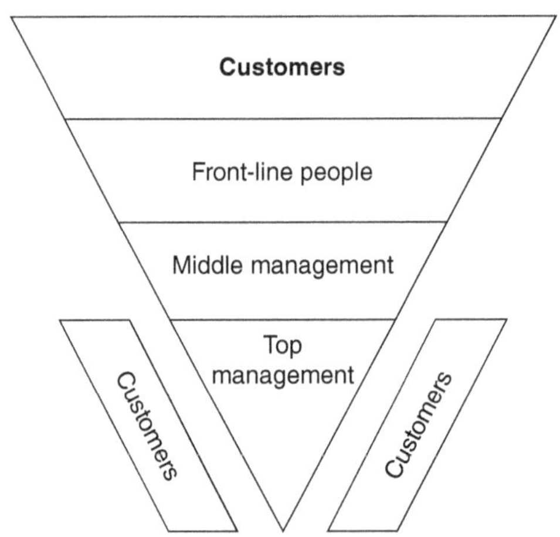

FIGURE 4.2 The customer-focused organization chart (Kotler, 1991 by permission).

	Traditional Organizations	Customer-Driven Organizations
Product and service planning	–Short-term focus –Reactionary management –Management by objectives planning process	–Long-term focus –Prevention-based management –Customer-driven strategic
Measures of performance	–Bottom-line financial results –Quick return on investment	–Customer satisfaction –Market share –Long-term profitability –Quality orientation –Total productivity
Attitudes toward customers	–Customers are irrational and a pain –Customers are a bottleneck to profitability –Hostile and careless –"Take it or leave it" attitude	–Voice of the customer is important –Professional treatment and attention to customers are required –Courteous and responsive –Empathy and respectful attitude
Quality of products and services	–Provided according to organizational requirements	–Provided according to customer requirements and needs
Marketing focus	–Seller's market –Careless about lost customers through customer satisfaction	–Increased market share and financial growth achieved
Process management approach	–Focus on error and defect detection	–Focus on error and defect prevention
Product and service delivery attitude	–It is OK for customers to wait for products and services	–It is best to provide fast time-to-market products and services
People orientation	–People are the source of problems and are burdens on the organization	–People are an organization's greatest resource
Basis for decision making	–Product-driven –Management by opinion	–Customer-driven –Management by data
Improvement strategy	–Crisis management –Management by fear and intimidation	–Continuous process improvement –Total process management
Mode of operation	–Career-driven and independent work –Customers, suppliers, and process owners have nothing in common	–Management-supported improvement –Teamwork between suppliers, process owners, and customers practiced

From Edosomwan (1993) by permission.

TABLE 4.1 Traditional Organizations versus Customer-Driven Organizations

The behavior of senior leaders carries tremendous symbolic meaning, which can quickly undermine the targeted message and destroy all credibility. Conversely, behavior that clearly demonstrates commitment to the vision can help spread the word that "They're serious this time." Leaders should expect to devote a minimum of 50 percent of their time to communication during the transition.

Measuring results. It is important to verify that you are delivering on promises to customers, shareholders, and employees. These measurements form the basis of the improvement efforts, and should include internal processes as well as external outcomes. Data must be available quickly to the people who use them and be easy to understand.

Rewarding employees. Employees should be treated like partners in the improvement effort and provided adequate and fair compensation for doing their jobs. Rewarding individuals with financial incentives can be manipulative, implying that the employee wouldn't do the job without the reward, which tends to destroy the very behavior you seek to encourage (Kohn, 1993). Recognizing exceptional performance or effort should be done in a way that encourages cooperation and team spirit, such as parties and public expressions of appreciation. Leaders should assure fairness: for example, management bonuses and worker pay cuts don't mix. Financial incentives should be fairly distributed throughout the organization, since most improvements are achieved due to the collective actions of the organization, rather than just a few people.

For too many organizations, the journey from a traditional to a customer-driven organization begins with recognition that a crisis is either upon the organization, or imminent. This wrenches the organization's leadership out of denial and forces them to abandon the status-quo. Their actions at this point define their success. The successful organization will establish a customer-focused vision, and develop plans to attain the vision, as outlined in Part II.

The common thread in the evolution of quality management is that attention to quality has moved progressively further up in the organizational hierarchy. Quality was first considered a matter for the line worker, then the inspector, then the supervisor, the engineer, the middle manager and, today, for upper management. Quality will continue to increase in importance, in tandem with customer relations. Ultimately, it is the customer's concern with quality that has been the driving force behind quality's increasing role in the organization. As Juran (1994) stated, the next century will be the century of quality.

PART II

Integrated Planning

Quality planning is the activity of developing the products and processes required to meet customers' needs. It involves a number of universal steps (Juran and DeFeo, 2010):

1. Define the customers.
2. Determine the customer needs.
3. Develop product and service features to meet customer needs.
4. Develop processes to deliver the product and service features.
5. Transfer the resulting plans to operational personnel.

As Juran intended and experience has shown, the term *universal* implies the activities are applied across any organization at various levels. The discussions in the following chapters are directed primarily at business-level planning to achieve profitability and organizational success through customer focus. The concepts will be similarly applied at the process and product levels in Part IV to develop and improve customer-focused products and services.

CHAPTER 5

Strategic Planning

Despite the inevitability of the future, it cannot be predicted. Nonetheless, long-range planning has valuable benefits, providing opportunity for managers to critically question (1) whether the effects of present trends can be extended into the future, (2) assumptions that today's products, services, markets, and technologies will be the products, services, markets, and technologies of tomorrow, and (3) perhaps most important, the usefulness of devoting their energies and resources to the defense of yesterday (Drucker, 1974).

Traditional strategic planning starts by answering two simple questions: "What *is* our business?" and "What *should* it be?"

Strategic planning is not forecasting, which Drucker (1974) pointedly noted: "is not a respectable human activity and not worthwhile beyond the shortest of periods." Strategic planning is necessary precisely because we cannot forecast the future. It deliberately seeks to upset the probabilities by innovations and organizational change.

Strategic planning is the continuous process of making present entrepreneurial decisions systematically and with the greatest knowledge of their futurity, organizing systematically the efforts needed to carry out these decisions, and measuring the results of these decisions against the expectations through organized, systematic feedback (Drucker, 1974).

Organizational Vision

The answer to these questions leads an organization to develop *value* and *mission* statements to explain the organization's broad (or sometimes quite specific) goals. The successful organization will outlive the people who are currently its members. Thus, the mission of the successful organization must provide vision for the long term, describing why the organization exists. No organization exists merely to "make a profit." Profits accrue to organizations that produce value in excess of their costs; that is, profits are an effect of productive existence, not a cause. Consider these examples:

1. Matsushita electric industrial company is one of the world's largest firms. Its stated mission is to eliminate poverty in the world by

making their products available to the people of the world at the lowest possible cost (Mintzberg and Quinn, 1991).

2. Henry Ford's mission was to provide low-cost transportation to the common man.

One might go a step further and ask why the organization was created to fulfill its mission. The answer, at least in the beginning, might lie in the values of the organization's founder. Henry Ford, for whatever reason, felt that it was important (i.e., valued) to provide the farmer with affordable and reliable motorized transportation. Furthermore, to elicit the cooperation of the members of the organization, the values of the organization must be compatible with the values of its members.

Organizational leaders are responsible for defining the organization's vision. Defining the vision requires developing a mental image of the organization at a future time. The future organization will more closely approximate the ideal organization, where "ideal" is defined as that organization which completely achieves the organization's values. How will such an organization "look"? What will its employees do? Who will be its customers? How will it behave toward its customers, employees, and suppliers? Developing a lucid image of this organization will help the leader see how she should proceed with her primary duty of transforming the present organization. Without such an image in her mind, the executive will lead the organization through a maze with a thousand dead ends. Conversely, with her vision to guide her, the transformation process will proceed on course. This is not to say that the transformation is ever "easy." But when there is a leader with a vision, it's as if the organization is following an expert scout through hostile territory. The path is clear, but the journey is still difficult.

When an individual has a vision of where he wants to go himself, he can pursue this vision directly. However, when dealing with an organization, simply having a clear vision is not enough. The leader must *communicate* the vision to the other members of the organization. Communicating a vision is a much different task than communicating instructions or concrete ideas.

Organizational visions that embody abstract values are necessarily abstract in nature. To effectively convey the vision to others, the leader must convert the abstractions to concretes. One way to do this is by living the vision. The leader demonstrates her values in every action she takes, every decision she makes, which meetings she attends or ignores, when she pays rapt attention and when she doodles absentmindedly on her notepad. Employees who are trying to understand the leader's vision will pay close attention to the behavior of the leader.

Another way to communicate abstract ideas is to tell stories. In organizations there is a constant flow of events. Customers encounter the organization through employees and systems, suppliers meet with engineers, literally thousands of events take place every day. From time to time an event occurs that captures the essence of the leader's vision. A clerk provides exceptional customer service, an engineer takes a risk and makes a mistake, a supplier keeps the line running through a mighty effort. These are concrete examples of what the leader wants the future organization to become. She should repeat these stories to others and publicly recognize the people who made the stories. She should also create stories of her own, even if it requires staging an event. There is nothing dishonest about creating a situation with powerful symbolic meaning and using it to communicate a vision. For example, Nordstrom has a story about a sales clerk who accepted a customer return of a defective tire. This story has tremendous symbolic meaning because Nordstrom doesn't sell tires! The story illustrates Nordstrom's policy of allowing employees to use their own best judgment in all situations, even if they make "mistakes," and of going the extra mile to satisfy customers. However, it is doubtful that the event ever occurred. This is irrelevant. When employees hear this story during their orientation training, the message is clear. The story serves its purpose of clearly communicating an otherwise confusing abstraction.

Strategy Development

After specifying the objectives of the business via the vision and mission statements, the next activity in strategic planning is to develop the strategies (i.e., the plan) to achieve these objectives. Yet planning must go beyond simply coming up with new things the business can do in the future. We must also ask of each present activity, product, process, or market, "If we weren't already doing this, would we start?" If the answer is "No," then the organization should develop plans to stop doing it, ASAP.

The planning aims to make organizational changes: changes in the way people work, changes in the systems to meet customers' future needs. The plan shows how to allocate scarce resources and builds accountability into the plan. Deadlines are necessary, as is feedback on progress and measurement of the final result.

A traditional basis of strategy formation is the comparison of internal Strengths and Weaknesses to external Opportunities and Threats (SWOT). As shown in Fig. 5.1, strategy is created at the intersection of an external appraisal of the threats and opportunities facing an organization in its environment, considered in terms of key factors for success, and an internal appraisal of the strengths and weaknesses of the organization itself, distilled into a set of distinctive competencies. Outside opportunities are exploited by inside strengths, while threats are avoided (or addressed) and weaknesses circumvented (or addressed). Opportunities and threats are identified by understanding customers and their markets. Internal strengths and weaknesses are evaluated through rigorous organizational assessments. (Each of these is discussed in detail in the chapters that follow.) Taken into consideration, both in the creation of the strategies and in their

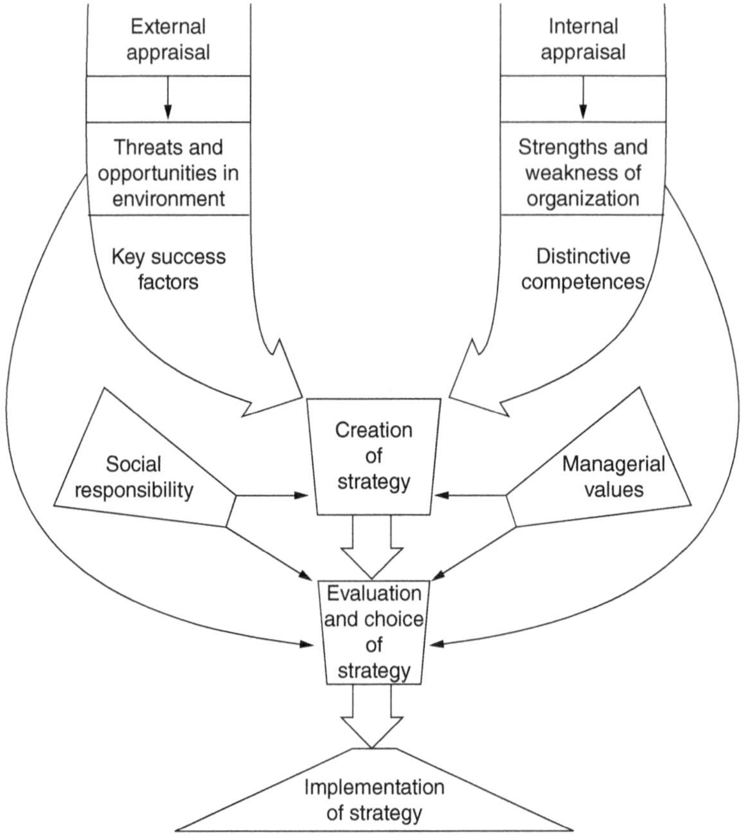

FIGURE 5.1 SWOT model of strategy formulation (Mintzberg, 1994).

subsequent evaluation, are the values of the leadership as well as its social responsibility.

Once a strategy has been chosen, it is implemented as a cross-functional improvement project using the improvement methods described in Part IV. The structured DMAIC improvement model, an improvement to Shewhart's Plan-Do-Study-Act, effectively builds buy-in among the stakeholder groups as well as the key middle management.

Projects are defined to achieve the vision established by the leadership team, with understanding of the customer needs and expectations (see Chap. 6). Organizational assessments, as well as customer data, provide the basis for the Measure stage of DMAIC. Key drivers of the outcomes are established in the Analyze stage. An improvement strategy is implemented in the Improve stage and continually verified in the Control stage.

Larger strategies are executed primarily at the executive level. The implementation of organization-wide initiatives in the Improve stage may be best broken into smaller projects aligned with a specific substrategy and executed at lower levels of the organization.

Strategic Styles

Strategic plans are necessarily specific to a particular organization, at a particular time in its life cycle. What works for a Fortune 100 firm in a developed market will not likely apply directly to a start-up in a niche market. Reeves et al. (2012) define four broad categories of strategic style:

1. *Classical.* This traditional approach to defining longer-term plans should be limited to organizations in fairly predictable, mature environments. Examples include oil exploration and production, air freight/logistics, beverages, utilities, paper products, and tobacco.
2. *Adaptive.* This flexible approach fosters experimentation to develop strategy within unpredictable environments in which you have little ability to change. Many of today's markets are continually changing due to competition, innovation, and/or economic uncertainty, making some aspects of a strategic plan obsolete or irrelevant after perhaps only a few months. In these environments, firms must become learning organizations, essentially integrating their strategic planning with operations to quickly develop and then track results of each strategic iteration. Examples include biotechnology, communications equipment, specialty retail, and computer hardware.
3. *Shaping.* When the environment is unpredictable, but you have some ability to change the environment, a shaping strategy is recommended. In this case, the organization develops a strategy that seeks to influence the market, such as through innovation. This may disrupt a stagnant market or a fragmented market, but in either case the objective is to redefine the market to the company's advantage. The authors cite Facebook as an example of successful deployment of the shaping strategy. Their ability to overtake an initially dominant competitor in MySpace was complemented by redefining the social media space with applications, such as games. Other examples include software development, airlines, catalog retail, consumer services, and wireless services.
4. *Visionary.* When an organization can both shape the environment and reliably predict the future, they can develop bold and decisive plans to create new markets or redefine an organization. Since the market is predictable, the organization can take the time and commit resources to develop and execute a complete plan. The authors cite examples such as XM satellite radio, UPS as "the enablers of global e-commerce," aerospace and defense, and food products.

Unfortunately, the authors noted that most organizations surveyed were using the visionary (40 percent) or the classical (35 percent) strategic styles, which rely on predictable markets, when assumptions of predictability were clearly unwarranted (Reeves et al., 2012).

Possibilities-Based Strategic Decisions

Clearly, defining and implementing organizational vision involves elements of creativity. Yet, a rigorous, scientific approach is necessary to ensure that a full breadth of options is explored and evaluated. Lafley et al. (2012) define the following seven steps to strategy making, which differ from traditional methods in clearly articulating possibilities:

1. *Frame a choice.* Define the issue using two or more mutually exclusive options. This moves the discussion from investigating issues to evaluating solutions and making decisions. It further ingrains the team with the notion that they have choices. It's often useful to include the status quo as an option, to explore the assumptions necessary in maintaining current practices, which often highlights the need for action.
2. *Generate possibilities.* Creatively brainstorm to develop additional options. At this point, options are not evaluated beyond general plausibility, but sufficient detail is necessary so the team can understand the nature of the option. Practically, the authors recommend three to five options.
3. *Specify conditions.* Define limitations of each option to describe the conditions under which the option would be strategically desirable. Note that this is not the time to argue merits of any option, or whether these necessary conditions exist now or could exist in the future. Instead, it is an opportunity to define the issues that would have to be evaluated in order to make the option attractive to the team. When conditions have been completely defined for an option, the team should be in agreement that, if all conditions were met, the option would be acceptable. If a given condition is desirable, but not necessary, it should be removed. The ultimate goal at this step is to understand the limitations of every option before analysis begins.
4. *Identify barriers.* Determine which conditions are least probable. The focus in this step is to identify the conditions that are most troublesome to the team members: Which conditions would you be most likely to be concerned about attaining?
5. *Design tests.* For each key barrier condition, construct tests that, when implemented, would convince the team that the conditions can be met. What is the standard of proof required for the team to

assert, with an acceptable degree of confidence, that the condition can be met? This may involve surveys of customers or suppliers, or honest discussions with key customers or suppliers, depending on the team's confidence in the condition. The authors suggest that the team members with greatest skepticism for a given option be tasked with developing the tests for that option.

6. *Conduct the tests.* Implement the tests defined in the preceding step. The authors recommend starting with the barrier condition that the team has the least confidence in, on the premise that the condition can be quickly dismissed without additional testing. They refer to this as the "lazy man's approach" and cite it as a key driver for reducing costs and effort of the analysis.
7. *Make the choice.* The option with the fewest key barriers will naturally surface as the preferred option by the team.

Lafley, a former chairman and CEO of Proctor & Gamble, cites the use of the possibilities-based approach to strategic decision-making in P&G's transformation of the Olay product line to a premium "prestige" brand, eventually accounting for $2.5 billion in annual sales. When considering the option of marketing a prestige product to younger clientele through mass-market channels, the following necessary conditions were defined (Lafley et al., 2012):

- *Industry segmentation.* A sufficiently large number of woman want to "fight the seven signs of aging."
- *Industry structure.* The emerging masstige (i.e., a prestige product in mass market) segment will be at least as structurally attractive as the current mass-market segment.
- *Channel.* Mass retailers will embrace the idea of creating a masstige experience to attract prestige customers.
- *Consumers.* A pricing sweet spot exists that will induce mass consumers to pay a premium and prestige shoppers to purchase in the mass channel.
- *Business model capabilities.* P&G can create prestige-like brand positioning, packaging, and in-store promotions in the mass channel; and P&G can build strong partnerships with mass retailers to create and exploit a masstige segment.
- *Costs.* P&G can create a prestige-like product with a cost structure that enables it to hit the pricing sweet spot.
- *Competitors.* Because of channel conflict, prestige competitors will not try to follow Olay into the masstige segment; and mass competitors will find it hard to follow because the lower price point is covered by the basic Olay Complete line.

In this example, Lafley cites the pricing sweet spot as the most challenging issue. Analysis focused on evaluating the price points at which consumers perceived prestige but were willing to purchase in the mass market. This particular challenge, and the analysis resulting from the possibility-based conditions, proved the most influential to ensuring success of the strategy.

The authors note the key differences in the possibilities-based approach compared to more traditional strategic planning:

1. Rather than asking "What should we do?" ask "What might we do?" Whereas the former leads to hasty decisions, the latter fosters introspective thought, which can then be scrutinized.
2. The specification of conditions leads the team to consider the assumptions necessary for the option to be desirable. This forces the team to imagine possibilities, rather than deal directly with the perceived limitations.
3. The team is essentially tasked with focusing on defining the right questions that lead them to the best decision, rather than trying to jump to the best solution quickly. The focus on inquiry is a fundamental aspect of the scientific approach to problem-solving.

Strategic Development Using Constraint Theory*

More than ever before, operational leaders are finding themselves in need of system-level tools to sustain the business success they've fought so hard to achieve. One such system-level tool is constraint management. Constraint management acknowledges that quality is but one important element in the business equation. Constraint management seeks to help managers at all levels of an organization maintain proper focus on the factors that are most critical to overall success: *system constraints*. In some systems, these might be quality related. In other systems, they may extend well beyond the traditional territory of quality.

There are many types of constraints. Some are not *physical* (e.g., lack of space, not enough resources, etc.). In many cases they derive from *policies*: the laws, regulations, rules, or procedures that determine what we can or can't do. Who hasn't heard it said, "That's the way we do things around here"? Or, alternatively, "That's *not* the way we do things around here." What you're hearing is the verbalization of a policy, possibly unwritten, but accepted as traditional practice nonetheless. When a policy of any kind inhibits what we need (or want) to accomplish, it, too, constitutes a constraint.

---

*Thanks to H. William Dettmer from Goal Systems International for writing this section.

The practice of constraint management tacitly recognizes that constraints limit what we can do in any circumstance, and it provides the vehicle to understand why this happens and what can be done about the constraints we face.

Constraint management is an outgrowth of the Theory of Constraints (TOC), a set of principles and concepts introduced by Eliyahu M. Goldratt, an Israeli physicist, in the 1980s in a book titled *The Goal* (Goldratt, 1986). These principles and concepts are a blend of both existing and new ideas. The new ideas build upon older ones to produce a robust, holistic approach to understanding and managing complex systems. To extend the theoretical principles and concepts into application, Goldratt developed three classes of tools, which will be described in more detail later. For now, the important point to remember is that TOC, and constraint management as a whole, constitutes a *systems management* methodology.

The Systems Approach

What do we mean by *systems management*? Throughout the twentieth century, management thought was largely activity oriented. In the early 1900s, Frederick Taylor's scientific management (Taylor, 1947) focused on dividing and subdividing work into discrete tasks or activities that could be closely monitored, measured, and "tweaked" to produce the most efficient performance from each activity. By the second half of the century, the focus had enlarged somewhat to encompass managing processes composed of several activities. At some level, these processes could become quite large and complex, such as a production process, a purchasing process, or a marketing and sales process. One way of dealing with complexity is to compartmentalize it—to cut it up into "manageable bites." Organizations typically do that by creating functional departments. Each department is responsible for some function that constitutes a part of the whole system. One could even say that these "parts" are actually individual processes. This is an orderly way to come to grips with the issue of complexity. Throughout the 1980s and early 1990s, the meteoric rise of the quality movement reinforced the idea that success lay in continuous refinement of processes. The ultimate objective became "Six Sigma," a level of defect-free performance unheard of 20 years before. Unquestionably, both commercial and noncommercial organizations needed this focus. Poor product quality (which is usually a result of faulty process quality) can bring down an organization faster than just about anything else. But many companies, despite herculean efforts and the expenditure of significant amounts of money, were disappointed to find that their payback wasn't what they expected it to be. The idea that "if you build a better mousetrap, the world will beat a path to your door" worked exceptionally well for the companies whose overriding constraint had been product quality. Yet for other companies, the strategy seemed to be somewhat underwhelming.

Despite the admonition to "consider your internal customer," many departments still behave as if they're in a "silo" by themselves. They pay lip service to the idea, but for a variety of reasons they don't practice it very well. Their focus remains inward, on individual measures of performance and efficiency. Most efforts are spent improving the links of the supply chain, with little effort devoted to the linkages, or interfaces between links, and the operation of the chain as a whole.

Systems Thinking

What these companies failed to appreciate was that a higher level of thinking was needed: *systems thinking*. Once the quality of individual processes is put reasonably well into line, other factors emerge to warrant attention. Consider the analogy of a football team.

Major professional sports spend a lot of time and money on process improvement, even though they probably don't look at it that way. A team owner can spend millions on a contract for a star quarterback. By applying natural talent, they expect to "improve the passing process." But in many cases, the touchdowns don't appear, despite the huge sums spent on star quarterbacks. At some point in the "process failure mode effects analysis," the coaches discover it's impossible for this highly valued quarterback to complete passes from flat on his back. They find that the offensive line needs shoring up. Or a good blocking back is needed, or a better game plan, or any number of other factors.

The point is that any organization, like a football team, succeeds or fails as a complete system, not as a collection of isolated, independent parts or processes. In the same way that a motion picture clip tells us much more about a situation than an instantaneous snapshot, systems thinking gives us a clearer picture of the whole organizational dynamic. In *The Fifth Discipline* (Senge, 1990), Peter Senge proposes that the only sustainable competitive advantage comes from transforming a company into a "learning organization." The keys to doing this, Senge maintains, are five basic disciplines that every organization striving for success must master: systems thinking, personal mastery, mental models, building a shared vision, and team learning. Guess which one he considers the most important. Though he numbers it fifth, he lists it first, and he titled his book after it.

System Optimization versus Process Improvement

If one "thinks system," the question inevitably arises: What do we do with process improvement? Do we ignore it now that we're thinking at a higher level? No, process improvement is *still* important. It constitutes the building blocks upon which system performance is based. But like the football team alluded to above, once you have a "star performer" at every position, you have a challenge of a different sort: coordinating and synchronizing the efforts of every component in the system to produce the best *system* result. In other words, once the ducks are in line, the task is to make them march in step together. We refer to this as *system optimization*.

How important is it to optimize the system, rather than its component parts? Deming himself answered that question in *The New Economics for Industry, Government, Education* (Deming, 1993, pp. 53, 100). He observed:

> *Optimization* is *the process of orchestrating the efforts of all components toward achievement of the stated aim. Optimization is management's job. Everybody wins with optimization. Anything less than optimization of the system will bring eventual loss to every component in the system. Any group should have as its aim optimization of the larger system that the group operates in. The obligation of any component is to contribute its best to the system, not to maximize its own production, profit, or sales, nor any other competitive measure. Some components may operate at a loss themselves in order to optimize the whole system, including the components that take a loss.*

This is a powerful indictment of the way most companies have been doing business since Frederick Taylor's time, not excluding the "quality enlightenment" era of the 1980s and 1990s. In essence, Deming said that maximizing local efficiencies everywhere in a system is not necessarily a good thing to do.

Systems as Chains

To express the concept of system constraints more simply, Goldratt has equated systems to chains (Goldratt, 1990, p. 53):

> *We are dealing here with "chains" of actions. What determines the performance of a chain? The strength of the chain is determined by the strength of its weakest link. How many weakest links exist in a chain? As long as statistical fluctuations prevent the links from being totally identical, there is only one weakest link in a chain.*

Goldratt goes on to suggest that there are as many constraints in a system as there are truly independent chains. Realistically, in most systems there aren't very many truly independent chains. The dictionary (Barnes and Noble, 1989) defines *system* as:

> *an assemblage or combination of things or parts forming a complex or unitary whole; the structure or organization, society, business ...*

Thomas H. Athey defines a system as any set of components that could be seen as working together for the overall objective of the whole (Athey, 1982, p. 12). The underlying theme in these definitions is an interrelatedness or interdependency. By definition, then, a "system" can't have too many truly independent chains. So if there aren't too many independent chains in a particular system—whether a manufacturing, service, or government system—at any given time, only a very few variables truly determine the performance of the system.

This idea has profound implications for managers. If only a very few variables determine system performance, the complexity of managers' jobs can be dramatically reduced. Look at it in terms of the Pareto rule, which suggests that only 20 percent of a system accounts for 80 percent of the problems within it. If this is a valid conclusion, managers should be able to concentrate most of their attention on that critical 20 percent. Goldratt's concept of chains and "weakest links" takes the Pareto concept a step further: the weakest link accounts for 99 percent of the success or failure of a system to progress toward its goal (Goldratt, 1990, p. 53).

Basic Constraint Management Principles and Concepts

Constraint management exhibits the theoretical foundation that Deming considered so important to effective management action (Deming, 1986). Theories can be either *descriptive* or *prescriptive*. A descriptive theory generally tells only why things are the way they are. It doesn't provide any guidance for what to *do* about the information it provides. An example of descriptive theory might be Newton's laws of gravitation, or Einstein's theories of relativity. Prescriptive theory describes, too, but it also guides through prescribed actions. Most management theories are prescriptive. The Deming philosophy prescribes through its 14 points (Deming, 1986). Ken Blanchard's *One-Minute Manager* (Blanchard, 1982) and Peter Senge's *The Fifth Discipline* (Senge, 1990) explain prescriptive theories in detail. Constraint management is prescriptive as well. It provides a common definition of a constraint, four basic underlying assumptions, and five focusing steps to guide management action.

Definition of a Constraint

Simply put, a constraint is *anything that limits a system* (company or agency) *in reaching its goal* (Goldratt, 1990, pp. 56–57). This is a very broad definition, because it encompasses a wide variety of possible constraining elements. Constraints could be physical (equipment, facilities, material, people), or they could be policies (laws, regulations, or the way we choose to do business—or choose *not* to do business). Frequently, policies cause physical constraints to appear.

Types of Constraints

Identifying and breaking constraints becomes a little easier if there is an orderly way of classifying them. From the preceding discussion, we know that system constraints can be considered either physical or policy. Within those two broad categories, there are seven basic types (Schragenheim and Dettmer, 2000, Chap. 4):

- *Market.* Not enough demand for a product or service.
- *Resource.* Not enough people, equipment, or facilities to satisfy the demand for products or services.
- *Material.* Inability to obtain required materials in the quantity or quality needed to satisfy the demand for products or services.
- *Supplier/vendor.* Unreliability (inconsistency) of a supplier or vendor, or excessive lead time in responding to orders.
- *Financial.* Insufficient cash flow to sustain an operation. For example, a company that can't produce more until payment has been received for work previously completed, because they might need that revenue to purchase materials for a firm order that's waiting.
- *Knowledge/competence.* Knowledge: Information or knowledge to improve business performance is not resident within the system or organization. Competence: People don't have the skills (or skill levels) necessary to perform at higher levels required to remain competitive.
- *Policy.* Any law, regulation, rule, or business practice that inhibits progress toward the system's goal.

Note: In most cases, a policy is most likely behind a constraint from any of the first six categories. For this reason, the Theory of Constraints assigns a very high importance to policy analysis, which will be discussed in more detail under "The Logical Thinking Process," below.

Not all of these types apply to all systems. Material and supplier/vendor constraints might not apply to service organizations. Market constraints are generally not relevant in not-for-profit systems, such as government agencies. But resource, financial, knowledge/competence, and policy constraints can potentially affect all types of organizations.

Four Underlying Assumptions

Constraint management is based on four assumptions about how systems function (Schragenheim and Dettmer, 2000, Chap. 2). These assumptions are:

1. Every system has a goal and a finite set of necessary conditions that must be satisfied to achieve that goal. Effective effort to improve system performance is not possible without a clear understanding and consensus about what the goal and necessary conditions are.
2. The sum of a system's local optima does not equal the global system optimum. In other words, the most effective system does not come from maximizing the efficiency of each system component individually, without regard to its interaction with other components.

3. Very few variables—maybe only one—limit the performance of a system at any one time. This is equivalent to the "weakest link" concept discussed earlier.
4. All systems are subject to logical cause and effect. There are natural and logical consequences to any action, decision, or event. For those events that have already occurred, these consequences can be visually mapped to aid in situation or problem analysis. For those decisions that have yet to occur, or which are contemplated, the outcomes of these actions, decisions, or events can be logically projected into the future and visually mapped as well.

All of the description and prescription contained in constraint management are predicated on these assumptions.

Goal and Necessary Conditions

The first assumption above holds that every system has a goal and a set of necessary conditions that must be satisfied to achieve that goal (Schragenheim and Dettmer, 2000, Chap. 2). The philosopher Friedrich Nietzsche once observed that by losing your goal, you have lost your way. Or another way of putting it: *if you don't know what the destination is, then any path will do.*

While this assumption is undoubtedly valid in most cases, there are obviously some organizations that have not expended the time or effort to clearly and unequivocally define what their goal is. And even if they have defined a goal, most have not gone the extra step to define the minimum necessary conditions, or critical success factors, for achieving that goal.

For example, most for-profit companies have something financial as their goal. Goldratt contends that the goal of for-profit companies is to "make more money, now and in the future" (Goldratt, 1990, p. 12). Another way of saying this is *profitability*. This, of course, would not be an appropriate goal for a government agency, such as the Department of Defense or Department of Education. Nonfinancial goals would have to be developed for such agencies. But it works quite well for most companies engaged in commercial business.

However, having profitability as a goal isn't enough. For any organization to be profitable, and for those profits to consistently increase, there is a discrete set of necessary conditions it must satisfy. Some of these will be unique to the industry that the company is in, others will be generic to all for-profit companies. But one thing that all organizations will have in common: there will be very few of these necessary conditions, maybe fewer than five.

Necessary conditions are critical success factors. They are actually required to achieve the goal. For instance, customer satisfaction is unquestionably essential to continued progress toward a financial goal. Employee satisfaction might be considered necessary for achieving the goal. Including necessary conditions like these as part of the goal hierarchy gives it credibility, identifying it as something that is not just temporary but must be satisfied throughout the lifetime of the organization. Figure 5.2 illustrates a typical goal/necessary condition hierarchy.

Necessary conditions differ from the goal. While the goal itself usually has no limit (it's normally worded in such a way that it's not likely ever to be fully realized), necessary conditions are more finite. They might be characterized as a "zero-or-one" situation: it's either there or it isn't (a "yes-or-no" state). For example, a for-profit organization might

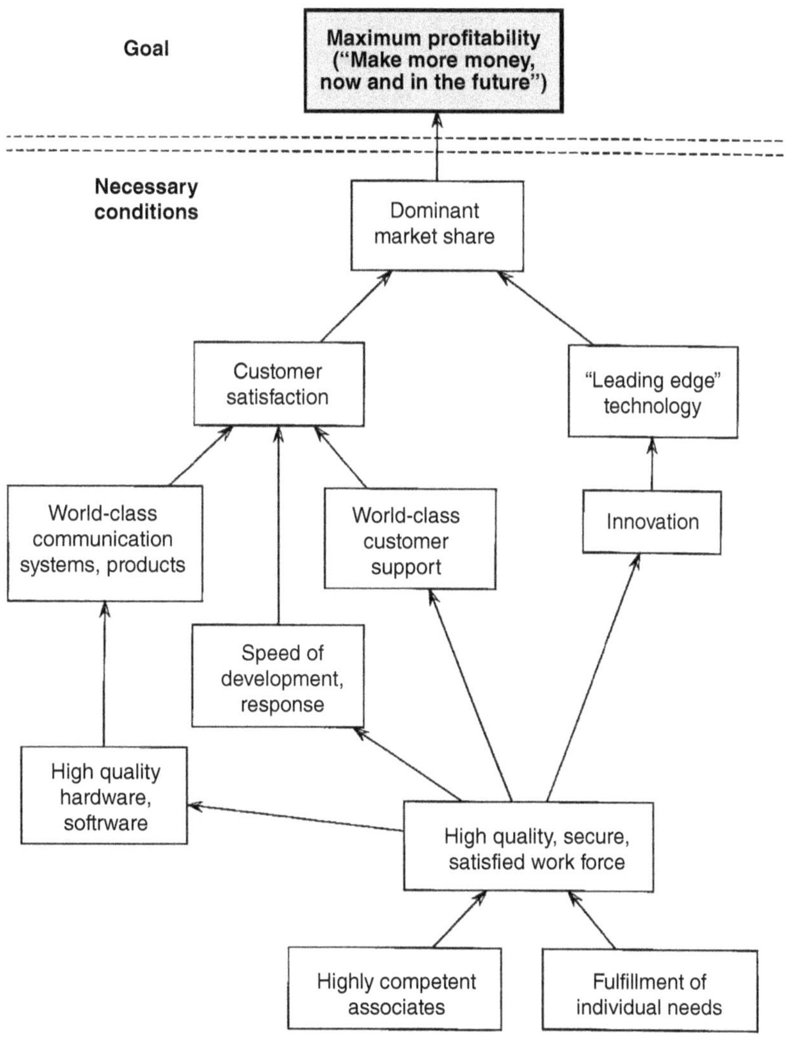

Figure 5.2 Hierarchy of goal and necessary conditions.

want to make as much money as it can—no limits. But employee security and satisfaction, as a necessary conditions, should be established at a well-defined minimum level. A for-profit company's goal can't expect to satisfy its employees without limit, but the organization should recognize the need to achieve a certain level of employee security and satisfaction as one minimum requirement for achieving the goal. This is not to say that all necessary conditions are zero-or-one in nature. Some, such as customer satisfaction, can be increased, and doing so can be expected to improve progress toward the goal. But even such variable necessary conditions have practical limits.

The importance of identifying a system's (organization's) goal and necessary conditions is that they become the standard by which all results are judged and all contemplated decisions are evaluated. Did yesterday's, or last month's, actions better satisfy a necessary condition or contribute to realizing the goal? If so, the organization knows that it's making progress in the right direction. Can we expect the decisions contemplated today, or next week, to advance the company toward its goal, or to satisfy a necessary condition? If so, the decision is a good one from the system perspective.

The Five Focusing Steps

Once the necessary conditions are established, constraint theory prescribes applying five "focusing steps" in order to continuously proceed toward satisfying those necessary conditions (Goldratt, 1986, p. 307).

Goldratt created the five focusing steps as a way of making sure management "keeps its eye on the ball"—what's really important to success: the system constraint. In one respect, these steps are similar to the Shewhart Cycle: Plan-Do-Check/Study-Act (Deming, 1986). They constitute a continuous cycle. You don't stop after just one "rotation." The five focusing steps (Schragenheim and Dettmer, 2000, Chap. 2) are:

1. *Identify.* The first step is to *identify the system's constraint.* What limits system performance now? Is it inside the system (a resource or policy) or is it outside (the market, material supply, a vendor … or another policy)? Once the system constraint is identified, if it can be broken without much investment, immediately do so, and revert to the first step again. If it can't be easily broken, proceed to the second step.
2. *Exploit. Decide how to exploit the system's constraint. "Exploit"* means to "get the most" out of the constraining element without additional investment. In other words, change the way you operate so that the maximum financial benefit is achieved from the constraining element. For example, if the system constraint is market demand (not enough sales), it means catering to the market so as to win more sales. On the other hand, if the constraint is an

internal resource, it means using that resource in the best way to maximize its marginal contribution to profit. This might mean process quality improvement, re-engineering the flow of work through the process, or changing the product mix. Exploitation of the constraint should be the kernel of tactical planning—ensuring the best performance the system can draw now. For this reason, the responsibility for exploitation lies with the line managers who must provide that plan and communicate it, so that everyone else understands the exploitation scheme for the immediate future.

3. *Subordinate.* Once the decision on how to exploit the constraint has been made, *subordinate everything else to that decision*. This is, at the same time, the most important and the most difficult of the five focusing steps to accomplish. Why is it so difficult? It requires everyone and every part of the system not directly involved with the constraint to subordinate, or "put in second place," their own cherished success measures, efficiencies, and egos. It requires everyone, from top management on down, to accept the idea that excess capacity in the system at most locations is not just acceptable—it's actually a good and necessary thing!

   Subordination formally relegates all parts of the system that are not constraints (referred to as "non-constraints") to the role of supporters of the constraint. This can create behavioral problems at almost all levels of the company. It's very difficult for most people to accept that they and/or their part of the organization aren't just as critical to the success of the system as any other. Consequently, most people at non-constraints will resist doing the things necessary to subordinate the rest of the system to the constraint. This is what makes the third step so difficult to accomplish.

   What makes the constraint more critical to the organization is its *relative weakness*. What distinguishes a non-constraint is its *relative strength*, which enables it to be more flexible. So the current performance of the organization really hinges on the weak point. While the other parts of the system could do more, because of that weak point *there is no point* in doing more. Instead, the key to better performance is wisely subordinating the stronger points so that the weak point can be exploited in full.

   Subordination actually redefines the objectives of every process in the system. Each process is supposed to accomplish a mission that's necessary for the ultimate achievement of the goal. But among processes there may be conflicting priorities, such as competition for the same resources. Subordinating non-constraints actually focuses the efforts of every process on truly supporting the goal of organization. It allows the constraint to be exploited in the best way possible.

Consider a raw material warehouse. What is its objective? The storing and releasing of material is needed as a "bridge" between the time materials arrive from the vendors and the time the same materials are needed on the production floor. When a specific work center is the constraint, any materials needed by that particular work center should be released precisely at the required time. If market demand is the only constraint, any order coming in should trigger material release.

However, even if no new orders enter the system, shop foremen often like to continue working, so as to keep their efficiency high. But if the non-constraints in a production system are properly subordinated, material should *not* be released. The material release process must be subordinated to the needs of the system constraint, not to arbitrary efficiency measurements. Maintaining the order in the warehouse is part of the subordination process. Release of materials not immediately needed for a firm order should be treated as a lower priority than the quick release of materials the constraint will soon need to fulfill a definite customer requirement.

Subordination serves to focus the efforts of the system on the things that help it to maximize its current performance. Actions that contradict the subordination rationale should be suppressed.

It's possible that, after completing the third step, the system constraint might be broken. If so, it should be fairly obvious. Output at the system level will usually take a positive jump, and some other part of the system might start to look like a "bottleneck." If this is the case, go back to the first step and begin the five focusing steps again. Identify which new factor has become the system constraint, determine how best to exploit that component and subordinate everything else.

4. *Elevate.* However, if, after completing Step 3, the original constraint is still the system constraint, at this point the best you can be assured of is that you're wringing as much productivity out of it as possible—it's not possible for the system to perform any better than it is without additional management action. In taking this action, it's necessary to proceed to the fourth step to obtain better performance from the system. That step is to *evaluate alternative ways to elevate the constraint* (or constraints, in the unlikely event that there is more than one). Elevate means to "increase capacity." If the constraint is an internal resource, this means obtaining more time for that resource to do productive work. Some typical alternatives for doing this might be to acquire more machines or people, or to add overtime or shifts until all 24 hours of the day are used. If the constraint is market demand (lack of sales), elevation might mean investing in an advertising campaign, or a new product introduction to boost sales. In any case, elevating invariably means "spend more money to make more money."

Notice the use of the word *evaluate* in this step. This word is emphasized for a good reason. From the preceding examples—buying more equipment or adding shifts, or overtime—it should be clear that there's more than one way to skin a cat. Some alternatives are less expensive than others. Some alternatives are more attractive for reasons that can't be measured directly in financial terms (e.g., being easier to manage). In any case, a choice on the means to elevate will usually be required, so jumping on the first option that you think of might not necessarily be a good idea.

One of the reasons to favor one elevation alternative over another is the identity of the next potential constraint. Constraints don't "go away," per se. When a constraint is broken, some other factor, either internal or external to the system, becomes the new system constraint—albeit at a higher level of overall system performance, but a constraint nonetheless. It's possible that the next potential constraint might be more difficult to manage than the one we currently have; it might reduce the margin of control we have over our system.

It's also possible these alternatives might drive the system constraint to different locations—one of which might be preferable to the other. Or it could be that dealing with the potential new constraint might require a much longer lead time than breaking the current constraint. In this case, if we decide to break the current constraint, we would want to get a "head start" on the tasks needed to exercise some control over the new constraint.

Ineffective Elevation: An Example

For example, one company involved in the manufacture of solid state circuit boards found its constraint to be the first step in its process: a surface-mount (gaseous diffusion) machine (Schragenheim and Dettmer, 2000, Chap. 2). Without considering which other resource might become the new constraint, they opted to purchase another surface-mount machine. This certainly relieved the original constraint. But the automated test equipment (ATE)—about eight steps down the production line—became the new constraint, and managing the constraint at this location was no easy task. It was more complex to schedule at that point, and it suffered more problems. Moreover, moving the constraint out of the ATE section was even more challenging. Buying more ATE was more expensive than buying additional surface-mount equipment. Finding qualified ATE operators was also more difficult.

In short, it took more time, effort, and money to manage or break the ATE constraint than it did to break the surface-mount constraint. Had the company been able to anticipate that ATE would become the system constraint, they could have chosen to either (1) leave the constraint where it was—at the surface-mount machine, or (2) begin long-lead time acquisition of ATE and ATE operators to boost the ATE section's capacity *before* increasing the surface-mount capacity. Doing so would have increased system performance, yet preserved the system constraint at a location that was far easier to manage.

Another important factor to consider is return on investment. Once the company described above broke the surface-mount constraint, there was potential to generate more Throughput, but how much? If the ATE's capacity was only slightly more than that of the original surface-mount machine, the company might have gained only a small increase in Throughput as a payback for the cost of the new surface-mount unit. This could become a definite disappointment.

As long as the next constraint poses a substantially higher limit than the existing one, it's probably safe to say that the company did the right thing. Even if exploiting the ATE is more difficult, the increase in Throughput might be worth the aggravation. The ATE could always be loaded a little less, and the company would still realize more money. What's the lesson here? Assessing the real return on investment from an *elevation* action requires an understanding of constraint theory, where the next constraint will be, and how much Throughput will increase before hitting the next constraint. So the "evaluate" part of the elevation step can be extremely important. It's important to know where the new constraint will occur, because it could affect our decision on how to elevate.

How to Determine Where the Next Constraint Will Be

The easiest way to do this is to apply the first three of the five focusing steps "in our heads," before actually elevating for the first time. In other words, identify the next most limiting factor, inside or outside the system, that will keep the whole system from achieving better performance after the current constraint is broken. Then determine what actions will be necessary to exploit that new constraint in the future, and how the rest of the system will have to act to subordinate itself to the exploitation of the new constraint.

Once this is done, the ramifications of each alternative to elevate should be obvious, and a better-informed decision is possible about which alternative to choose—and it might not be the obvious choice, or the cheapest one!

5. *Go back to Step 1, but beware of "inertia."* Even if the *exploit* and *subordinate* steps don't break the system constraint, the *elevate* step very likely will, unless a conscious decision is made to curtail

elevation actions short of that point. In either case, after the *subordinate* or *elevate* steps it's important to go back to the first step (*identify*) to verify where the new system constraint is, or to determine that it has not migrated away from the original location. Sometimes a constraint moves, not as a result of intentional actions, but as a result of a change in the environment. For instance, a change in preferences of the market might drive a company to change its product mix to such an extent that the constraint moves elsewhere. While such external changes don't happen very frequently, it's worth the effort to go back to the first step from time to time, just to verify that what we believe to be the constraint still is, in fact, the system's limiting factor.

The warning about inertia says: "Don't become complacent." There are two reasons for this. First, when the constraint moves, the actions or policies we put into place to *exploit* and *subordinate* the rest of the system to the "old" constraint may no longer be the best things to do for the benefit of the whole system. If we don't re-evaluate where the new system constraint is, this deficiency would never be noticed. Second, there is often a tendency to say, "Well, we've solved that problem. There's no need to revisit it again." But today's solution eventually becomes tomorrow's historical curiosity. An organization that's too lazy (or distracted by other demands for its attention) to revisit old solutions can be sure that eventually—probably sooner, rather than later—it won't be getting the best possible performance from its system.

Tools of Constraint Management

Success or failure in any endeavor often relies on the selection and proper use of the right tools. Constraint management is no exception. While the five focusing steps are effective guidelines for the tactical and strategic management of any kind of system, in specific situations the nature of constraints and the problems associated with them call for different tools and procedures. Exploiting a constraint would be done differently in a service environment than in a production process. Subordination would be different in a heavy manufacturing company that produces standardized products than it would be in a small job shop. Wouldn't it be useful to have an aid that could point us toward the right constraint management actions for each situation?

The Logical Thinking Process

With so many different kinds of constraints, and with policy constraints underlying most of them, how can we identify what specific changes we should be working on? Many of these constraints aren't easy to identify. Often, they're not physical, or they're not easy to measure. They sometimes extend beyond the boundaries of production processes alone, although they still affect manufacturing, and sometimes—especially if they're policies—they pervade the whole organization.

To facilitate the analysis of complex systems, Goldratt created a logical thinking process. The thinking process is composed of six logic diagrams, or "trees." (Dettmer, 1997, 1998). It was specifically designed to analyze the policies of an organization and determine which one(s) might constitute a constraint to better performance.

This thinking process is unique from one perspective: it's one of the few (maybe the only) problem-solving methodologies that goes beyond problem identification and solution generation, and into solution verification and implementation planning. The components of the thinking process include:

1. *The Current Reality Tree (CRT).* Designed to help identify the system constraint, especially when that constraint is a policy of some kind. Figure 5.3 shows an example of a typical CRT.
2. *The "Evaporating Cloud" (EC).* A kind of conflict resolution diagram. Helps create breakthrough solutions to resolve hidden, underlying conflicts that tend to perpetuate the constraint. Figure 5.4 illustrates a typical EC.
3. *The Future Reality Tree (FRT).* Tests and validates potential solutions. Provides logical verification that a proposed solution will actually deliver the desired results. Figure 5.5 depicts an FRT.
4. *The Negative Branch (NB).* Actually a subset of the FRT. Helps identify and avoid any new, devastating effects that might result from the solution. Figure 5.6 represents a notional example of an NB, and how it might have been used to anticipate the disastrous consequences of a very high-profile decision. Notice that this example underscores the fact that application of the thinking process tools is not confined to commercial business situations alone.
5. *The Prerequisite Tree (PRT).* Helps to surface and eliminate obstacles to implementation of a chosen solution. Also time-sequences the actions required to achieve the objective. Figure 5.7 shows a typical PRT.
6. *The Transition Tree (TT).* Can facilitate the development of step-by-step implementation plans. Also helps explain the rationale for the proposed actions to those responsible for implementing them. This can be especially important when those charged with executing a plan are not the same people who developed it. Figure 5.8 contains a typical TT. Either the TT or the PRT can form the basis of a project activity network for implementation of change.

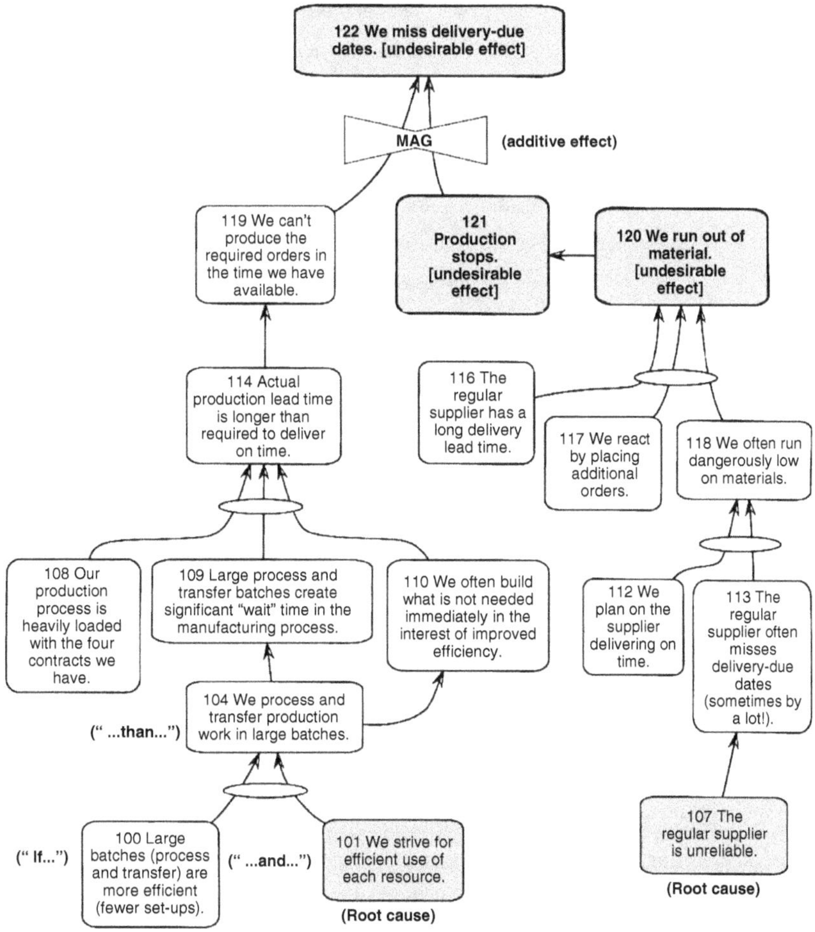

FIGURE 5.3 Current Reality Tree—manufacturing example (adapted from Schragenheim and Dettmer, 2000).

Four of these trees—the Current Reality Tree, Future Reality Tree, Negative Branch and Transition Tree are cause-and-effect trees. They're read using "If … then …." The Evaporating Cloud and Prerequisite Tree are referred to as "necessary condition" trees. They're read a little differently, using "In order to have … we must …."

These tools are specifically designed to help answer the three major questions inherent in the first three of the five focusing steps:

- *What* to change?
- What to change *to*?
- How to *cause* the change?

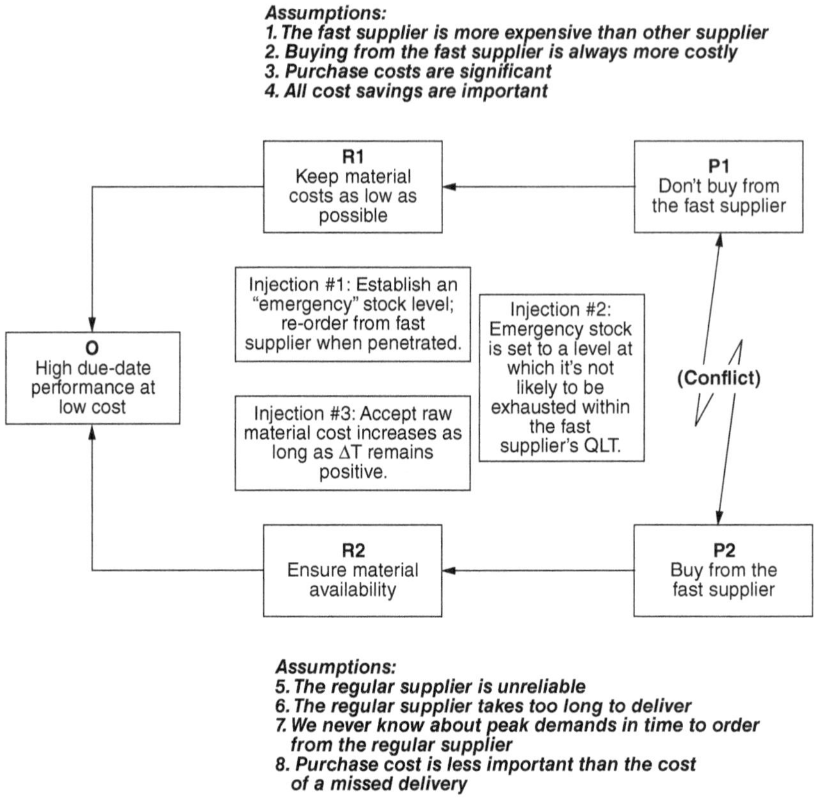

FIGURE 5.4 "Evaporating Cloud"—unreliable supplier (adapted from Schragenheim and Dettmer, 2000).

"Drum-Buffer-Rope" Production Scheduling

Probably the best-known of the constraint management tools developed by Goldratt is called "Drum-Buffer-Rope" (DBR). The origin of this name dates back to the analogy Goldratt and Cox used in *The Goal* (Goldratt, 1986) to describe a system with dependencies and statistical fluctuations. The analogy was a description of a Boy Scout hike. The drum was the pace of the slowest Boy Scout, which dictated the pace for the others. The buffer and rope were additional means to ensure all the Boy Scouts walked at approximately the pace of the slowest boy.

Goldratt and Fox, in *The Race* (Goldratt and Fox, 1986), describe in detail the manufacturing procedure that stems from the concepts of a drum, buffer, and rope originally introduced through the Boy Scout hike. The DBR method provides the means for synchronizing an entire manufacturing process with "weakest link" in the production chain. Figure 5.9 illustrates the DBR concept.

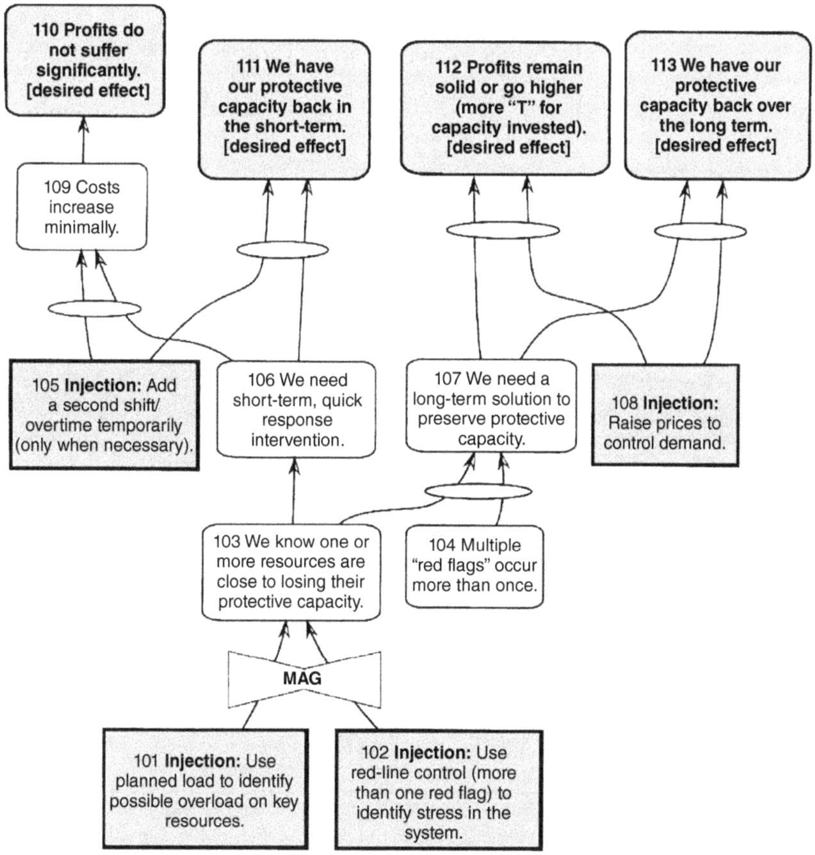

Figure 5.5 Future Reality Tree—"red-line control" (manufacturing control) (adapted from Schragenheim and Dettmer, 2000).

- *The "Drum."* In a manufacturing or service company, the "drum" is the schedule for the resource or work center with the most limited capacity: the Capacity Constrained Resource (CCR). The reason the CCR is so important is that it determines the maximum possible output of the entire production system. It also represents the whole system's output, since the system can't produce any more than its least-capable resource.
- The "buffer" and the "rope" ensure that this resource is neither starved for work nor overloaded (causing backlogs). In constraint management, buffers are composed of *time*, not *things*.
- *The "Buffer."* Starvation can result from upstream process variability, which might delay the transfer of work-in-process beyond its expected time. To ensure a CCR is not starved for work, a buffer time is established to protect against variability. This is a period of time in advance of the scheduled "start processing" time that a

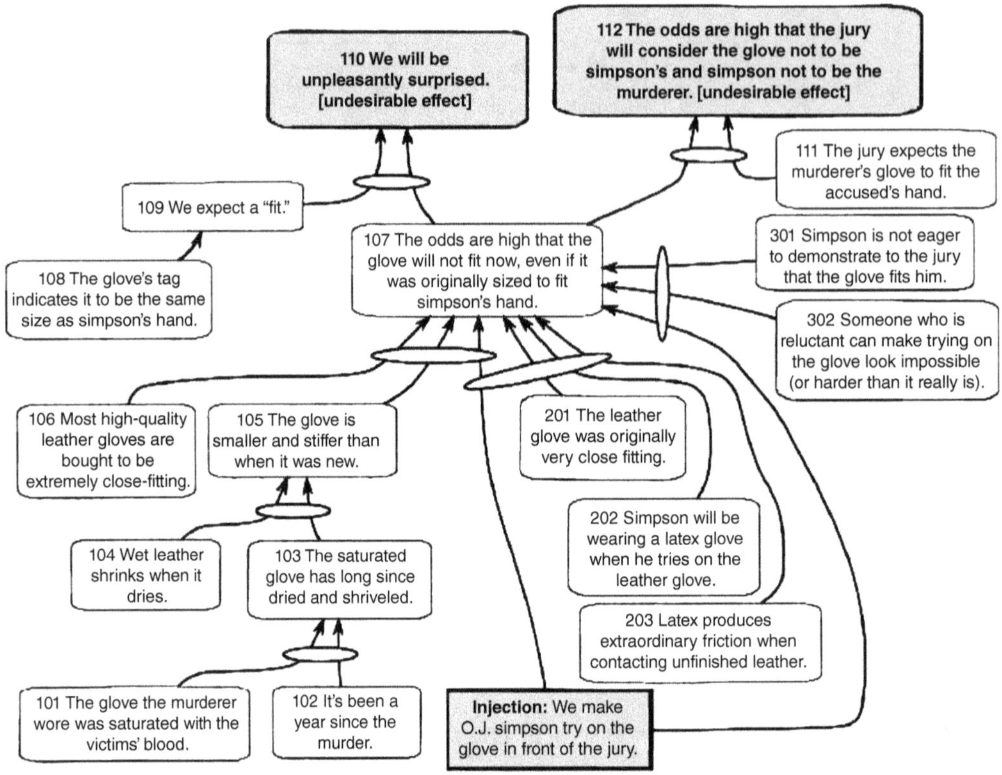

FIGURE 5.6 Los Angeles District Attorney's Negative Branch—"If it does not fit, you must acquit ..." (adapted from Dettmer, 1998).

particular job arrives at the CCR. For example, if the CCR schedule calls for this valuable resource to begin processing a particular work order at 3:00 PM on Tuesday, the material for that job might be released early enough to allow all preceding processing steps to be completed at 3:00 PM on Monday (a full work day ahead of the time required). The buffer time serves to "protect" the most valuable resource from having no work to do—a serious failing, since the output of this resource is equivalent to the output of the entire system.

- It should be noted that only critical points in a service or production process are protected by buffers. (Refer to Fig. 5.9.) These critical points are the capacity-constrained resource, any subsequent process step where assembly with other parts occurs, and the shipping schedule. Because the protection against variability is concentrated only at the most critical places (and eliminated everywhere else in the process), actual lead time can be shortened considerably, sometimes by 50 percent or more, without

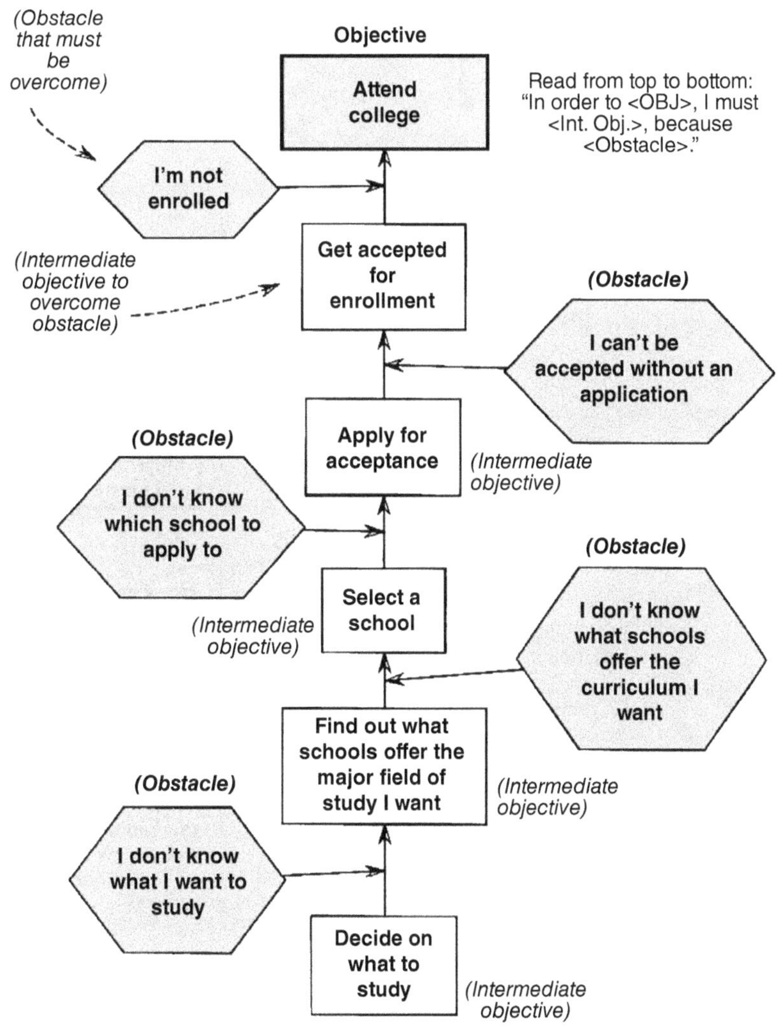

Figure 5.7 Prerequisite Tree—example (adapted from Dettmer, 1997).

compromising due-date reliability. Shorter lead times and higher delivery reliability are important service characteristics that customers often look for.

- *The "Rope."* The rope is constraint management's safeguard against overloading the CCR. In essence, it's a material release schedule that prevents work from being introduced into the system at a rate faster than the CCR can process it. The rope concept is designed to prevent the backlog of work at most points in the system (other than the planned buffers at the critical protected points). This is important because work-in-process queues are one of the chief causes of long delivery lead times.

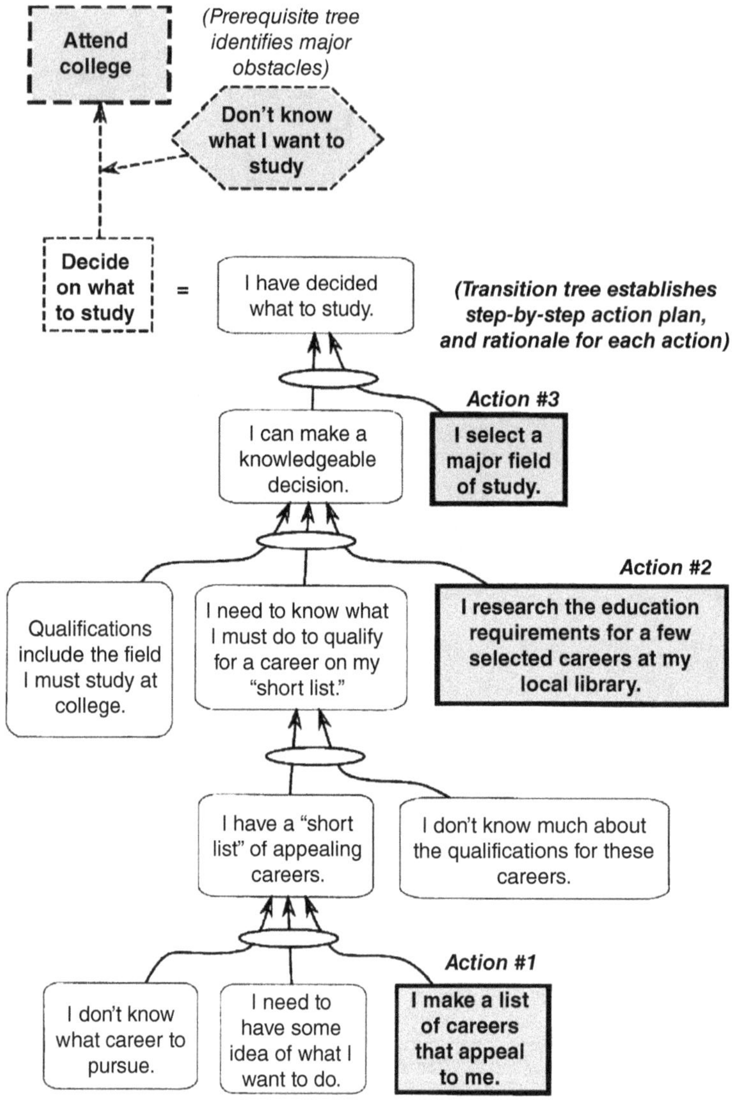

Figure 5.8 Transition Tree—example (adapted from Dettmer, 1997).

When the entire Drum-Buffer-Rope concept is applied, delivery reliability of 100 percent is not an unreasonable target, and actual lead time reductions of 70 percent are common. Bethlehem Steel's Sparrows Point plant increased delivery reliability from 49 percent to nearly 100 percent while they reduced actual lead times from 16 weeks to about four weeks (Dettmer, 1998, Chap. 1).

Critical Chain

Another valuable asset in the constraint management toolbox is called Critical Chain. Also the title of a book by Goldratt (Goldratt, 1997), the

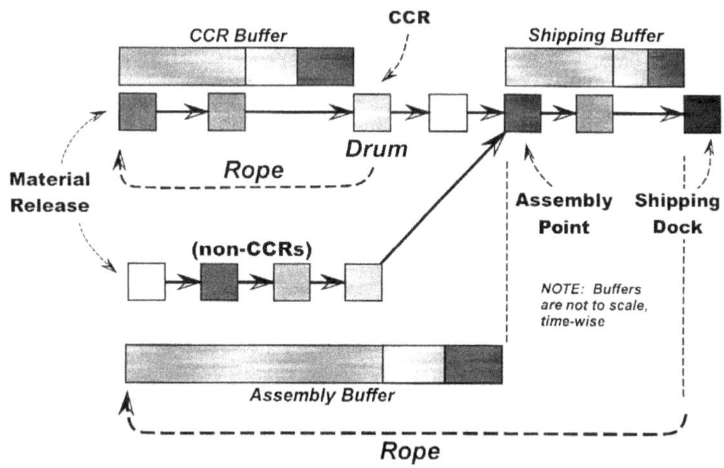

FIGURE 5.9 Basic DBR concept (adapted from Schragenheim and Dettmer, 2000).

Critical Chain concept provides an effective way to schedule project activities by effectively accommodating uncertainty and resolving simultaneous needs (contentions) for the same resource. Critical Chain constitutes the application to one-time projects of the same principles that DBR applies to repetitive production. The result of applying Critical Chain scheduling and resource allocation is a higher probability of completing projects on time, and, in some cases, actually shortening total project duration. Originally applied to the management of a single project, the Critical Chain method has been expanded to multiproject environments, based on the concept of the "drum," described in Drum-Buffer-Rope, above.

Since projects aren't quite the same as repetitive production, some differences in employing Critical Chain project planning are inevitable. But the concepts are much the same as those of DBR. What distinguishes Critical Chain from PERT/CPM and other traditional project management approaches?

First, Critical Chain recognizes and accounts for some human behavioral phenomena that traditional project management methods don't (Leach, 2000; Newbold, 1998). These phenomena include:

1. The tendency of technical professionals to "pad" their time estimates for individual tasks, in an effort to protect themselves from late completion.
2. The so-called "student syndrome"—waiting until the last minute to begin work on a task with a deadline.
3. Parkinson's law (ensuring that an activity consumes every bit of the estimated time, no matter how quickly the associated tasks can actually be completed).

4. Multitasking—the tendency of management to assign people more than one deadline activity simultaneously. Multitasking can create a devastating effect. Project personnel switch back and forth between several tasks, causing "drag" in all of them. The result is that other resources that depend on these task completions for their inputs are delayed. Delays "cascade" when several simultaneous projects are involved.

The delays that result from this kind of behavior are caused by a flawed management assumption: *The only way to ensure on-time completion of a project is to ensure that EVERY activity will finish on time.* This common management belief prompts management to attribute unwarranted importance to meeting scheduled completion time for each separate activity in a project. People then pad their estimates of task times to ensure they can get everything done in time. But at the same time, they try to not finish much earlier than their own inflated estimation, so as not to be held to shorter task times on subsequent projects. All of these "human machinations" cause a vicious circle, dragging projects out longer and longer, but the reliability of meeting the original schedules isn't improved.

Second, to solve this problem, Critical Chain takes most of the protective time out of each individual activity and positions some of it at key points in the project activity network: at convergence points and just ahead of project delivery. Since accumulating protection on an entire chain is much more effective than protecting every activity, only half of the aggregated "protective pad" extracted from individual activities is put back in at the key locations. The rest can contribute to earlier project completion. In traditional project execution, if protective time in a specific activity isn't used, it's lost forever—unusable by later activities that might need more protection than they were originally assigned. This formerly "lost time" is, in many cases, usable in Critical Chain.

Third, Critical Chain devotes more attention to the availability of critical resources when they're needed for specific activities. Leveling the resources on any single project is mandatory. The Critical Chain is really the longest sequence in the project that considers *both* dependent, sequential activity links *and* resource links. The critical path reflects only the sequential linking of dependent tasks.

The key elements of Critical Chain Project Management include:

- *The Critical Chain.* The set of tasks that determines project duration, considering both task precedence *and* resource dependencies (Newbold, 1998). The longest sequence of dependent activities is the "critical path" in a PERT/CPM approach. But when the duration of this sequence is adjusted for optimum resource availability (resource leveling), the whole definition of the critical path becomes

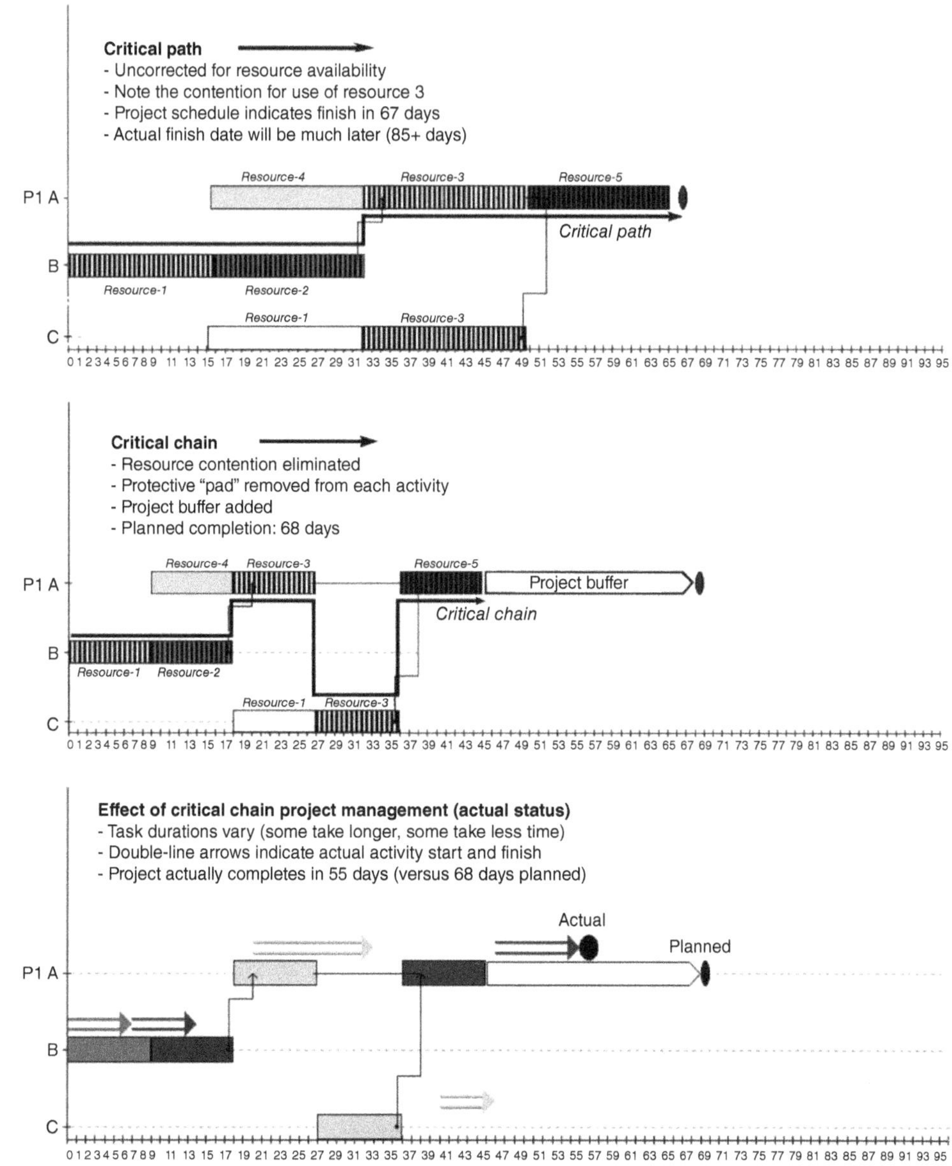

FIGURE 5.10 Critical Chain versus Critical Path.

irrelevant. What results is a different time duration, based on resource usage. This is the Critical Chain. (Refer to Fig. 5.10.) Because the Critical Chain now constitutes the constraint that determines the earliest date that a project can finish, it's crucial to monitor progress along the Critical Chain, because it reflects the progress of the entire project.

- *The project buffer.* A project buffer is established at the end of the final activity on the Critical Chain, and before the required delivery date. It's designed to protect against extreme variability and uncertainty that may impact the Critical Chain.
- *Feeding buffers.* Each activity (or sequence of activities) that feeds the Critical Chain is buffered with some reasonable amount of time to protect against variability in that particular task, or sequence of tasks. Feeding buffers protect the Critical Chain from delays occurring in activities that are not on the Critical Chain. Variability is *all* that these buffers protect against. They don't protect against multitasking: the tendency of organizations to assign technical personnel tasks for different projects or obligations simultaneously. In fact, one of the determinants of success or failure in applying Critical Chain is the willingness of management to move away from multitasking.
- *Buffer management.* A means of control that, at any time in the project life, affords project managers the opportunity to determine how much of various buffers have been used and to take action as soon as the project appears to be in danger of exceeding its scheduled time. Task completion delays will cause "penetration" into buffers, but warning of this happening comes so much earlier that it's often possible to prevent schedule overruns sooner, with considerably less (or less extreme) corrective action. Monitoring the buffers, especially the project buffer, results in a higher probability that the project will complete on time.
- *The drum.* This concept, similar to that of DBR, applies only to multiproject situations. The drum concept requires choosing one of the most heavily loaded resources as the "drum" and staggering multiple projects according to the availability of that resource.

The ultimate effects of Critical Chain project management are higher delivery due date reliability, more frequent earlier delivery, less "crashing," and conservation of project costs. Buffer management, in particular, provides much better focused information on current project status.

Constraint Management Measurements

One of the unique contributions of the Theory of Constraints to the management body of knowledge is the measurements used to assess progress toward the system's goal. Goldratt recognized some deficiencies inherent in traditional measurement systems and conceived of a different—more reliable—way to measure results and evaluate decisions.

Dilemma: System or Process?

The measurement issue harks back to the earlier discussion about systems versus processes and the fallacy of assuming that the sum of local efficiencies is the system optimum. Traditional rationale maintains that achieving the highest possible productivity in every discrete function of the system equates to good management. Productivity is typically represented as the ratio of outputs to inputs. These inputs and outputs are sometimes expressed in financial terms. Managers often spend inordinate time chasing higher productivity for their own departments, without much concern for whether the whole system benefits or not. This underscores the heart of the problem: How can we be sure that the decisions we make day-to-day truly benefit the system as a whole. In other words, how can local decisions be related to the global performance of the company?

This is not necessarily an easy question to answer. Consider yourself a production manager for a moment. A sales manager comes to you and asks you to interrupt your current production run (that is, break a setup) to process a small but urgent order for a customer. How will doing what the sales manager wants affect the company's bottom line? What will it cost to break the setup (both financially and to the production manager's productivity figures)? How much will it benefit the company? Or the production department? These are not easy questions to answer, yet throughout many companies people are called upon to make such decisions daily.

New Financial Measures

Assuming that a company's goal is to make more money, Goldratt conceived of three simple financial measures to ensure that local decisions line up effectively with this goal. These measures are easy to apply by anyone at virtually any level of a company: *Throughput*, *Inventory* or *Investment*, and *Operating Expense* (Goldratt, 1990, pp. 19–51).

Throughput (T) is defined as the rate at which a system generates money through sales (Goldratt, 1990, p. 19). Another way to think about it is the marginal contribution of sales to profit. Throughput can be assessed for the entire company over some period of time, or it can be broken out by product line, or even by individual unit of product sold. Mathematically, Throughput equates to sales revenue minus variable cost.

$$T = SR - VC$$

Inventory (I) is defined as all the money the system invests in purchasing things it intends to sell (presumably after adding some value to them) (Goldratt, 1990, p. 23). Because Goldratt's concept of Inventory includes fixed assets, such as equipment, facilities, and real estate, the term "I" has come to represent "investment" as well, rather than just "inventory" alone.

Inventory/Investment certainly includes the materials that the company will turn into finished products or services. But it also includes the assets of a company, which are eventually sold off at depreciated or scrap value and replaced with new assets. This is even true of factory buildings themselves. However, for day-to-day decisions, most managers consider "I" to represent the consumable inventory of materials that will be used to produce finished products or services.

Operating Expense (OE) is defined as all the money the system spends turning Inventory into Throughput. (Goldratt, 1990, p. 29). Notice that overhead is not included in the Throughput formula (or the definition). Overhead, and most other kinds of fixed costs, are included in Operating Expense. Constraint management measurements deliberately segregate fixed costs from the Throughput calculation for a valid reason: allocating fixed costs to units of product sold produces a distorted concept of actual product costs in most day-to-day situations.

For example, let's say you're a small manufacturer of precision-machined parts. You're working on an order for 100 units of a particular part for an original equipment manufacturer. In the middle of this run, the customer calls and asks you to add 10 more units to the order. This will increase your production time by 53 minutes. How much more have these additional 10 units cost you? As long as you're not backlogged with work, the cost of the extra units is the value of the raw materials alone! You didn't pay any more in salary to the machine operator (he or she works by the hour, not by the piece). In most cases you wouldn't pay any more in electricity costs. You have to turn the lights on for business for the whole day anyway. And the cost of the general manager's company car didn't change just because you produced 10 additional units of the product. The real cost of the increase in production volume is limited primarily to the cost of materials alone.

Labor costs are also considered an Operating Expense, because in almost all cases they are paid by some fixed unit of time, not by the individual unit of product produced. We pay people by the hour, week, month, or year, whether they are actively producing a product for sale or not. Moreover, the capacity to produce a product or service (the resources: people, facilities, equipment, etc.) is obtained in "chunks." It's really difficult to hire six-tenths of a person, or to buy three-quarters of a machine.

So the expenditure of cost for capacity usually comes in sizeable increments—a step function. Products, on the other hand, are normally priced and sold by the unit—smaller steps, perhaps, but closer to a continuous function. All this makes it difficult to attach an accurate allocation of fixed costs to a unit of product. Which means that those who do so obtain a distorted impression of product costs—not a good basis from which to form operating decisions.

For daily management decisions, which we'd like to be able to relate to the system's goal, T, I, and OE are much more useful than the traditional organizational success measures of net profit (NP), return on investment (ROI), and cash flow (CF). Yet there has to be a connection between the two types of financial measures. And here it is:

$$NP = T - OE$$
$$ROI = (T - OE)/I$$
$$CF = T - OE \pm \Delta I$$

Net profit is the difference between Throughput and Operating Expense. Return on investment is net profit (Throughput minus Operating Expense) divided by Inventory/Investment. And cash flow is net profit (Throughput minus Operating Expense) plus-or-minus the change in Inventory.

If Throughput is increased, net profit increases, even if Operating Expense remains the same. If Operating Expense is reduced, net profit also increases (as long as Throughput at least remains constant). If Inventory is reduced, ROI increases, even if there are no changes to Throughput or Operating Expense. These measures relate to operational management decisions much better than net profit and return on investment, keeping daily decisions more in line with the system's goal than abstract efficiency measures such as machine utilization, units produced per day/week, etc.

For example, here's how a manager might use T, I, and OE to evaluate a decision he or she is contemplating (Dettmer, 1998, p. 33):

- Will the decision result in a better use of the worst-constrained resource (i.e., more units of product available to sell in the same or less time)?
- Will it make full use of the worst-constrained resource?
- Will total sales revenue increase because of the decision?
- Will it speed up delivery to customers?
- Will it provide a characteristic of product or service that our competitors don't have (e.g., speed of delivery)?
- Will it win repeat or new business for us?
- Will it reduce scrap or rework?
- Will it reduce warranty or replacement costs?
- Will we be able to divert some people to do other work (work we couldn't do before) that we can charge customers for? If so, the decision will improve Throughput.

- Will we need less raw material or purchased parts?
- Will we be able to keep less material on hand?
- Will it reduce work-in-process?
- Will we need less capital facilities or equipment to do the same work? If the answer is "yes," the decision will reduce Inventory or Investment.
- Will overhead go down?
- Will payments to vendors decrease?

If so, the decision will decrease Operating Expense.

Let's return to the example mentioned earlier about the production manager faced with a decision to break a setup in response to a sales manager's request. Using conventional reasoning, an efficiency-oriented production manager would see his productivity figures suffering because of the time lost to doing the new setup. Inserting this urgent order would also disrupt a formal schedule, slipping every subsequent order's scheduled starting and finishing times.

But a constraint-oriented manager would look at it a little differently. His or her first question would be, "Am I internally constrained by a shortage of resources?" The answer should be fairly obvious, because the manager would already be aware of the size of the backlog, if any. If the answer is "no," the production process has excess capacity to be able to accommodate the urgent order without delaying scheduled work. The cost of the additional order is only the raw materials used and possibly the loss of materials from a unit of the job currently on the machine when the setup is broken. The machine operators' time doesn't cost any more. They're paid by the hour, whether they're doing setups or producing products. The direct increase in net profit to the company by agreeing to the sales manager's request would be the Throughput (sales revenue minus variable costs) for the new order. And it's possible that providing this expedited service to a customer might win more repeat business in the future. So as long as the manufacturing process is not internally constrained, the production manager cognizant of constraint theory would probably accept this new job, while a traditional manager might tell the sales person, "Get in line! We operate on a first-in, first-out basis." And this manager's "numbers" would look good at the end of the month—but what would that decision have done for the company?

The use of T, I, and OE for making management decisions is not a replacement for generally accepted accounting procedures (GAAP). Those are required, and will probably always be, for external reporting purposes—annual reports to stockholders, securities and exchange filings, tax reporting, etc. But Throughput, Inventory, and Operating Expense are considerably easier for most line managers to understand and use in gauging the financial effects of their daily decisions.

The Strategic Implications of T, I, and OE

What makes Throughput, Inventory/Investment, and Operating Expense even more beneficial is the strategic implications of their application by senior managers and executives. Of course, every company executive wants to improve net profit, return on investment, and cash flow. However, in standard accounting T, I, and OE are "embedded" in these terms and often difficult to single out, even though they are better managed separately.

Consider the bar graph in Fig. 5.11. It shows three bars representing T, I, and OE. Most companies devote extraordinary effort to reducing costs (both fixed and variable). The cost-of-quality concept emphasizes this as a justification for pursuing quality in the first place. Lean manufacturing does, too. The same is true of reducing inventory. In fact, in many corporate strategies, cost saving features high on the priority list.

The graph tells us a different story, however. While there might well be savings to be had in these areas (OE and I reductions), these savings have a point of diminishing returns. There's a practical level below which neither Inventory nor Operating Expense can be reduced without hurting a company's ability to generate Throughput. Beyond that point, you're not cutting "fat" anymore, you're cutting "muscle." And the truth is that most managers can't really say where that point lies. Moreover, many companies have been engaged in cost reduction efforts for so long that all of the OE (and even the variable costs of Throughput) have largely been wrung out of the system. To improve financial performance, there's nowhere else to go . . .

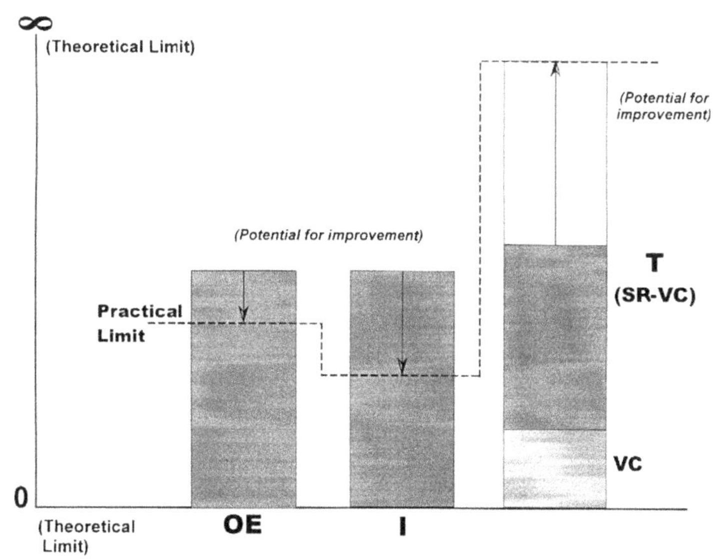

FIGURE 5.11 Limits to T, I, and OE. Why chasing cost reduction is nonproductive (*Source:* Schragenheim and Dettmer, 2000).

except to Throughput. And look at the Throughput bar in Fig. 5.10. The improvement potential may not be infinite, but from a practical perspective there's a lot more potential to improve profitability by increasing T than there ever will be by reducing I and OE. So what does it make sense to prioritize: rearranging the deck chairs on the Titanic, or steering away from the iceberg? In summary, not many companies can "save their way" to prosperity.

Using T, I, and OE is variously referred to as Throughput accounting, constraint accounting, or Throughput-based decision support. They aren't rigorous enough to replace GAAP for standard accounting needs. But they're simpler and usually more effective for decision making than traditional management accounting. More details on how to use T, I, and OE may be found in Goldratt (1990), Noreen et al. (1995), Corbett (1998), and Smith (1999).

Summary and Conclusion

To summarize, constraint management:

- Is a systems management methodology.
- Separates the "critical few" from the "trivial many."
- Emphasizes attention on the critical few factors that determine system success.

It's based on four underlying assumptions:

1. Every system has a goal and a finite set of necessary conditions that must be satisfied to achieve that goal.
2. The sum of a system's local optima does not equal the global system optimum.
3. Very few variables—maybe only one—limit the performance of a system at any one time.
4. All systems are subject to logical cause and effect.

The Theory of Constraints, as conceived by Goldratt, is embodied in the five focusing steps (Identify, Exploit, Subordinate, Elevate, Repeat/Inertia), a set of financial progress measurements (Throughput, Inventory, and Operating Expense), and three generic tools (the Logical Thinking Process, Drum-Buffer-Rope, and Critical Chain) that can be applied in a variety of organizational situations.

CHAPTER 6

Understanding Customer Expectations and Needs

Noritaki Kano modeled the relationship between customer satisfaction and quality shown in Fig. 6.1. The Kano model shows that there is a basic level of quality that customers assume the product will have. For example, all automobiles have windows and tires. If asked, customers don't even mention the basic quality items; they take them for granted. However, if this quality level *isn't* met, the customer will be dissatisfied; note that the entire "basic quality" curve lies in the lower half of the chart, representing dissatisfaction. Thus, providing basic quality is insufficient to create a satisfied customer.

The expected quality line represents those expectations that customers explicitly consider, for example, the length of time spent waiting in line at a checkout counter. The model shows that customers will be dissatisfied if their quality expectations are not met; satisfaction increases as more expectations are met.

The exciting quality curve lies entirely in the satisfaction region. This is the effect of innovation. Exciting quality represents *unexpected* quality items. The customer receives more than he or she expected. For example, Cadillac pioneered a system where the headlights stay on long enough for the owner to walk safely to the door.

Competitive pressure will constantly raise customer expectations. Today's exciting quality is tomorrow's basic quality. Firms that seek to lead the market must innovate constantly. Conversely, firms that seek to offer standard quality must constantly research customer expectations to determine the currently accepted quality levels. It is not enough to track competitors since expectations are influenced by outside factors as well. At one time, now relatively ubiquitous features like stereo radios or more recently airbags in automobiles were considered luxury items.

Clearly, understanding the customers' current expectations and excitements is a continuing activity. Unfortunately, broad markets typically have many types of customers, with different needs and expectations.

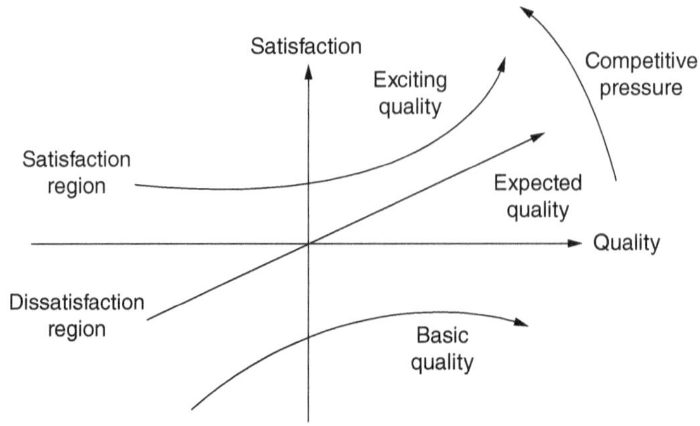

Figure 6.1 Kano model showing relationship between customer satisfaction and quality levels.

Customer Classifications

Customers can be classified in several ways to assist in categorizing their needs. One distinction commonly made is between external and internal customers. The primary determinant of which type of customer is being discussed is whether the relationship involves the transfer of resources across the organizational boundary.

- *External customer.* An external customer transaction is one that involves the exchange of the organization's product or service for money or other valuable consideration that the organization does not already possess. The transaction may result in transfer of ownership of a resource, or in the granting of specific rights to use a resource, as with a lease arrangement.
- *Internal customer.* In an internal customer-supplier transaction ownership or right of use of the resource being exchanged does not transfer outside of the organization as a result of the exchange. However, control of a resource may be transferred between individuals or groups within the organization. With internal customer transactions there is no transfer of outside assets (e.g., cash) into the organization from outside the organization.
- *End user.* An end user is the internal or external customer who actually uses the product or service. Even though end users are the focus of the design effort for the product or service, they nonetheless may not have much consideration in the actual delivery of the product or service.
- *Transfer agents.* A transfer agent acts as an intermediary between the producer and the end user. Transfer agents include the purchase

decision maker, source of payment, people who evaluate the product's capabilities, people involved in transporting the product from the manufacturer, people who install the equipment, and any other person involved in the process of acquiring the product and preparing it for use by the end user. As a general rule, transfer agents perform functions that are transparent to the end user; for instance, the user of an automobile doesn't really know or care how the vehicle was transported from the manufacturer to the dealer. This presents opportunities to reduce costs without affecting perceived quality. Yet transfer agents often play a major role in end user satisfaction. A rude, obnoxious, or ill-informed salesperson can create customer dissatisfaction that no amount of product quality will ever overcome. Because of their closer contact with producers and greater buying power, transfer agents often have greater voice than end users, and can act as an information filter between producers and end users. The transfer agents' interests may be different from the interests of the end user. The combination of these factors can create a disconnect between end user and producer that leads to a failure to meet the needs of the end user. If the gap becomes wide enough, it can create an opening that an alert competitor can use to gain market share at the expense of the often perplexed producer.

- *Maintenance customers.* These are people who repair or replace products that have defects or other problems that make them unsuitable for service, as well as those who perform routine preventive maintenance on products. This category of customer often plays a major role in end user satisfaction, for better or for worse. Maintenance customers are a potential goldmine of product improvement ideas. They are often more familiar with the actual performance of the product than the original designers, and may also have intimate knowledge of competitive products. Especially in larger companies, designers may be relatively isolated from other designers, with little understanding of how their design piece would fit with the rest of the design. Maintenance customers work with complete products under actual field conditions. This, combined with intimate end user contact, makes their input invaluable for design improvement. Adding maintenance customers to the product design team can produce big payoffs.

In practice, these categories may become somewhat ambiguous, as people fill different role at different times. The key to keeping the correct perspective is to understand that the unifying criterion is the satisfaction of the end user. All activities in the process are being performed to achieve this final result.

Customer Identification and Segmentation

Customer identification and segmentation is part of the strategic marketing activity. The basis of customer identification and segmentation is that the firm cannot be all things to all people. There are many possible reasons this might be so: the customers' buying requirements vary, customers are geographically dispersed, the competition can serve some market needs better than you, etc. If relatively homogenous groups of customers can be identified, it will be easier to target products and services that meet these needs.

A three-step strategy is recommended:

1. Identify homogenous market segments. Identify segmentation variables and segment the market. Develop profiles of the resulting segments.
2. Decide which market segments you wish to enter. Evaluate the attractiveness of each segment. Select the target segment(s).
3. Develop a product or service positioned to appeal to the special requirements of the selected market segments. Identify possible positioning concepts for each target segment. Select, develop, and communicate the chosen positioning concept.

Market segmentation is typically conducted by evaluating existing and prospective customer motivations, attitudes, and behavior through interviews, focus groups, and surveys. Product/service attributes are evaluated, including their importance ratings. Brand awareness and brand ratings are studied, as are product-usage patterns. Due to the strategic importance of these studies, extreme care must be taken to ensure that samples are representative of the market as a whole and of the segments being studied. These studies belong to the enumerative class of statistical studies (see Chap. 9). Professional assistance is a good investment, as advanced statistical techniques are often employed, such as cluster analysis, factor analysis, or regression analysis. Once identified, clusters are profiled by attitudes, demographics, etc. Cluster labels are determined by studying the dominant characteristic of customers within the cluster.

From this analysis, target segment(s) are determined. The strategy used for selecting the target segment(s) may form a basis for the development of the strategic quality plan.

Key segmentation concepts are illustrated in Fig. 6.2.

- *No market segmentation* (A). Here we have a completely undifferentiated set of customers. When this situation exists, each customer is viewed as being no different from any other customer. *Mass marketing* is used. Uniform quality is the goal.

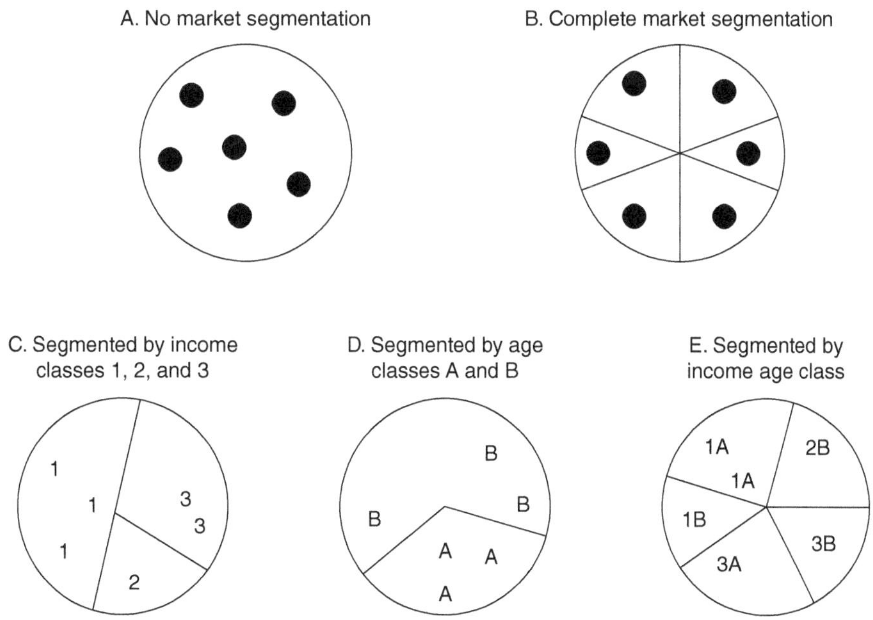

FIGURE 6.2 Segmentation concepts.

- *Complete market segmentation* (B). Each customer is viewed as possessing a unique set of requirements. *Customized marketing* is used. Quality requirements are set by each customer's individual demands.
- *Segmentation by a single criterion* (C, D). A single differentiating factor (e.g., income or age) is used. A different marketing approach is used for each level of the differentiating factor. Quality requirements vary for each level of the differentiating factor.
- *Segmentation by multiple criteria* (E). Two or more differentiating factors are used. For example, customers might be classified by both income and age. Again, marketing strategy and quality requirements may vary for each segment.

If a segment is small relative to the entire market, and if the competition in the segment involves few or no competitors, it is said to be a market "niche." Market niches are often deliberately selected for special products or services for which the company possesses a competitive advantage.

While it may be simple to conceptualize quality requirements that vary by customer segment, implementing these different requirements in a single company is problematic. Generally, firms have only one quality *policy*, and requirements tend to flow from the policy. Also, customers tend to attach images to specific brand names, and the reputation for quality plays a role in the customer's image. Thus, when companies wish to appeal to widely divergent customer types they sometimes find it necessary to set up completely separate entities. For example, Toyota created Lexus as a separate entity to manufacture and market its luxury car line.

While segments are often selected based on demographic criteria (e.g., age, income), this is not the only way to segment customers. Of particular interest to quality managers is segmentation by product attributes. Quality can be considered a product attribute and marketed directly to selected customer segments. More often, quality is marketed indirectly, as is the case with luxury versus economy automobiles. The quality manager should ensure that the quality requirements *of the customer segment* are being met by the product or service offered.

The basis of segmentation is to identify groups of customers with similar likes and dislikes. In other words, demographics and other differences are surrogates for customer preferences. Three broad patterns of preferences are illustrated in Fig. 6.3.

- *Homogeneous preferences* indicate a market where all consumers have roughly the same preferences. There are no *natural segments* as far as the two attributes are concerned. We would predict that competing brands would be similar and located near the center.
- *Diffused preferences* indicate a market where consumer preferences vary a great deal. Again, there are no natural segments. The center of the space minimizes the sum of consumer dissatisfaction. However, if several competitors exist, we would predict that they offer dissimilar products to match consumer preferences.

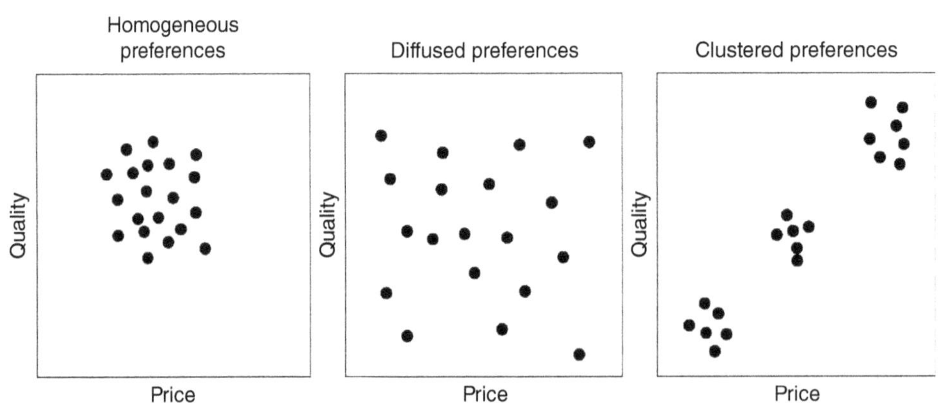

FIGURE 6.3 Preference segment patterns.

- *Clustered preferences* represent *natural market segments*. When distinct preference clusters exist, we predict that competing products would be dissimilar between clusters, and similar within the clusters. An example would be luxury, economy, and sporty automobiles.

Collecting Data on Customer Expectations and Needs

Since an organization exists to serve its customers, it is essential that the organization continually evaluate its ability to serve the customers. As discussed, different customers are likely to have different expectations, partly due to market differences, as well as their perceptions of expected quality. Since these factors are likely to change over time, as shown in the Kano model, it is essential the organization develop and maintain systems to acquire this data, and use it for strategic planning.

The primary objective of this data collection is the evaluation of the customers' perception of the firm's product and service quality and its impact on customer satisfaction. To be effective, the communication will provide sufficient detail to help the firm determine how they could improve quality and satisfaction, as part of their continuous improvement efforts.

There are several primary strategies commonly used to obtain information from or about customers: customer service and support, surveys, case studies, and field experiments.

Data is collected for sample surveys from a sample of a universe to estimate the characteristics of the universe, such as their range or dispersion, the frequency of occurrence of events, or the expected values of important universe parameters. A given survey is applied to one or more statistical populations, and results analyzed using the enumerative statistical tools described in Chap. 9. When survey results are collected at regular intervals, the results can be analyzed using the analytical tools of statistical process control, as described in Chap. 9 to obtain information on the underlying process. There is inherent benefit to the analytical approach, as small, routine, periodic surveys can provide more relevant and timely data than infrequent enumerative studies. Without the information available from time-ordered series of data, it will not be possible to learn about processes that produce changes in customer satisfaction or perceptions of quality.

A case study is an analytic description of the properties, processes, conditions, or variable relationships of either single or multiple units under study. Sample surveys and case studies are usually used to answer descriptive questions ("How do things look?") and normative questions ("How well do things compare with our requirements?"). A field experiment seeks the answer to a cause-and-effect question ("Did the change result in the desired outcome?"). Complaint and suggestion systems typically provide all customers with an easy-to-use method of providing favorable or unfavorable feedback to management. Due to selection bias, these methods do not provide statistically valid information. However, because they are a census rather than a sample, they provide opportunities for individual customers to have their say. These are moments of truth that can be used to increase customer loyalty. They also provide anecdotes that have high face validity and are often a source of ideas for improvement.

Customer Service and Support

A key source of customer information is your sales, service, and support staff who communicate with customers on a daily basis, sometimes referred to as a customer contact workers (CCW). The nature of the contact can be face-to-face encounters, telephone communication, or written correspondence such as emails or online chats. To customers, these workers are "the organization," so their importance to customer satisfaction is obvious. Yet in traditional organizations CCWs are often among the least experienced and lowest paid employees of the firm, resulting in customer dissatisfaction well documented in the literature.

CCWs are placed in "boundary positions" (where the organization meets the outside world), creating a number of stresses not experienced by other members of the organization. As outsiders, customers make demands upon the CCW. The CCW may not be able to meet the customer demands due to organizational policies and restrictions. Furthermore, the CCW must present the organization to the customer in a positive manner that doesn't unduly anger or annoy the customer, while retaining a customer perspective within the organization. Given the low status of many CCW positions, with little ability to exercise control over internal policy, there is a significant challenge with these divided loyalties, resulting in stress and confusion, often to both the CCW and the customer.

This situation is best addressed by giving the worker the authority to act in a wide range of customer situations. This empowerment results in the greatest customer satisfaction. In essence, empowerment turns the organizational hierarchy upside down. Carlzon (1987) views each customer contact as a "moment of truth" where the commitment to serve the customer is put to the test. Carlzon maintains that the purpose of management is to design organizational systems to ensure that moments of truth are properly handled from the customer's perspective. (Deming would agree.) Carlzon uses the analogy of a soccer coach, who can neither dribble down the field nor provide constant and immediate instructions to players on shooting, passing, or defense. Rather, the coach's responsibility is to develop the players' skills and empower them to exercise judgment in the use of those skills. Without empowerment, the CCW has little power to influence the outcome of the moment of truth.

Empowerment requires discretion and awareness of consequence, both of which can be greatly enhanced through proper training. Firms with excellent reputations for customer service typically provide their CCWs with extensive up-front training prior to placing them on the "front line." Continuous classroom and on-the-job training of CCWs accounts for an additional 1 to 5 percent of the CCW's working hours. The best firms provide training that is formal, rigorous, and ongoing. It may involve scenarios and role-playing, but must include relevant (sometimes in-depth) coverage of the cross-functional processes impacted by the potential decisions associated with the CCW's role.

Despite the advantages of training, there are some traits of effective CCWs that cannot be provided through training. Developing effective customer service systems requires a selection process designed to identify candidates with the proper psychological traits for the job, such as patience, ability to handle stress in a positive and congenial manner, proper communication skills, and so on. Often, recommendations and references from current CCWs or previous employers will be beneficial. Various psychological tests can also be used, such as personality profiles or tests of a person's ability to detect non-verbal cues. Many firms use structured interviews to determine how candidates respond to various situations, particularly to complaints.

Effective recovery from complaints is an important element in customer satisfaction and retention. Although complaints arise from a variety of sources, the most prevalent are product defects, errors in service, untimely service, poor communication, and inadequate company systems and processes, even during the complaint process itself. Not all problems result in complaints. It has been shown that the percentage of defects or service failures that actually result in a complaint is related to the seriousness of the problem and the price paid for the product or service. Minor problems are reported less than major problems; problems are reported more often when higher prices are paid. Since only a portion of the customers who have experienced a given problem will take the time to report it, systems should be designed to document the complaint with the intent to review and ascertain the true extent of the problem. This feedback can be useful in identifying opportunities and threats to meeting customer and market expectations.

Given a suitable company policy and procedure for complaints, complaint processing becomes largely a communications matter between the CCW and the customer. The most important activity in the process is to listen to the customer. Listening skills should be routinely taught to all CCWs; these skills are absolutely essential to complaint processors. The CCW should attempt to understand the nature, magnitude, and potential impact of the complaint on the customer. They should agree with the customer that the complaint is valid, and offer the company's apology for the inconvenience. CCWs should scrupulously avoid arguing or becoming defensive. Every effort should be made to get the facts. If possible, an immediate solution should be offered. If resolving the problem cannot be accomplished on the spot, the customer should be told precisely what steps will be taken to resolve the problem. During the resolution process the customer should be kept informed of the status of the complaint.

Many complaints are not covered under legally binding agreements, such as warranties. Instead, the emphasis is on maintaining good relations with the customer by honoring implicit "guarantees." In these situations, the spirit of the understanding between the company and its customer is more important than the letter of a legal agreement. The message conveyed to the customer should be "You are a very important person to us." After all, it's the truth.

Although it may not appear intuitive, research has shown that complaints are an excellent opportunity to gain customer loyalty. Customers who are satisfied with the way in which their complaints are handled are *more likely* to patronize a firm in the future than customers who had no complaints! Proper complaint handling, from the customer's perspective, usually involves receiving a courteous, quick, and fair response to the immediate issue involved, for example, replacing a defective item. Making certain that this happens routinely should not be left to chance. Complaint handling, like everything else, is a process. As such, it should be designed to accomplish its goal (a satisfied customer), tested to ensure that it is properly designed, and continuously improved.

While complaint handling is important, it shouldn't be forgotten that complaints are undesirable events that should be prevented. Complaint prevention should be an ongoing activity, which requires data collection to understand the nature of its occurrence. As with all process data, the statistical control chart (described in Chap. 9) is the proper tool for analysis, providing differentiation between sporadic sources of process variation and systematic causes of variation. This distinction is critical for identifying the nature of the opportunity or threat, as well as the proper response.

While general sales and customer service principles are often transaction oriented, the customer sales and service should always be designed and conducted with the larger concept of relationship management in mind. Relationship management is most appropriate for those customers who can most affect the company's future. For many companies, the top 20 percent of customers account for a disproportionate share of the company's sales. Contact with such customers should not be limited to sales calls. These key accounts should be monitored to ensure that their special needs are being addressed. The relationship should be viewed as a partnership, with the supplier proactively offering suggestions as to how the customer might improve their sales, quality, etc. For larger accounts, a formal relationship management program should be considered.

Kotler (1991, pp. 679–680) presents the following steps for establishing a relationship management program within a company:

Identify the key customers meriting relationship management. The company can choose the 5 or 10 largest customers and designate them for relationship management. Additional customers can be added who show exceptional growth or who pioneer new developments in the industry, and so on.

Assign a skilled relationship manager to each key customer or customer group. The salesperson who is currently servicing the customer should receive training in relationship management or be replaced by someone who is more skilled in relationship management. The relationship manager should have characteristics that match or appeal to the customer.

Develop a clear job description for relationship managers. It should describe their reporting relationships, objectives, responsibilities, and evaluation criteria. The relationship manager is responsible for the client, is the focal point for all information about the client, and is responsible for mobilizing company services to the client. Each relationship manager will have only one or a few relationships to manage.

Appoint an overall manager to supervise the relationship managers. This person will develop job descriptions, evaluation criteria, and resource support to increase relationship managers' effectiveness.

Each relationship manager must develop long-range and annual customer-relationship plans. The annual relationship plan will state objectives, strategies, specific actions, and required resources.

Figure 6.4 lists a number of relationship factors.

Surveys

Survey development consists of the following major tasks (GAO, 1986, p. 15):

- Initial planning of the questionnaire
- Developing the measures
- Designing the sample
- Developing and testing the questionnaire
- Producing the questionnaire
- Preparing and distributing mailing materials
- Collecting data
- Reducing the data to forms that can be analyzed
- Analyzing the data

Figure 6.5 shows a typical timetable for the completion of these tasks.

THINGS AFFECTING RELATIONSHIPS

Good Things	*Bad Things*
Initiate positive phone calls	Make only callbacks
Make recommendations	Make justifications
Can do in language	Accommodative language
Use phone	Use correspondence
Show appreciation	Wait for misunderstandings
Make service suggestions	Wait for service requests
Use "we" problem-solving language	Use "owe us" legal language
Get to problems	Only respond to problems
Use jargon/shorthand	Use long-winded communication
Personality problems aired	Personality problems hidden
Talk of "our future together"	Talk about making good on the past
Routinize responses	Fire drill/emergency responsiveness
Accept responsibility	Shift blame
Plan the future	Rehash the past

FIGURE 6.4 Actions affecting buyer-seller relationships. (Levitt, 1983.)

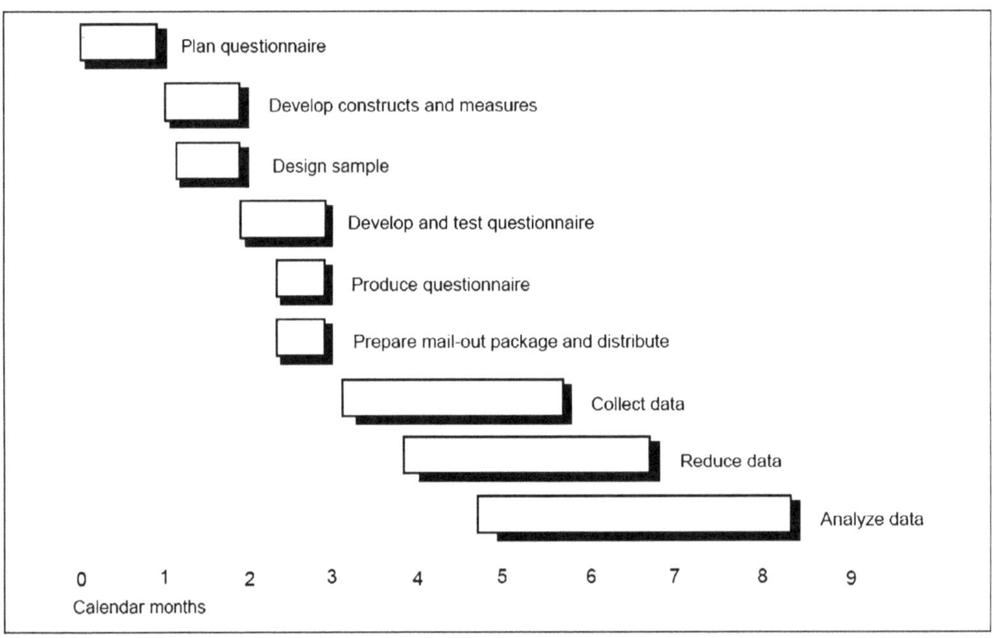

FIGURE 6.5 Typical completion times for major questionnaire tasks (GAO, 1986).

Guidelines for Developing Questions

The axiom that underlies the guidelines shown below is that the question writer(s) must be thoroughly familiar with the respondent group and must understand the subject matter from the perspective of the respondent group. This is often problematic for the quality professional when the respondent group is the customer; methods for dealing with this situation are discussed below. There are eight basic guidelines for writing good questions:

1. Ask questions in a format that is appropriate to the questions' purpose and the information required.
2. Make sure the questions are relevant, proper, and qualified as needed.
3. Write clear, concise questions at the respondent's language level.
4. Give the respondent a chance to answer by providing a comprehensive list of relevant, mutually exclusive responses from which to choose.
5. Ask unbiased questions by using appropriate formats and item constructions and by presenting all important factors in the proper sequence.
6. Get unbiased answers by anticipating and accounting for various respondent tendencies.
7. Quantify the response measures where possible.
8. Provide a logical and unbiased line of inquiry to keep the reader's attention and make the response task easier.

The above guidelines apply to the *form* of the question. Using the critical incident technique to develop good question *content* is described below.

Response Types

There are several commonly used types of survey responses.

- *Open-ended questions.* These are questions that allow the respondents to frame their own response without any restrictions placed on the response. The primary advantage is that such questions are easy to form and ask using natural language, even if the question writer has little knowledge of the subject matter. Unfortunately, there are many problems with analyzing the answers received to this type of question. This type of question is most useful in determining the scope and content of the survey, not in producing results for analysis or process improvement.

- *Fill-in-the-blank questions.* Here the respondent is provided with directions that specify the units in which the respondent is to answer. The instructions should be explicit and should specify the answer units. This type of question should be reserved for very specific requests, for instance, "What is your age on your last birthday? (age in years)."
- *Yes/no questions.* Unfortunately, yes/no questions are very popular. Although they have some advantages, they have many problems and few uses. Yes/no questions are ideal for dichotomous variables, such as defective or not defective. However, too often this format is used when the measure spans a range of values and conditions, for example, "Were you satisfied with the quality of your new car (yes/no)?" A yes/no response to such questions contains little useful information.
- *Ranking questions.* The ranking format is used to rank options according to some criterion, for example, importance. Ranking formats are difficult to write and difficult to answer. They give very little real information and are very prone to errors that can invalidate all the responses. They should be avoided whenever possible in favor of more powerful formats and formats less prone to error, such as rating. When used, the number of ranking categories should not exceed five.
- *Rating questions.* With this type of response, a rating is assigned on the basis of the score's absolute position within a range of possible values. Rating scales are easy to write, easy to answer, and provide a level of quantification that is adequate for most purposes. They tend to produce reasonably valid measures. Here is an example of a rating format:

  For the following statement, check the appropriate box:
  The workmanship standards provided by the purchaser are
  - *Clear*
  - *Marginally adequate*
  - *Unclear*
- *Guttman format.* In the Guttman format, the alternatives increase in comprehensiveness; that is, the higher-valued alternatives include the lower-valued alternatives. For example,

  Regarding the benefit received from training in quality improvement:
  - *No benefit identified*
  - *Identified benefit*
  - *Measured benefit*
  - *Assessed benefit value in dollar terms*
  - *Performed cost/benefit analysis*

- *Likert and other intensity scale formats.* These formats are usually used to measure the strength of an attitude or an opinion. For example,

  Please check the appropriate box in response to the following statement: "The quality auditor was knowledgeable."
  - *Strongly disagree*
  - *Disagree*
  - *Neutral*
  - *Agree*
  - *Strongly agree*

  Intensity scales are very easy to construct. They are best used when respondents can agree or disagree with a statement. A problem is that statements must be worded to present a single side of an argument. We know that the respondent agrees, but we must infer what he believes. To compensate for the natural tendency of people to agree, statements are usually presented using the converse as well, for instance, "The quality auditor was not knowledgeable."

  When using intensity scales, use an odd-numbered scale, preferably with five categories. If there is a possibility of bias, order the scale in a way that favors the hypothesis you want to disprove and handicaps the hypothesis you want to confirm. This will confirm the hypothesis with the bias against you—a stronger result. If there is no bias, put the most undesirable choices first.

- *Semantic differential format.* In this format, the values that span the range of possible choices are not completely identified; only the end points are labeled. For example,

  Indicate the number of times you initiated communication with your customer in the past month.

few								many
(2 or less)	■	■	■	■	■	■	■	(20 or more)

  The respondent must infer that the range is divided into equal intervals. The range seems to work better with seven categories rather than the usual five.

  Semantic differentials are very useful when we do not have enough information to anchor the intervals between the poles. However, they are very difficult to write well, and if not written well the results are ambiguous.

Survey Development Case Study

This actual case study involves the development of a mail survey at a community hospital. The same process has been used by the author to develop customer surveys for clientele in a variety of industries.

The study of service quality and patient satisfaction was performed at a 213-bed community hospital in the southwestern United States. The hospital is a nonprofit, publicly funded institution providing services to the adult community; pediatric services are not provided. The purpose of the study was to:

- Identify the determinants of patient quality judgments.
- Identify internal service delivery processes that impacted patient quality judgments.
- Determine the linkage between patient quality judgments and intent-to-patronize the hospital in the future or to recommend the hospital to others.

To conduct the study, the author worked closely with a core team of hospital employees, and with several ad hoc teams of hospital employees. The core team included the nursing administrator, the head of the Quality Management Department, and the head of nutrition services.

The team decided to develop their criteria independently. It was agreed that the best method of getting information was directly from the target group, in-patients. Due to the nature of hospital care services, focus groups were not deemed feasible for this study. Frequently, patients must spend a considerable period of time convalescing after being released from a hospital, making it impossible for them to participate in a focus group soon after discharge. While the patients are in the hospital, they are usually too sick to participate. Some patients have communicable diseases, which makes their participation in focus groups inadvisable.

Since memories of events tend to fade quickly (Flanagan, 1954, p. 331), the team decided that patients should be interviewed within 72 hours of discharge. The target patient population was, therefore, all adults treated as in-patients and discharged to their homes. The following groups were not part of the study: families of patients who died while in the hospital, patients discharged to nursing homes, and patients admitted for psychiatric care.

The team used the Critical Incident Technique (CIT) to obtain patient comments. The CIT was first used to study procedures for selection and classification of pilot candidates in World War II (Flanagan, 1954). A bibliography assembled in 1980 listed over seven hundred studies about or using the CIT (Fivars, 1980). Given its popularity, it is not surprising that the CIT has also been used to evaluate service quality.

CIT consists of a set of specifically defined procedures for collecting observations of human behavior in such a way as to make them useful in addressing practical problems. Its strength lies in carefully structured data collection and data classification procedures that produce detailed information not available through other research methods. The technique, using either direct observation or recalled information collected via interviews, enables researchers to gather firsthand patient-perspective information. This kind of self-report preserves the richness of detail and the authenticity of personal experience of those closest to the activity being studied. Researchers have concluded that the CIT produces information that is both reliable and valid.

This study attempted to follow closely the five steps described by Flanagan as crucial to the CIT: (1) establishment of the general aim of the activity studied, (2) development of a plan for observers or interviewers, (3) collection of data, (4) analysis (classification) of data, and (5) interpretation of data.

Establishment of the General Aim of the Activity Studied The general aim is the purpose of the activity. In this case the activity involves the whole range of services provided to in-patients in the hospital. This includes every service activity between admission and discharge. From the service provider's perspective the general aim is to create and manage service delivery processes in such a way as to produce a willingness by the patient to utilize the provider's services in the future. To do this the service provider must know which particular aspects of the service are remembered by the patient.

Our general aim was to provide the service provider with information on what patients remembered about their hospital stay, both pleasant and unpleasant. This information was to be used to construct a new patient survey instrument that would be sent to recently discharged patients on a periodic basis. The information obtained would be used by the managers of the various service processes as feedback on their performance, from the patient's perspective.

Interview Plan Interviewers were provided with a list of patients discharged within the past 3 days. The discharge list included all patients. Nonpsychiatric patients who were discharged to "home" were candidates for the interview. Home was defined as any location other than the morgue or a nursing home. Interviewers were instructed to read a set of predetermined statements. Patients to be called were selected at random from the discharge list. If a patient could not be reached, the interviewer would try again later in the day. One interview form was prepared per patient. To avoid bias, 50 percent of the interview forms asked the patient to recall unpleasant incidents first and 50 percent asked for pleasant incidents first. Interviewers were instructed to record the patient responses using the patient's own words.

Collection of Data Four interviewers participated in the data collection activity; all were management-level employees of the hospital. Three of the interviewers were female, and one was male. The interviews were conducted when time permitted during the interviewer's normal, busy workday. The interviews took place during September 1993. Interviewers were given the instructions recommended by Hayes (1992, pp. 14–15) for generating critical incidents.

A total of 36 telephone attempts were made and 23 patients were reached. Of those reached, three spoke only Spanish. In the case of one of the Spanish-speaking patients, a family member was interviewed. Thus, 21 interviews were conducted, which is slightly greater than the 10 to 20 interviews recommended by Hayes (1992, p. 14). The 21 interviews produced 93 critical incidents.

Classification of Data The Incident Classification System required by CIT is a rigorous, carefully designed procedure with the end goal being to make the data useful to the problem at hand while sacrificing as little detail as possible (Flanagan, 1954, p. 344). There are three issues in doing so: (1) identification of a general framework of reference that will account for all incidents, (2) inductive development of major area and subarea categories that will be useful in sorting the incidents, and (3) selection of the most appropriate level of specificity for reporting the data.

The critical incidents were classified as follows:

- Each critical incident was written on a 3 × 5 card, using the patient's own words.
- The cards were thoroughly shuffled.
- Ten percent of the cards (10 cards) were selected at random, removed from the deck and set aside.
- Two of the four team members left the room while the other two grouped the remaining 83 cards and named the categories.
- The 10 cards originally set aside were placed into the categories found in step 3.
- Finally, the two members not involved in the initial classification were told the names of the categories. They then took the reshuffled 93 cards and placed them into the previously determined categories.

The above process produced the following dimensions of critical incidents:

- Accommodations (5 critical incidents)
- Quality of physician (14 critical incidents)
- Care provided by staff (20 critical incidents)
- Food (26 critical incidents)
- Discharge process (1 critical incident)
- Attitude of staff (16 critical incidents)
- General (11 critical incidents)

Interpretation of Data Inter-judge agreement, the percentage of critical incidents placed in the same category by both groups of judges, was 93.5 percent. This is well above the 80 percent cutoff value recommended by experts. The setting aside of a random sample and trying to place them in established categories is designed to test the comprehensiveness of the categories. If any of the withheld items were not classifiable it would be an indication that the categories do not adequately span the patient satisfaction space. However, the team experienced no problem in placing the withheld critical incidents into the categories.

Ideally, a critical incident has two characteristics: (1) it is *specific* and (2) it describes the service provider in *behavioral terms* or the service product with *specific adjectives* (Hayes, 1992, p. 13). Upon reviewing the critical incidents in the General category, the team determined that these items failed to have one or both of these characteristics. Thus, the 11 critical incidents in the General category were dropped. The team also decided to merge the two categories "Care provided by staff" and "Attitude of staff" into the single category "Quality of staff care." Thus, the final result was a five-dimension model of patient satisfaction judgments: Food, Quality of physician, Quality of staff care, Accommodations, and Discharge process.

A rather obvious omission in the above list is billing. This occurred because the patients had not yet received their bill within the 72-hour time frame. However, the patient's bill was explained to the patient prior to discharge. This item is included in the Discharge process dimension. The team discussed the billing issue and it was determined that billing complaints do arise after the bills are sent, suggesting that billing probably is a satisfaction dimension. However, the team decided not to include billing as a survey dimension because (1) the time lag was so long that waiting until bills had been received would significantly reduce the ability of the patient to recall the details of their stay, (2) the team feared that the patients' judgments would be overwhelmed by the recent receipt of the bill, and (3) a system already existed for identifying patient billing issues and adjusting the billing process accordingly.

Survey Item Development As stated earlier, the general aim was to provide the service provider with information on what patients remembered about their hospital stay, both pleasant and unpleasant. This information was then to be used to construct a new patient survey instrument that would be sent to recently discharged patients on a periodic basis. The information obtained would be used by the managers of the various service processes as feedback on their performance, from the patient's perspective.

The core team believed that accomplishing these goals required that the managers of key service processes be actively involved in the creation of the survey instrument. Thus, ad hoc teams were formed to develop survey items for each of the dimensions determined by the critical incident study. The teams were given brief instruction by the author in the characteristics of good survey items. Teams were required to develop items that, in the opinion of the core team, met five criteria: (1) relevance to the dimension being measured, (2) concise, (3) unambiguous, (4) one thought per item, and (5) no double negatives. Teams were also shown the specific patient comments that were used as the basis for the categories and informed that these comments could be used as the basis for developing survey items.

Writing items for the questionnaire can be difficult. The process of developing the survey items involved an average of three meetings per dimension, with each meeting lasting approximately 2 hours. Ad hoc teams ranged in size from 4 to 11 members. The process was often quite tedious, with considerable debate over the precise wording of each item.

The core team discussed the scale to be used with each ad hoc team. The core team's recommended response format was a five point Likert-type scale. The consensus was to use a five point *agree-disagree* continuum as the response format. Item wording was done in such a way that agreement represented better performance from the hospital's perspective.

In addition to the response items, it was felt that patients should have an opportunity to respond to open-ended questions. Thus, the survey also included general questions that invited patients to comment in their own words. The benefits of having such questions is well known. In addition, it was felt that these questions might generate additional critical incidents that would be useful in expanding the scope of the survey.

The resulting survey instrument contained 50 items and three open-ended questions.

Survey Administration and Pilot Study The survey was to be tested on a small sample. It was decided to use the total design method (TDM) to administer the survey (Dillman, 1983). Although the total design method is exacting and tedious, Dillman indicated that its use would ensure a high rate of response. Survey administration would be handled by the Nursing Department.

TDM involves rather onerous administrative processing. Each survey form is accompanied by a cover letter, which must be hand-signed in blue ink. Follow up mailings are done one, three, and seven weeks after the initial mailing. The three- and seven-week follow-ups are accompanied by another survey and another cover letter. No "bulk processing" is allowed, such as the use of computer-generated letters or mailing labels. Dillman's research emphasizes the importance of viewing the TDM as a completely integrated approach (Dillman, 1983, p. 361).

Because the hospital in the study is small, the author was interested in obtaining maximum response rates. In addition to following the TDM guidelines, he recommended that a $1 incentive be included with each

	Delivered	Returned	%
Surveys	92	45	49%
Surveys with $1 incentive	47	26	55%
Surveys without $1 incentive	45	19	42%

Table 6.1 Pilot Patient Survey Return Information

survey. However, the hospital administrator was not convinced that the additional $1 per survey was worthwhile. It was finally agreed that to test the effect of the incentive on the return rate $1 would be included in 50 percent of the mailings, randomly selected.

The hospital decided to perform a pilot study of 100 patients. The patients selected were the first 100 patients discharged to home. The return information is shown in Table 6.1.

Although the overall return rate of 49 percent is excellent for normal mail-survey procedures, it is substantially below the 77 percent average and the 60 percent "minimum" reported by Dillman. As possible explanations, the author conjectures that there may be a large Spanish-speaking constituency for this hospital. As mentioned above, the hospital is planning a Spanish version of the survey for the future.

The survey respondent demographics were analyzed and compared to the demographics of the nonrespondents to ensure that the sample group was representative. A sophisticated statistical analysis was performed on the responses to evaluate the reliability and validity of each item. Items with low reliability coefficients or questionable validity were reworded or dropped.

Focus Groups

The focus group is a special type of group in terms of purpose, size, composition, and procedures. A focus group is typically composed of seven to ten participants who are unfamiliar with each other. These participants are selected because they have certain characteristic(s) in common that relate to the topic of the focus group.

The researcher creates a permissive environment in the focus group that nurtures different perceptions and points of view, without pressuring participants to vote, plan, or reach consensus. The group discussion is conducted several times with similar types of participants to identify trends and patterns in perceptions. Careful and systematic analysis of the discussions can provide clues and insights as to how a product, service, or opportunity is perceived.

A focus group can thus be defined as a carefully planned discussion designed to obtain perceptions on a defined area of interest in a permissive, nonthreatening environment. The discussion is relaxed, comfortable, and often enjoyable for participants as they share their ideas and perceptions. Group members influence each other by responding to ideas and comments in the discussion.

In quality management, focus groups are useful to:

- Gather insight useful in the strategic planning process.
- Generate ideas for survey questionnaires.
- Develop needs assessment, e.g., training needs.
- Test new program ideas.
- Determine customer decision criteria.
- Recruit new customers.

The focus group is a socially oriented research procedure. The advantage of this approach is that members stimulate one another, which may produce a greater number of comments than would individual interviews. If necessary, the researcher can probe for additional information or clarification. Focus groups produce results that have high face validity, that is, the results are in the participant's own words rather than in statistical jargon. The information is obtained at a relatively low cost, and they can be obtained very quickly.

There is less control in a group setting than with individual interviews. When group members interact, it is often difficult to analyze the resulting dialogue. The quality of focus group research is highly dependent on the qualifications of the interviewer. Trained and skilled interviewers are hard to find. Group-to-group variation can be considerable, further complicating the analysis. Finally, focus groups are often difficult to schedule.

CHAPTER 7

Benchmarking

Benchmarking is a popular method for developing requirements and setting goals. In more conventional terms, benchmarking can be defined as measuring your performance against that of best-in-class companies, determining how the best-in-class achieve those performance levels, and using the information as the basis for your own company's targets, strategies, and implementation. Benchmarking involves research into the best practices at the industry, firm, or process level. Benchmarking goes beyond a determination of the "industry standard"; it breaks the firm's activities down to process operations and looks for the best-in-class for a particular operation. For example, Xerox corporation studied the retailer LL Bean to help them improve their parts distribution process.

The benefits of competitive benchmarking include:

- Creating a culture that values continuous improvement to achieve excellence
- Enhancing creativity by devaluing the not-invented-here syndrome
- Increasing sensitivity to changes in the external environment
- Shifting the corporate mind-set from relative complacency to a strong sense of urgency for ongoing improvement
- Focusing resources through performance targets set with employee input
- Prioritizing the areas that need improvement
- Sharing the best practices between benchmarking partners

Benchmarking is based on learning from others, rather than developing new and improved approaches. Since the process being studied is there for all to see, a firm will find that benchmarking cannot give them a sustained competitive advantage. Although helpful, benchmarking should never be the *primary* strategy for improvement.

Competitive analysis is an approach to goal setting used by many firms. This approach is essentially benchmarking confined to one's own industry. Although common, competitive analysis virtually guarantees second-rate quality because the firm will always be following their competition. If the entire industry employs the approach it will lead to industry-wide stagnation, establishing opportunities for outside innovators.

Camp (1989) lists the following steps for the benchmarking process:

1. Planning
   1.1. Identify what is to be benchmarked
   1.2. Identify comparative companies
   1.3. Determine data collection method and collect data
2. Analysis
   2.1. Determine current performance "gap"
   2.2. Project future performance levels
3. Integration
   3.1. Communicate benchmark findings and gain acceptance
   3.2. Establish functional goals
4. Action
   4.1. Develop action plans
   4.2. Implement specific actions and monitor progress
   4.3. Recalibrate benchmarks
5. Maturity
   5.1. Leadership position attained
   5.2. Practices fully integrated into process

The first step in benchmarking is determining what to benchmark. To focus the benchmarking initiative on critical issues, begin by identifying the process outputs most important to the customers of that process (i.e., the key quality characteristics). This step applies to every organizational function, since each one has outputs and customers. The Quality Function Deployment (QFD) customer needs assessment, discussed in Chap. 15, is a natural precursor to benchmarking activities.

Getting Started with Benchmarking

The essence of benchmarking is the acquisition of information. The process begins with the identification of the process that is to be benchmarked. The process chosen should be one that will have a major impact on the success of the business. The rules used for identifying candidates for business process re-engineering can also be used here (see Chap. 2).

Once the process has been identified, contact a business library and request a search for the information relating to your area of interest. The library will identify material from a variety of external sources, such as magazines, journals, special reports, etc. You should also conduct research using the internet and other electronic networking resources. However, be prepared to pare down what will probably be an extremely large list of candidates (e.g., an internet search on the word *benchmarking* produced nearly 20 million hits). Don't forget your organization's internal resources. If your company has an "intranet," use it to conduct an internal search. Set up a meeting with people in key departments, such as R&D. Tap the expertise of those in your company who routinely work with customers, competitors, suppliers, and other "outside" organizations. Often your company's board of directors will have an extensive network of contacts.

The search is, of course, not random. You are looking for the best of the best, not the average firm. There are many possible sources for identifying the elites. One approach is to build a compendium of business awards and citations of merit that organizations have received in business process improvement. Sources to consider are *Industry Week*'s Best Plant's Award, National Institute of Standards and Technology's Malcolm Baldrige Award, *USA Today* and the Rochester Institute of Technology's Quality Cup Award, European Foundation for Quality Management Award, Occupational Safety and Health Administration (OSHA), Federal Quality Institute, Deming Prize, Competitiveness Forum, *Fortune* magazine, and United States Navy's Best Manufacturing Practices, to name just a few. You may wish to subscribe to an "exchange service" that collects benchmarking information and makes it available for a fee. Once enrolled, you will have access to the names of other subscribers—a great source for contacts.

Don't overlook your own suppliers as a source for information. If your company has a program for recognizing top suppliers, contact these suppliers and see if they are willing to share their "secrets" with you. Suppliers are predisposed to cooperate with their customers; it's an automatic door-opener. Also contact your customers. Customers have a vested interest in helping you do a better job. If your quality, cost, and delivery performance improve, your customers will benefit. Customers may be willing to share some of their insights as to how their other suppliers compare with you. Again, it isn't necessary that you get information about direct competitors. Which of your customer's suppliers are best at billing? Order fulfillment? Customer service? Keep your focus at the process level and there will seldom be any issues of confidentiality. An advantage to identifying potential benchmarking partners through your customers is that you will have a referral that will make it easier for you to start the partnership.

Another source for detailed information on companies is academic research. Companies often allow universities access to detailed information for research purposes. While the published research usually omits reference to the specific companies involved, it often provides comparisons and detailed analysis of what separates the best from the others. Such information, provided by experts whose work is subject to rigorous peer review, will often save you thousands of hours of work.

After a list of potential candidates is compiled, the next step is to choose the best three to five targets. A candidate that looked promising early in the process might be eliminated later for any number of reasons, including poor performance, a lack of commitment to sharing information or practices, low availability, or questionable value of information (Vaziri, 1992).

As the benchmarking process evolves, the characteristics of the most desirable candidates will be continually refined. This occurs as a result of a clearer understanding of your organization's key quality characteristics and critical success factors and an improved knowledge of the marketplace and other players. This knowledge and the resulting actions tremendously strengthen an organization.

Why Benchmarking Efforts Fail

The causes of failed benchmarking projects are the same as those for other failed projects (DeToro, 1995):

- *Lack of sponsorship.* A team should submit to management a one- to four-page benchmarking project proposal that describes the project, its objectives, and potential costs. If the team can't gain approval for the project or get a sponsor, it makes little sense to proceed with a project that's not understood or appreciated or that is unlikely to lead to corrective action when completed.
- *Wrong people on team.* Individuals involved in benchmarking should own or work in the process under review. It's useless for a team to address problems in business areas that are unfamiliar or where the team has no control or influence.
- *Teams don't understand their work completely.* If the benchmarking team didn't map, flowchart, or document its work process, and if it didn't benchmark with organizations that also documented their processes, there can't be an effective transfer of techniques. The intent in every benchmarking project is for a team to understand how its process works and compare it with another company's process at a detailed level. The exchange of process steps is essential for improved performance.
- *Teams take on too much.* Broad issues quickly become unmanageable and must be broken into smaller, more manageable projects that can be approached logically. A suggested approach is to create a functional flowchart of an entire area, such as production or marketing, and identify its processes. Criteria can then be used to select a subprocess to be benchmarked that would best contribute to the organization's objectives.
- *Lack of long-term management commitment.* Since managers aren't as familiar with specific work issues as their employees, they tend to

underestimate the time, cost, and effort required to successfully complete a benchmarking project. Managers should be informed that, while it's impossible to know the exact time it will take for a typical benchmarking project, a rule of thumb is that a team of four or five individuals requires a third of their time for 5 months to complete a project.

- *Focusing on metrics rather than processes.* Some firms focus their benchmarking efforts on performance targets (metrics) rather than processes. Knowing that a competitor has a higher return on assets doesn't mean that its performance alone should become the new target (unless an understanding exists about how the competitor differs in the use of its assets and an evaluation of its process reveals that it can be emulated or surpassed).
- *Not positioning benchmarking within a larger strategy.* Benchmarking is one of many total quality management tools—such as problem solving, process improvement, and process re-engineering—used to shorten cycle time, reduce costs, and minimize variation. Benchmarking is compatible with and complementary to these tools, and they should be used together for maximum value.
- *Misunderstanding the organization's mission, goals, and objectives.* All benchmarking activity should be launched by management as part of an overall strategy to fulfill the organization's mission and vision by first attaining the short-term objectives and then the long-term goals.
- *Assuming every project requires a site visit.* Sufficient information is often available from the public domain, making a site visit unnecessary. This speeds the benchmarking process and lowers the cost considerably.
- *Failure to monitor progress.* Once benchmarking has been completed for a specific area or process benchmarks have been established and process changes implemented, managers should review progress in implementation and results.

The best way of addressing these issues is to prevent their occurrence through carefully planning and managing the project from the outset. This list can be used as a checklist to evaluate project plans; if the plans don't clearly preclude these problems, then the plans are not complete.

CHAPTER 8

Organizational Assessment

Organizational assessments provide an understanding of the current state of the organization relative to specific goals or objectives related to the three main stakeholder groups: customers, shareholders, and employees.

Assessing Quality Culture

Juran and Gryna (1993) define an organization's *quality culture* as the opinions, beliefs, traditions, and practices concerning quality. While sometimes difficult to quantify, an organization's culture has a profound effect on the quality produced by that organization. Without an understanding of the cultural aspects of quality, significant and lasting improvements in quality levels are unlikely.

Two of the most common means of assessing organization culture are the focus group and the written questionnaire. The areas addressed generally cover attitudes, perceptions, and activities within the organization that impact quality. Because of the sensitive nature of cultural assessment, anonymity is usually necessary. The organization should develop its own set of questions, as the generation of questions is an education in itself. The critical-incident technique is a common method for developing questions that produce favorable results. The critical-incident technique involves selecting a small representative sample ($n \approx 20$) from the group you wish to survey and ask open-ended questions, such as:

> "Which of our organization's beliefs, traditions, and practices have a beneficial impact on quality?"
>
> "Which of our organization's beliefs, traditions, and practices have a detrimental impact on quality?"

The questions are asked by unbiased interviewers and the respondents are guaranteed anonymity. Although interviews are usually conducted in person or by phone, written responses are sometimes obtained. The order in which the questions are asked (beneficial/detrimental) is randomized to avoid bias in the answers. Interviewers are instructed not to prompt the respondent in any way. It is important that the responses be recorded verbatim, using the respondent's own words. Participants are urged to provide as many responses as they can; a group of 20 participants will typically produce 80 to 100 responses.

The responses typically provide a great deal of information. When grouped into categories, the categories may be examined to glean additional insight into the common themes. The responses and categories can be used to develop valid survey items (see Chap. 6) or to prepare focus-group questions. The follow-up activity is why so few people are needed at this stage—statistical validity is obtained during the survey stage.

The results of the quality culture survey will be used to understand the perceived strengths and weaknesses of the current quality initiative. It may identify sources of resistance to change, as well as frustrations with the status quo. Business-level improvement projects may be developed to focus on specifically changing some aspect of the quality culture, or more generally on transforming the quality effort to a more customer-focused approach.

Organizational Metrics

Organizational metrics, sometimes called Key Performance Indicators (KPI), are developed to understand the overall health of an organization. They provide the fundamental element of balanced scorecards and dashboards, which are used to quickly show how well the organization is performing relative to the past, a target, or both.

The choice of metric is important only so far as the metric is used to guide behavior or establish strategy. Poorly chosen metrics may lead to suboptimal behavior if they lead people away from the organization's goals instead of toward them. Joiner (1994) suggests three system-wide measures of performance: overall customer satisfaction, total cycle time, and first-pass quality. An effective metric for quantifying first-pass quality is total cost of poor quality (see Cost of Quality section). Once chosen, the metrics must be communicated to the members of the organization. To be useful, the employee must be able to influence the metric through his or her performance, and it must be clear precisely how the employee's performance influences the metric.

Rose (1995) lists the following attributes of good metrics:

- They are customer centered and focused on indicators that provide value to customers, such as product quality, service dependability, and timeliness of delivery, or are associated with internal work processes that address system cost reduction, waste reduction, coordination and teamwork, innovation, and customer satisfaction.
- They measure performance across time, which shows trends rather than snapshots.

- They provide direct information at the level at which they are applied. No further processing or analysis is required to determine meaning.
- They are linked with the organization's mission, strategies, and actions. They contribute to organizational direction and control.
- They are collaboratively developed by teams of people who provide, collect, process, and use the data.

Traditional KPI are established within four broad categories (Kaplan and Norton, 1992): customer-based, internal process, learning and growth, and financial. These may be evaluated as follows (Pyzdek and Keller, 2010):

Customer. Customers generally consider four broad categories in evaluating a supplier: Quality, Timeliness, Performance and Service, and Value. The customer communication methods outlined in the previous section will provide the means to understand the relative importance the customer base places on these categories, as well as their general expectations. On the basis of this feedback, internal goals may be defined, and operational metrics established from the goals. An example is provided in Table 8.1.

Internal process. These are the metrics perhaps most familiar to the operational personnel; however, in this case those operational metrics that strongly align with the strategic objectives are best suited. Joiner's recommendation of total cycle time (i.e., time to process the order) and first-pass quality are relevant indicators of internal process performance. Process cycle efficiency, calculated as the value-added time divided by the total lead time, or Overall Equipment Effectiveness are relevant Lean-focused metrics for evaluating internal performance and resource utilization.

Goal	Candidate Metrics
We will cut the time required to introduce a new product from 9 to 3 months	Average time to introduce a new product for most recent month or quarter Number of new products introduced in most recent quarter
We will be the best in the industry for on-time delivery	Percentage of on-time deliveries Best in industry on-time delivery percentage divided by our on-time delivery percentage Percentage of late deliveries
We will intimately involve our customers in the design of our next major product	Number of customers on design team(s) Number of customer suggestions incorporated in new design

Table 8.1 Defining Operational Metrics from Goals

Learning and growth. Metrics in this category might focus on the total deliverables (in dollars saved) from continuous improvement projects, new product or service development times, improvements in employee perspective or quality culture, revenue or market share associated with new product, and so on.

Financial. Many suitable financial metrics are available and widely tracked, including revenue, profitability, market share, and so on. Cost of quality, discussed in the next section, is also recommended.

Cost of Quality

The history of evaluating the cost of quality (sometimes referred to as the *cost of poor quality*) dates to the first edition of *Juran's QC Handbook* in 1951. Today, quality cost accounting systems are part of every modern organization's quality improvement strategy. Indeed, quality cost accounting and reporting are part of many quality standards. Quality cost systems help management plan for quality improvement by identifying opportunities for greatest return on investment. However, the quality manager should keep in mind that quality costs address only half of the quality equation. The quality equation states that quality consists of doing the right things and not doing the wrong things. "Doing the right things" implies developing product and service features that satisfy or delight the customer. "Not doing the wrong things" means avoiding defects and other behaviors that cause customer *dissatisfaction*. Quality costs address only the latter aspect of quality. It is conceivable that a firm could drive quality costs to zero and still go out of business.

The fundamental principle of the cost of quality is that any cost that would not have been expended if quality were perfect is a cost of quality. This includes such obvious costs as scrap and rework, but it also includes many costs that are far less obvious, such as the cost of reordering to replace defective material. Service businesses also incur quality costs; for example, a hotel incurs a quality cost when room service delivers a missing item to a guest. Specifically, quality costs are a measure of the costs specifically associated with the achievement or nonachievement of product or service quality—including all product or service requirements established by the company and its contracts with customers and society. Requirements include marketing specifications, end-product and process specifications, purchase orders, engineering drawings, company procedures, operating instructions, professional or industry standards, government regulations, and any other document or customer needs that can affect the definition of product or service. More specifically, quality costs are the total of the cost incurred by (a) investing in the *prevention* of nonconformances to requirements, (b) *appraising* a product or service for conformance to requirements, and (c) *failure* to meet requirements (see Fig. 8.1).

Prevention Costs
The costs of all activities specifically designed to prevent poor quality in products or services. Examples are the costs of new product review, quality planning, supplier capability surveys, process capability evaluations, quality improvement team meetings, quality improvement projects, and quality education and training.

Appraisal Costs
The costs associated with measuring, evaluating, or auditing products or services to assure conformance to quality standards and performance requirements. These include the costs of incoming and source inspection/test of purchased material; in process and final inspection/test; product, process, or service audits; calibration of measuring and test equipment; and the costs of associated supplies and materials.

Failure Costs
The costs resulting from products or services not conforming to requirements or customer/user needs. Failure costs are divided into internal and external failure cost categories.

Internal Failure Costs
Failure costs occurring prior to delivery or shipment of the product, or the furnishing of a service, to the customer. Examples are the costs of scrap, rework, re-inspection, retesting, material review, and down grading.

External Failure Costs
Failure costs occurring after delivery or shipment of the product, and during or after furnishing of a service, to the customer. Examples are the costs of processing customer complaints, customer returns, warranty claims, and product recalls.

Total Quality Costs
The sum of the above costs. It represents the difference between the actual cost of a product or service, and what the reduced cost would be if there were no possibility of substandard service, failure of products, or defects in their manufacture.

FIGURE 8.1 Quality costs—general description (Campanella, 1990, by permission).

For most organizations, quality costs are hidden costs. Unless specific quality cost identification efforts have been undertaken, few accounting systems include provision for identifying quality costs. Because of this, unmeasured quality costs tend to increase. Poor quality impacts companies in two ways: higher cost and lower customer satisfaction. The lower satisfaction creates price pressure and lost sales, which results in lower revenues. The combination of higher cost and lower revenues eventually brings a crisis that may threaten the very existence of the company. Rigorous cost of quality measurement is one technique for preventing such a crisis from occurring. Figure 8.2 illustrates the hidden cost concept.

The goal of any quality cost system is to reduce quality costs to the lowest practical level. Juran and Gryna (1988) present these costs graphically as shown in Fig. 8.3. In the figure it can be seen that the cost of failure declines as conformance quality levels improve toward perfection, while the cost of appraisal plus prevention increases. There is some "optimum" target quality level where the sum of prevention, appraisal, and failure

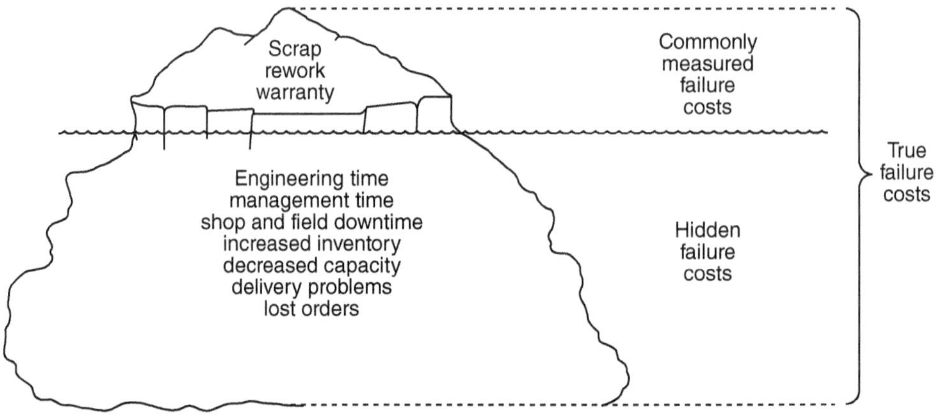

FIGURE 8.2 Hidden cost of quality and the multiplier effect (Campanella, 1990, by permission).

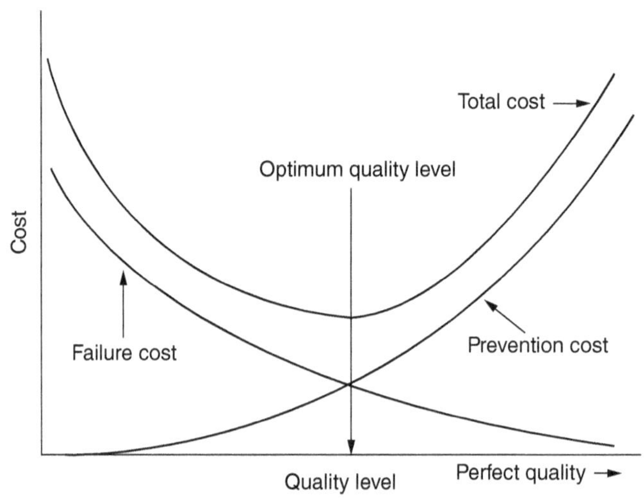

FIGURE 8.3 Classical model of optimum quality costs (Juran, 1988, by permission).

costs is at a minimum. Efforts to improve quality to better than the optimum level will result in increasing the total quality costs.

Juran acknowledged that in many cases the classical model of optimum quality costs is flawed. It is common to find that quality levels can be economically improved to literal perfection. For example, millions of stampings may be produced virtually error free from a well-designed and well-constructed stamping die. The classical model created a mind-set that perfection was not cost effective. The new model of optimum quality cost incorporates the possibility of zero defects and is shown in Fig. 8.4.

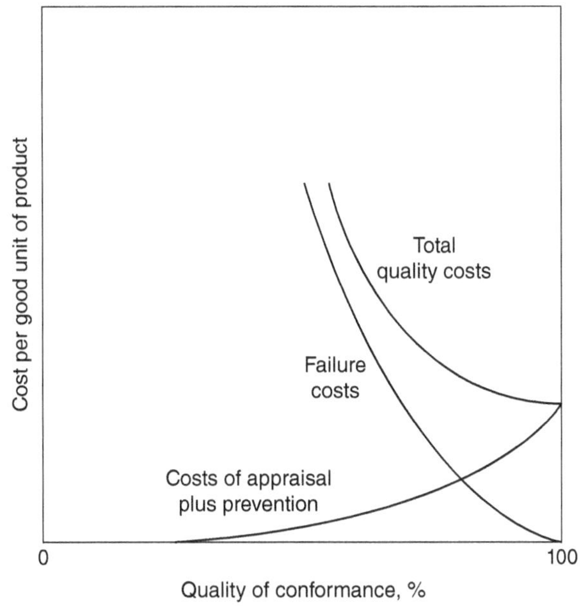

Figure 8.4 New model of optimum quality costs (Juran, 1988, by permission).

Quality costs are lowered by identifying the root causes of quality problems and taking action to eliminate these causes. The tools and techniques described in Chap. 5 are useful in this endeavor. KAIZEN, reengineering, and other continuous improvement approaches are commonly used.

As a general rule, quality costs increase as the detection point moves further up the production and distribution chain. The lowest cost is generally obtained when nonconformances are prevented in the first place. If nonconformances occur, it is generally least expensive to detect them as soon as possible after their occurrence. Beyond that point there is loss incurred from additional work that may be lost. The most expensive quality costs are from nonconformances detected by customers. In addition to the replacement or repair loss, a company loses customer goodwill and reputation. In extreme cases, litigation may result, adding even more cost and loss of goodwill.

Another advantage of early detection is that it provides more meaningful feedback to help identify root causes. The time lag between production and field failure makes it very difficult to trace the occurrence back to the process state that produced it. While field failure tracking is useful in *prospectively* evaluating a "fix," it is usually of little value in *retrospectively* evaluating a problem.

The accounting department bears *primary* responsibility for accounting matters, including cost of quality systems. The quality department's role in development and maintenance of the cost of quality system is to provide guidance and support to the accounting department.

The cost of quality system is an integrated subsystem of the larger cost accounting system. Terminology, format, etc. should be consistent between the cost of quality system and the larger system to speed the learning process and reduce confusion. The ideal cost of quality accounting system will simply aggregate quality costs to enhance their visibility to management and facilitate efforts to reduce them. For most companies, this task falls under the jurisdiction of the controller's office.

Quality cost measurement need not be accurate to the penny to be effective. The purpose of measuring such costs is to provide broad guidelines for management decision making and action. The very nature of cost of quality makes such accuracy impossible. In some instances it will only be possible to obtain periodic rough estimates of such costs as lost customer goodwill, cost of damage to the company's reputation, etc. These estimates can be obtained using special audits, statistical sampling, and other market studies. These activities can be jointly conducted by teams of marketing, accounting, and quality personnel. Since these costs are often huge, these estimates must be obtained. However, they need not be obtained every month. Annual studies are usually sufficient to indicate trends in these measures.

Quality cost management helps firms establish priorities for corrective action. Without such guidance, it is likely that firms will misallocate their resources, thereby getting less than optimal return on investment. If such experiences are repeated frequently, the organization may even question or abandon their quality cost reduction efforts. The most often used tool in setting priorities is Pareto analysis. Typically employed at the outset of the quality cost reduction effort, Pareto analysis is used to evaluate failure costs to identify those "vital few" areas in most need of attention. Documented failure costs, especially external failure costs, almost certainly understate the true cost and are highly visible to the customer. Pareto analysis is combined with other quality tools, such as control charts and cause-and-effect diagrams, to identify the root causes of quality problems. Of course, the analyst must constantly keep in mind the fact that most costs are hidden. Pareto analysis cannot be effectively performed until the hidden costs have been identified. Analyzing only those data easiest to obtain is an example of the GIGO (garbage-in, garbage-out) approach to analysis.

After the most significant failure costs have been identified and brought under control, appraisal costs are analyzed. Are we spending too much on appraisal in view of the lower levels of failure costs? Here quality cost analysis must be supplemented with risk analysis to assure that failure and appraisal cost levels are in balance. Appraisal cost analysis is also used to justify expenditure in prevention costs.

Prevention costs of quality are investments in the discovery, incorporation, and maintenance of defect prevention disciplines for all operations affecting the quality of product or service (Campanella, 1990). As such, prevention needs to be applied correctly and *not* evenly across the board. Much improvement has been demonstrated through reallocation of prevention effort from areas having little effect to areas where it really pays off; once again, the Pareto principle in action. Examples of categorized quality costs are provided in Table 8.2.

Analyzing quality costs requires a suitable base, so that the quality cost is analyzed as a percent of an appropriate base: Generally, a suitable base is related to quality costs in a meaningful way, well known to the managers who will review the quality cost reports, and a measure of business volume in the area where quality cost measurements are to be applied.

Several bases are often necessary to get a complete picture of the relative magnitude of quality costs. Some commonly used bases are (Campanella, 1990, p. 26):

- A labor base (such as total labor, direct labor, or applied labor)
- A cost base (such as shop cost, operating cost, or total material and labor)
- A sales base (such as net sales billed, or sales value of finished goods)
- A unit base (such as the number of units produced, or the volume of output)

While actual dollars spent is usually the best indicator for determining where quality improvement projects will have the greatest impact on profits and where corrective action should be taken, unless the production rate is relatively constant, it will not provide a clear indication of quality cost improvement trends. Since the goal of the cost of quality program is improvement over time, it is necessary to adjust the data for other time-related changes such as production rate, inflation, etc. Total quality cost compared to an applicable base results in an index that may be plotted and analyzed using statistical control charts.

For long-range analyses and planning, net sales is the base most often used for presentations to top management (Campanella, 1990, p. 24). If sales are relatively constant over time, the quality cost analysis can be performed for relatively short spans of time. In other industries this figure must be computed over a longer time interval to smooth out large swings in the sales base. For example, in industries such as shipbuilding or satellite manufacturing, some periods may have no deliveries, while others have large dollar amounts. It is important that the quality costs incurred be related to the sales for the same period. Consider the sales as the "opportunity" for the quality costs to happen.

Prevention Costs	
Marketing/customer/user	Marketing research; customer/user perception surveys/clinics; contract/document review
Product/service/design development	Design quality progress reviews; design support activities; product design qualification test; service design qualification; field tests
Purchasing	Supplier reviews; supplier rating; purchase order tech data; supplier quality planning
Operations (manufacturing or service)	Operations process validation; operations quality planning (including design and development of quality measurement and control equipment); operations support quality planning; operator quality education; operator SPC/process control
Quality administration	Administrative salaries; administrative expenses; quality program planning; quality performance reporting; quality education; quality improvement; quality audits; other prevention costs
Appraisal Costs	
Purchasing	Receiving or incoming inspections and tests; measurement equipment; qualification of supplier product; source inspection and control programs
Operations (manufacturing or service)	Planned operations inspections, tests, audits (including checking labor; product or service quality audits; inspection and test materials); set-up inspections and tests; special tests; process control measurements; laboratory support; measurement equipment (including depreciation allowances, measurement equipment expenses, maintenance and calibration labor); outside endorsements and certifications
External appraisal costs	Field performance evaluation; special product evaluations; evaluation of field stock and spare parts
Review of tests and inspection data	Including miscellaneous quality evaluations
Internal Failure Costs	
Product/service design (internal)	Design corrective action; rework due to design changes; scrap due to design changes
Purchasing	Purchased material reject disposition costs; purchased material replacement costs; supplier corrective action; rework of supplier rejects; uncontrolled material losses
Operations (product or service)	Material review and corrective action costs (including disposition costs; troubleshooting or failure analysis costs (operations); investigation support costs; operations corrective action); operations rework and repair costs; re-inspection/retest costs; extra operations; scrap costs (operations); downgraded end product or service; internal failure labor losses
Other internal failure costs	Including miscellaneous quality evaluations
External Failure Costs	Complaint investigation/customer or user service; returned goods; retrofit costs; recall costs; warranty claims; liability costs; penalties; customer/user goodwill; lost sales; other external failure costs

TABLE 8.2 Cost of Quality Examples (Hagan, 1990)

Some examples of cost of quality bases are (Campanella, 1990):

- Internal failure costs as a percent of total production costs
- External failure costs as an average percent of net sales
- Procurement appraisal costs as a percent of total purchased material cost
- Operations appraisal costs as a percent of total productions costs
- Total quality costs as a percent of production costs

PART III

Process Control

CHAPTER 9

Quantifying Process Variation

The successful quality system will address several aspects of process and product control. Effective controls may be established only after understanding the fundamental nature of the variation to be controlled.

Descriptive Statistics

Typically, descriptive statistics are computed to describe properties of empirical distributions, that is, distributions of data from samples. There are three areas of interest: the distribution's location or central tendency, its dispersion, and its shape. The analyst may also want some idea of the magnitude of possible error in the statistical estimates. Table 9.1 describes some of the more common descriptive statistical measures.

Figures 9.1 through 9.4 illustrate distributions with different descriptive statistics.

Many readers may be familiar with these statistics from their college curriculum. They are used in defining confidence intervals and to accept or reject statistical tests via hypothesis testing. These tests are somewhat dependent on the specific distribution of the population from which the samples are taken.

Enumerative and Analytic Studies

Deming (1975) defines enumerative and analytic studies as follows:

> *Enumerative study.* A study in which action will be taken on the universe.
>
> *Analytic study.* A study in which action will be taken on a process to improve performance in the future.

The term *universe* is somewhat synonymous with *population*: the entire group of interest. For example, the expected voters in a specific election might constitute a population of interest.

Sample Statistic	Description	Equation/Symbol
Measures of location		
Population mean	The center of gravity or centroid of the distribution	$\mu = \frac{1}{N}\sum_{i=1}^{N} x_i$ where N is the population size and x is an observation
Sample mean	The center of gravity or centroid of a sample from a distribution	$\bar{x} = \frac{1}{n}\sum_{i=1}^{n} x_i$ where n is the sample size and x is an observation
Median	The 50/50 split point. Precisely half of the data set will be above the median, and half below it.	$\tilde{x}$
Mode	The value that occurs most often. If the data are grouped, the mode is the group with the highest frequency.	None
Measures of dispersion		
Range	The distance between the sample extreme values	R = Largest – Smallest
Population standard deviation	A measure of the variation around the mean, in the same units as the original data	$\sigma_x = \sqrt{\frac{\sum_{j=1}^{N}(x_j - \bar{\bar{x}})^2}{N}}$
Sample standard deviation	A measure of the variation around the mean, in the same units as the original data	$s_x = \sqrt{\frac{\sum_{j=1}^{n}(x_j - \bar{x})^2}{n-1}}$
Measures of shape		
Skewness	A measure of asymmetry. Zero indicates perfect symmetry; the normal distribution has a skewness of zero. Positive skewness indicates that the "tail" of the distribution is more stretched on the side above the mean. Negative skewness indicates that the tail of the distribution is more stretched on the side below the mean.	$k = \frac{[(N-1)(N-2)]}{N}\sum_{j=1}^{N}\frac{(x_j - \bar{\bar{x}})^3}{\sigma_x^3}$
Kurtosis	Kurtosis is a measure of flatness of the distribution. Heavier tailed distributions have larger kurtosis measures. The normal distribution has a kurtosis of 3.	$\text{Kurtosis} = \left[\frac{N(N+1)}{(N-1)(N-2)(N-3)}\sum_{j=1}^{N}\frac{(x_j - \bar{\bar{x}})^4}{\sigma_x^4}\right] - \frac{3(N+1)^2}{(N-2)(N-3)}$

Table 9.1 Common Descriptive Statistics

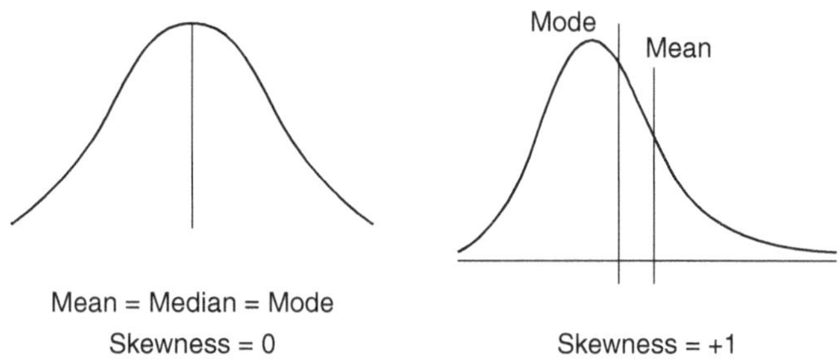

Figure 9.1 Illustration of mean, median, and mode.

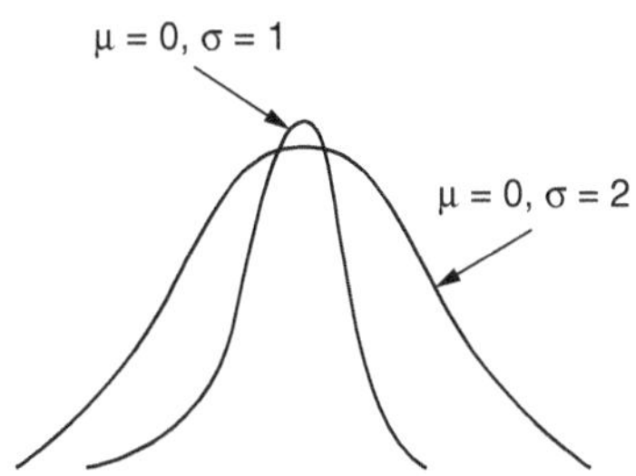

Figure 9.2 Illustration of sigma.

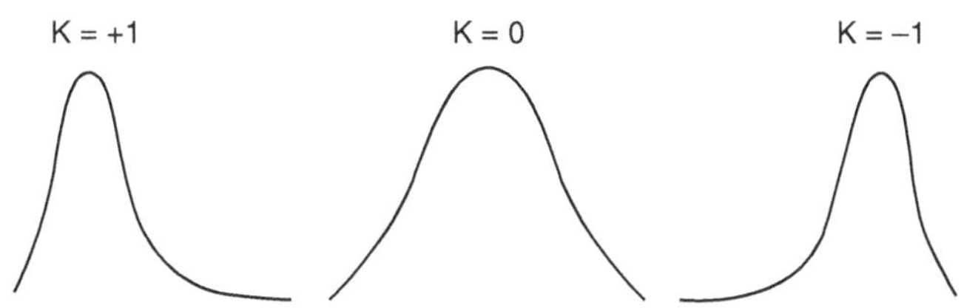

Figure 9.3 Illustration of skewness.

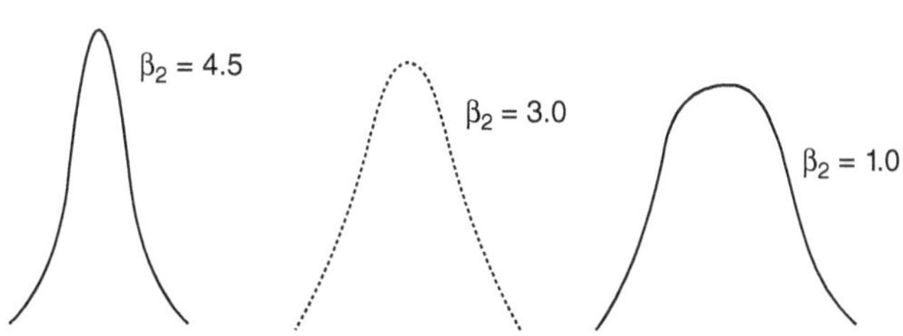

Figure 9.4 Illustration of kurtosis.

Consideration	Enumerative Study	Analytic Study
Aim	Parameter estimation	Prediction
Focus	Universe	Process
Method of access	Counts, statistics	Models of the process (e.g., flow charts, causes and effects, mathematical models)
Major source of uncertainty	Sampling variation	Extrapolation into the future
Uncertainty quantifiable?	Yes	No
Environment for the study	Static	Dynamic

TABLE 9.2 Important Aspects of Analytic Studies

In an analytic study the focus is on a *process* and how to improve it. The focus is the *future*. Thus, unlike enumerative studies, which make inferences about the universe actually studied, analytic studies are interested in a universe that has yet to be produced. Table 9.2 compares analytic studies with enumerative studies (Provost, 1988).

With regard to the analysis of processes, Deming (1986) comments:

> Analysis of variance, t-tests, confidence intervals, and other statistical techniques taught in the books, however interesting, are inappropriate because they provide no basis for prediction and because they bury the information contained in the order of production.

In organizations, processes are carried out as repeatable activities, carried out time and time again. The element of time is lost in the enumerative tools of confidence intervals and hypothesis testing. While useful for analyzing short-term data from a planned experiment, for example, these enumerative tools pool the variation that occurs over time into a single estimate of sample variation.

This is a persistent and unfortunate problem with the use of histograms. Apparently, most practitioners learn that histograms are useful to graphically show the shape of the data, which is fundamentally true. Unfortunately, the shape of the data and the expected shape of the process are completely different if the process is not stable. An example of this will be shown shortly.

Acceptance Sampling

Acceptance sampling is a traditional quality control technique that is applied to discrete lots or batches of a product. (A *lot* is a collection of physical units; the term *batch* is usually applied to chemical materials). The lot or batch is typically presented to the inspection department by either a supplier or a production department. The inspection department then inspects a sample from the lot or batch and, based on the results of the inspection, they determine the acceptability of the lot or batch. Acceptance sampling schemes generally consist of three elements:

- *The sampling plan.* How many units should be inspected? What is the acceptance criteria?
- *The action to be taken on the current lot or batch.* Actions include accept, sort, scrap, rework, downgrade, return to vendor, etc.
- *Action to be taken in the future.* Future action includes such options as switching to reduced or tightened sampling, switching to 100 percent inspection, shutting down the process, etc.

Acceptance sampling methods are generally based on ANSI/ASQ Z1.4 (formerly MIL-STD 105E), or variants of the plan known as Dodge-Romig Sampling tables.

These acceptance sampling plans have absolutely no place in a modern quality organization. They should be soundly rejected by the quality professional. The sampling plans are fundamentally flawed in assuming the samples are from a homogenous population (i.e., characterized by a single statistical distribution), when there is no evidence that the samples have been drawn from a stable process (the only situation under which the samples will be from a single distribution). When applied to an unstable process, the reliability of the acceptance sampling plan is misstated, and in fact unpredictable (since, by definition, the output of an unstable process is unpredictable).

If a process is in continuous production, Deming (1986) showed it is better to use a p chart (a control chart discussed in the next section) for process control than to apply an acceptance sampling plan. Based on the stable p chart you can determine the process average fraction defective, from which you can determine whether to sort the output or ship it by applying Deming's all-or-none rule:

$$\text{If } p < K_1/K_2 \text{ then ship, otherwise sort}$$

where K_1 is the cost of inspecting one piece and K_2 is the cost of shipping a defective, including lost customer goodwill. For example, if $K_1 = \$1$ and $K_2 = \$100$ then output from a process with an average fraction defective of less than 1 percent would be shipped without additional inspection; if the process average were 1 percent or greater, the output would be sorted. Note that this discussion does *not* apply to critical defects or critical defectives. Sampling for critical defects or defectives is done only to confirm that a previous 100 percent inspection or test was effective.

As Deming's rule shows, the alternatives are no inspection, 100 percent inspection, or sampling and analysis using a statistical process control chart. Supplier programs must emphasize the need for process control as a condition of sale, as discussed later in the chapter.

Statistical Control Charts

In his landmark 1931 book, *Economic Control of Quality of Manufacturing*, Shewhart described the following experiment:

> Write the letter "a" on a piece of paper. Now make another a just like the first one, then another and another until you have a series of a's (a, a, a, ...). You try to make all the a's alike but you don't; you can't. You are willing to accept this as an empirically established fact. But what of it? Let us see just what this means in respect to control. Why can we not do a simple thing like making all the a's exactly alike? Your answer leads to a generalization which all of us are perhaps willing to accept. It is that there are many causes of variability among the a's: the paper was not smooth, the lead in the pencil was not uniform, and the unavoidable variability in your external surroundings reacted upon you to introduce variation in the a's. But are these the only causes of variability in the a's? Probably not.
>
> We accept our human limitations and say that likely there are many other factors. If we could but name all the reasons why we cannot make the a's alike, we would most assuredly have a better understanding of a certain part of nature than we now have. Of course, this conception of what it means to be able to do what we want to do is not new; it does not belong exclusively to any one field of human thought; it is commonly accepted.
>
> The point to be made in this simple illustration is that we are limited in doing what we want to do; that to do what we set out to do, even in so simple a thing as making a's that are alike, requires almost infinite knowledge compared with that which we now possess. It follows, therefore, since we are thus willing to accept as axiomatic that we cannot do what we want to do and cannot hope to understand why we cannot, that we must also accept as axiomatic that a controlled quality will not be a constant quality. Instead, a controlled quality must be a variable quality. This is the first characteristic.
>
> But let us go back to the results of the experiment on the a's and we shall find out something more about control. Your a's are different from my a's; there is something about your a's that makes them yours and something about my a's that makes them mine. True, not all of your a's are alike. Neither are all of my a's alike. Each group of a's varies within a certain range and yet each group is distinguishable from the others. This distinguishable and, as it were, constant variability within limits is the second characteristic of control.

Shewhart goes on to define *control*:

> A phenomenon will be said to be controlled when, through the use of past experience, we can predict, at least within limits, how the phenomenon may be expected to vary in the future. Here it is understood that prediction within limits means that we can state, at least approximately, the probability that the observed phenomenon will fall within the given limits.

The critical point in this definition is that control is not defined as the complete absence of variation. Control is simply a state where all variation is predictable variation. In all forms of prediction there is an element of chance. Any unknown cause of variation is called a chance cause. If the influence of any particular chance cause is very small, and if the number of chance causes of variation is very large and relatively constant, we have a situation where the variation is predictable within limits. You can see from our definition above that a system such as this qualifies as a controlled system. Deming uses the term "common cause" rather "chance cause," and we will use Deming's term "common cause" as it is most prevalent in use.

An example of such a controlled system might be the production and distribution of peaches. If you went into an orchard to a particular peach tree at the right time of the year, you would find a tree laden with peaches (with any luck at all). The weights of the peaches will vary. However, if you weighed every single peach on the tree you would probably notice that there was a distinct pattern to the weights. In fact, if you drew a small random sample of, say, 25 peaches, you could probably predict the weights of those peaches remaining on the tree. This predictability is the essence of a controlled phenomenon. The number of common causes that account for the variation in peach weights is astronomical, but relatively constant. A constant system of common causes results in a controlled phenomenon.

Deming demonstrated the principles behind SPC with his Red Bead experiment, which he regularly conducted during his seminars. In this experiment, he used a bucket of beads or marbles. Most of the beads were white; however, a small percentage (about 10 percent) of red beads were thoroughly mixed with the white beads in the bucket. Students volunteered to be process workers, who would dip a sample paddle into the bucket and produce a day's "production" of 50 beads for the White Bead Company. Another student would volunteer to be an inspector. The inspector counted the number of white beads in each operator's daily production. The white beads represented usable output that could be sold to White Bead Company's customers, while the red beads were scrap. These results were then reported to a manager, who would invariably chastise operators for a high number of red beads. If the operator's production improved on the next sample, she was rewarded; if the production of white beads went down, more chastising.

Of course, most of the observers in the audience would chuckle heartily at management's actions, given that each production lot was merely a dip into a bucket that held a fixed percentage of red beads. That was the beauty of the demonstration, and Deming would draw the analogy to general business processes, where management would chastise employees for process variation that was largely out of their control. That is, most operational employees have no involvement with the design of their process, the qualifications of their suppliers, the specifications, and so on.

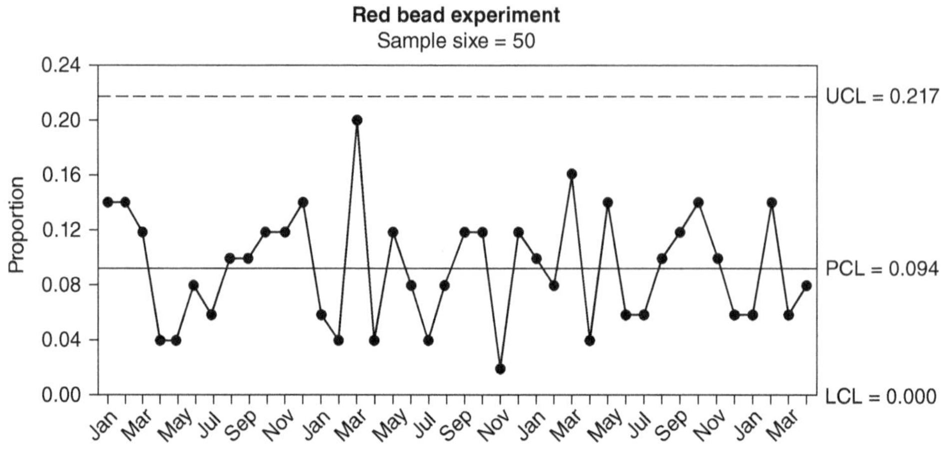

FIGURE 9.5 Example control chart of Deming's red bead experiment.

A control chart of the typical Red Bead "error rate" is shown in Fig. 9.5. The control chart shows variation between the process observations: each dip into the bucket yielded a different percentage of red beads. From the perspective of process analysis, has the process changed?

The control chart includes lines labeled UCL (an acronym for upper control limit) and LCL (for lower control limit). These control limits are calculated based on the statistics of the data, and provide the expected bounds of the process. The control limits in this example indicate that between 0 and 11 red beads (0 and 22 percent) should be expected in each sample of 50 beads.

Thus, the control limits for this stable process provide a means of predicting future performance of the process. Consider if this were the error rate of a key process. Predicting its future performance would aid the budgeting process, or the general allocation of resources needed to meet operational requirements given the waste, or perhaps just provide an economic justification for process improvement.

Figure 9.5 also demonstrates a fundamental premise of process analysis: variation exists in processes, just as Shewhart described in his writing example. The values plotted in the figure vary from approximately 3 to 20 percent. If this data represented the error rate from a key process in your organization, would someone question why the process "jumped" to 20 percent? Would they insist that someone investigate and determine what happened to make the process error rate increase so drastically, when the two prior months were 4 percent and 5 percent? When it fell the following month to 0.04, would they congratulate people for their effort, or smugly congratulate themselves for putting out the fire?

Any of these reactions should be greeted with the same amusement as in Deming's Red Bead experiment. The control chart makes clear that the process has not changed: it is stable; it is in statistical control. No one changed the bucket: it is the same bucket, with the same percentage of red beads.

Needless to say, not all phenomena arise from constant systems of common causes. At times the variation is caused by a source of variation that is not part of the constant system. These sources of variation were termed *assignable causes* by Shewhart; Deming calls them *special causes* of variation. Experience indicates that special causes of variation can usually be found and eliminated.

The basic rule of statistical process control is:

> Variation from common cause is indicative of the system. Changing the amount of common cause variation requires fundamental changes to the system itself. Special causes of variation can be identified by the control chart, and often quickly eliminated.

The control chart in Fig. 9.5 illustrates the need for statistical methods to determine the type of variation. The control charts answer the question: *Are these variations built into the system, or are they indicative of a change to the system?*

Variation between the control limits designated by the UCL and LCL lines are considered variation from the common cause system. Any variability beyond these limits is associated with special causes of variation. Any system exhibiting only common cause variation is considered *statistically controlled*. It must be noted that the control limits are not defined by management or customers: they are calculated from the data using statistical theory. A control chart is a practical tool that provides an operational definition of a special cause. That is, we cannot determine which causes of variation in the process are special causes until the control chart identifies them as such.

There are two broad categories of control charts: those for use with variable or continuous data (e.g., measurements) and those for use with attributes data (e.g., counts).

The basis of all control charts is the *rational subgroup*. Each plotted point on the chart contains the data from a single subgroup. Subgroups are considered rational subgroups if "all of the items are produced under conditions in which only random effects are responsible for the observed variation" (Nelson, 1988). We often form rational subgroups using consecutive items from the process, or items that are representative of the process during a short time period. This reduces the likelihood of special causes of variation occurring within the subgroup. The within-subgroup variation is used to define the control limits, which provide our estimate of the common cause variation in the process (i.e., the longer-term between-subgroup variation), so it is important to exclude special causes of variation in this estimate.

Defining the control limits (i.e., the expected future variation in the plotted statistic) using the within-subgroup variation implies that short-term variation is used to predict the expected longer-term variation. This is a perfect definition for process stability: if short-term variation can be used to predict longer-term variation, then the process is stable (i.e., in statistical control). Note the sharp contrast between this approach and the *random sampling* approach used for enumerative statistical methods, where short-term variation is pooled with longer-term variation to calculate a sample standard deviation.

Rational subgroups have the following properties (Keller, 2011b):

1. *The observations within the subgroup are independent, which implies that none of the observations influences or results from any other.* When observations are dependent on one another, the process has autocorrelation or serial correlation (these terms mean the same thing), which can cause the within-subgroup variation to be small relative to the between-subgroup variation. Examples of processes influenced by autocorrelation include:
   - *Chemical processes.* The temperature in a batch of beer is likely to be dependent on the temperature 5 minutes earlier. The autocorrelation diminishes over time, so the temperature an hour later may be less dependent.
   - *Service processes.* The wait time (i.e., time in queue) of a given customer at the grocery store checkout is likely to be somewhat dependent on the wait time of the customer immediately ahead, and perhaps the customer two or three places ahead. The last customer in line cannot be serviced until the others are completed, so their wait times have dependence.
   - *Discrete part manufacturing.* When feedback controls are used to control an automated process, this causes dependence since the process is adjusting based on these prior measurements.
2. *The observations within a subgroup are from a single stable process.* It has been mentioned that subgroups are often formed over a small time interval to limit the possibility for special causes to creep into the subgroup. If the subgroup contains the output of multiple-stream processes, the within-subgroup variation is likely to be larger than the longer-term between-subgroup variation. Examples of this include multiple cavity molding, multiple head filling stations, or samples from the teller station and the loan officer at the bank.
3. *The subgroups are formed from time-ordered data collection.* The x-axis of the control chart is time-ordered, so that the subgroups on the right of the chart represent a time period later than the subgroups to their left. Rational subgroups cannot be formed from a set of

random data from the process, or a box of parts shipped from your supplier, since the time sequence of the data has been lost.

Variable Control Charts

In SPC, the mean, range, and standard deviation are the statistics most often used for analyzing measurement data. Control charts are used to monitor these statistics. An out-of-control point for any of these statistics is an indication that a special cause of variation is present and that an immediate investigation should be made to identify the special cause.

Average and Range Control Charts

Average charts (usually called $\overline{X}$ charts in reference to the symbol used to designate the averages in Table 9.1) are statistical tools used to evaluate the central tendency of a process over time. Range charts are statistical tools used to evaluate the dispersion or spread of a process over time.

Average charts answer the question "Has a special cause of variation caused the central tendency of this process to change over the time period observed?" Range charts answer the question "Has a special cause of variation caused the process distribution to become more or less erratic?" Average and range charts can be applied to any continuous variable like weight, size, cycle time, error rate, and so on, subject to the conditions necessary for rational subgroups.

A predefined subgroup size is defined for the given process. Typical subgroup sizes are three or five observations in the subgroup.

The average and range are computed for each subgroup separately, then plotted on the control chart. Each subgroup's statistics are compared with the control limit, and patterns of variation between subgroups are analyzed.

Subgroup Equations for Average and Range Charts

$$\overline{X} = \frac{\text{sum of subgroup measurements}}{\text{subgroup size}}$$

$$R = \text{largest in subgroup} - \text{smallest in subgroup}$$

Table 9.3 contains 25 subgroups of five observations each. The average and range for each subgroup are shown in the table.

Control Limit Equations for Average and Range Charts Control limits for both the average and the range charts are computed such that it is highly unlikely that a subgroup average or range from a stable process would fall outside of the limits. All control limits are set at plus and minus three standard deviations from the centerline of the chart. Thus, the control limits for subgroup averages are plus and minus three standard deviations of the mean from the grand average; the control limits for

Sample 1	Sample 2	Sample 3	Sample 4	Sample 5	Average	Range	Sigma
110	93	99	98	109	101.8	17	7.396
103	95	109	95	98	100.0	14	6.000
97	110	90	97	100	98.8	20	7.259
96	102	105	90	96	97.8	15	5.848
105	110	109	93	98	103.0	17	7.314
110	91	104	91	101	99.4	19	8.325
100	96	104	93	96	97.8	11	4.266
93	90	110	109	105	101.4	20	9.290
90	105	109	90	108	100.4	19	9.607
103	93	93	99	96	96.8	10	4.266
97	97	104	103	92	98.6	12	4.930
103	100	91	103	105	100.4	14	5.550
90	101	96	104	108	99.8	18	7.014
97	106	97	105	96	100.2	10	4.868
99	94	96	98	90	95.4	9	3.578
106	93	104	93	99	99.0	13	6.042
90	95	98	109	110	100.4	20	8.792
96	96	108	97	103	100.0	12	5.339
109	96	91	98	109	100.6	18	8.081
90	95	94	107	99	97.0	17	6.442
91	101	96	96	109	98.6	18	6.804
108	97	101	103	94	100.6	14	5.413
96	97	106	96	98	98.6	10	4.219
101	107	104	109	104	105.0	8	3.082
96	91	96	91	105	95.8	14	5.718

Table 9.3 Data for Average Control Charts

the subgroup ranges are plus and minus three standard deviations of the range from the average range. These control limits are quite robust with respect to non-normality in the process distribution.

To facilitate calculations, constants are used in the control limit equations. The table in Appendix 1 provides control chart constants for subgroups of 25 or less.

Control Limit Equations for Range Charts

$$\bar{R} = \frac{\text{sum of subgroup ranges}}{\text{number of subgroups}}$$

$$\text{LCL} = D_3\bar{R}$$

$$\text{UCL} = D_4\bar{R}$$

The control limits are calculated from the data in Table 9.3 as follows:

$$\bar{R} = \frac{\text{sum of subgroup ranges}}{\text{number of subgroups}} = \frac{369}{25} = 14.76$$

$$\text{LCL}_R = D_3\bar{R} = 0 \times 14.76 = 0$$

$$\text{UCL}_R = D_4\bar{R} = 2.115 \times 14.76 = 31.22$$

Since it is not possible to have a subgroup range less than zero, the LCL is not shown on the control chart for ranges.

Control Limit Equations for Averages Charts Using $\bar{R}$

$$\bar{\bar{X}} = \frac{\text{sum of subgroup averages}}{\text{number of subgroups}}$$

$$\text{LCL} = \bar{\bar{X}} - A_2\bar{R}$$

$$\text{UCL} = \bar{\bar{X}} + A_2\bar{R}$$

The control limits are calculated from the data in Table 9.3 as follows:

$$\bar{\bar{X}} = \frac{\text{sum of subgroup averages}}{\text{number of subgroups}} = \frac{2487.5}{25} = 99.5$$

$$\text{LCL}_{\bar{X}} = \bar{\bar{X}} - A_2\bar{R} = 99.5 - 0.577 \times 14.76 = 90.97$$

$$\text{UCL}_{\bar{X}} = \bar{\bar{X}} + A_2\bar{R} = 99.5 + 0.577 \times 14.76 = 108.00$$

The completed average and range control charts are shown in Fig. 9.6.

The charts in Fig. 9.6 show a process in statistical control, so the limits of variability for this process are predictable (using the control limits). Since the process is in statistical control, we can also make predictions of our ability to meet customer requirements, as discussed in the upcoming section on Process Capability.

Average and Standard Deviation Control Charts

Average and standard deviation control charts are conceptually identical to average and range control charts. The difference is that the subgroup standard deviation is used to measure dispersion rather than the subgroup range. The subgroup standard deviation is statistically more efficient than the subgroup range for subgroup sizes greater than two. This efficiency advantage increases as the subgroup size increases, most dramatically when subgroup size is 10 or larger. In those cases, the standard deviation (or sigma chart)

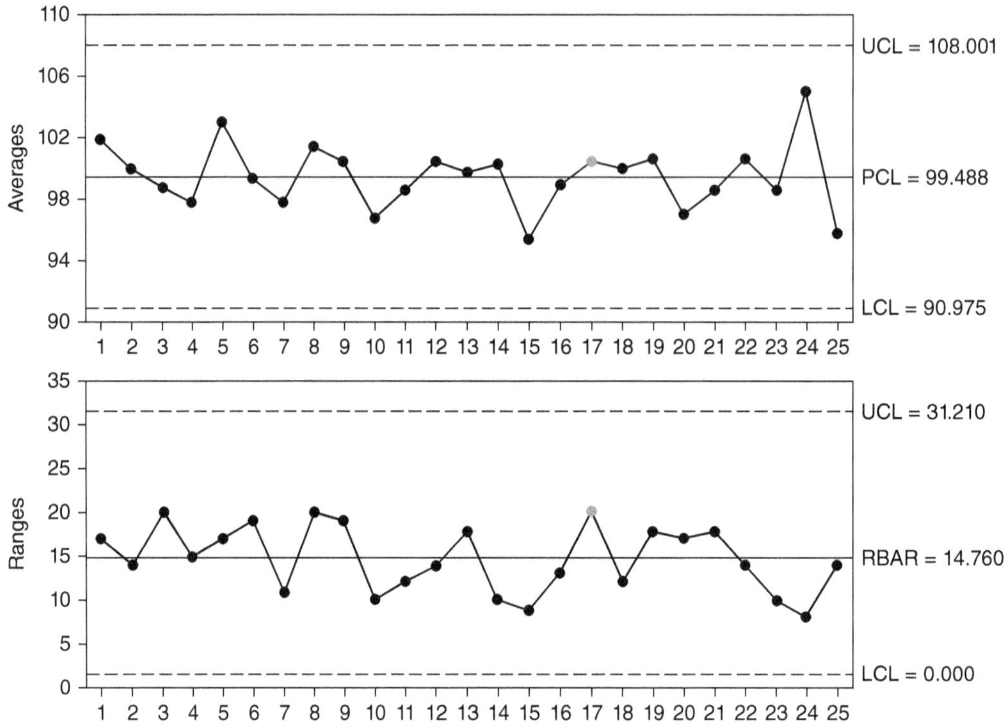

FIGURE 9.6 Completed average and sigma control charts.

should be used instead of the range chart. Although the range is easier to compute and easier for most people to understand, if the analyses are to be interpreted by statistically knowledgeable personnel and calculations are not a problem, the standard deviation chart is preferred for all subgroup sizes. In the vast majority of applications, software will be used to analyze the process. The software will automatically calculate all the control limits, and apply additional run test rules that will be discussed shortly. The MS Excel software used for these examples costs only a few hundred dollars.

Subgroup Equations for Average and Sigma Charts

$$\bar{X} = \frac{\text{sum of subgroup measurements}}{\text{subgroup size}}$$

$$s = \sqrt{\frac{\sum_{i=1}^{n}(x_i - \bar{X})^2}{n-1}}$$

The standard deviation, s, is computed separately for each subgroup, using the subgroup average rather than the grand average (i.e., the average of the subgroup averages, the centerline of the averages chart). This is an important point: using the grand average would introduce special cause variation if the process were out of control, thereby underestimating the process capability, perhaps significantly.

The calculated standard deviation for each subgroup is shown in Table 9.3.

Control Limit Equations for Average and Sigma Charts Control limits for both the average and the sigma charts are computed such that it is highly unlikely that a subgroup average or sigma from a stable process would fall outside of the limits. All control limits are set at plus and minus three standard deviations from the centerline of the chart. Thus, the control limits for subgroup averages are plus and minus three standard deviations of the mean from the grand average. The control limits for the subgroup sigmas are plus and minus three standard deviations of sigma from the average sigma.

These control limits are quite robust with respect to non-normality in the process distribution.

To facilitate calculations, constants are used in the control limit equations. The table in Appendix 1 provides control chart constants for subgroups of 25 or less.

Control Limit Equations for Sigma Charts Based on S-Bar

$$\bar{s} = \frac{\text{sum of subgroup sigmas}}{\text{number of subgroups}}$$

$$\text{LCL} = B_3\bar{s}$$

$$\text{UCL} = B_4\bar{s}$$

To illustrate the calculations and to compare the range method with the standard deviation results, the data used in the previous example will be re-analyzed using the subgroup standard deviation rather than the subgroup range.

The control limits are calculated from the Table 9.3 data as follows:

$$s = \frac{\text{sum of subgroup sigmas}}{\text{number of subgroups}} = \frac{155.45}{25} = 6.218$$

$$\text{LCL}_s = B_3\bar{s} = 0 \times 6.218 = 0$$

$$\text{UCL}_s = B_4\bar{s} = 2.089 \times 6.218 = 12.989$$

Since it is not possible to have a subgroup sigma less than zero, the LCL is not shown on the control chart for sigma.

Control Limit Equations for Averages Charts Based on S-Bar

$$\bar{\bar{X}} = \frac{\text{sum of subgroup averages}}{\text{number of subgroups}}$$

$$\text{LCL} = \bar{\bar{X}} - A_3\bar{s}$$

$$\text{UCL} = \bar{\bar{X}} + A_3\bar{s}$$

Using the data in Table 9.3:

$$\bar{\bar{X}} = \frac{\text{sum of subgroup averages}}{\text{number of subgroups}} = \frac{2487.5}{25} = 99.5$$

$$\text{LCL}_{\bar{X}} = \bar{\bar{X}} - A_3\bar{s} = 99.5 - 1.427 \times 6.218 = 90.63$$

$$\text{UCL}_{\bar{X}} = \bar{\bar{X}} + A_3\bar{s} = 99.5 + 1.427 \times 6.218 = 108.37$$

The completed average and sigma control charts are shown in Fig. 9.7. Note that the control limits for the averages chart are only slightly different from the limits calculated using ranges.

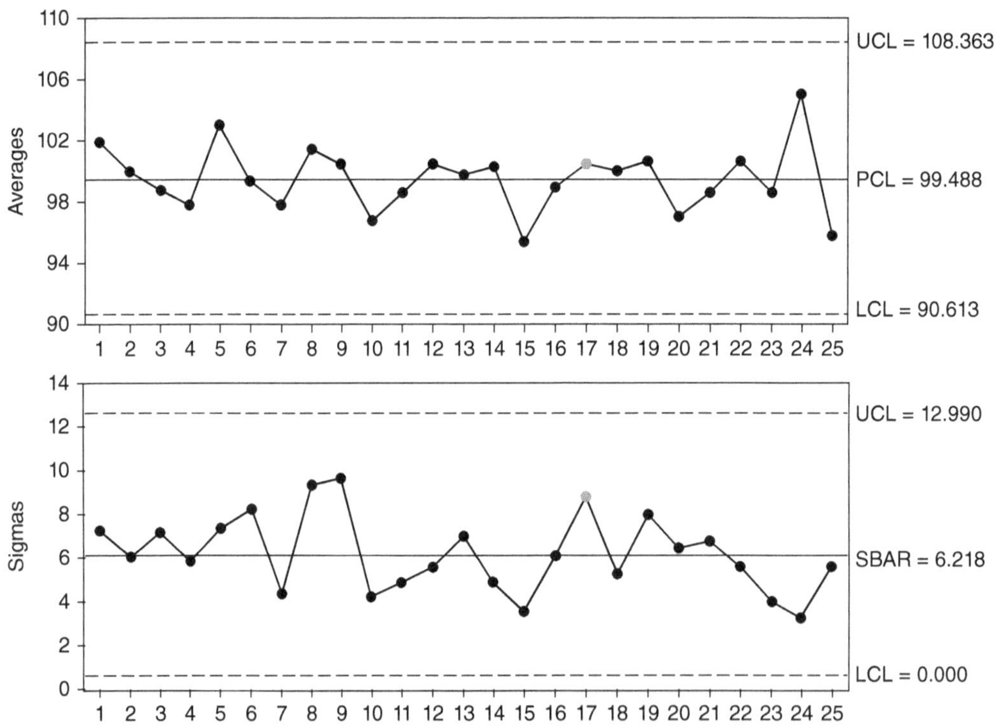

FIGURE 9.7 Completed average and sigma control charts. Note that the conclusions reached are the same as when ranges were used.

Control Charts for Individual Measurements (X Charts)

Individuals control charts (sometimes called *X charts*) are statistical tools used to evaluate the central tendency of a process over time. Individuals control charts are used when it is not feasible to use averages for process control. There are many possible reasons why average control charts may not be desirable: observations may be expensive to get (e.g., destructive testing), output may be too homogeneous over short time intervals (e.g., pH of a solution), the production rate may be slow and the interval between successive observations long, and so forth. Control charts for individuals are often used to monitor batch processes, such as chemical processes, where the within-batch variation is so small relative to between-batch variation that the control limits on a standard $\overline{X}$ chart would be too close together. Range charts (more strictly *moving range charts*) are used in conjunction with individuals charts to help monitor dispersion.

Calculations for Moving Ranges Charts As with average and range charts, the range is computed as shown above,

$$R = \text{largest in subgroup} - \text{smallest in subgroup}$$

Here, the range is calculated as the absolute value of the difference between a consecutive pair of process measurements, which meets the requirement of a rational subgroup in estimating short-term variation. The range control limit is computed as was described for averages and ranges charts, using the D_4 constant for subgroups of two, which is 3.267. That is,

$$LCL = 0 \text{ (for } n = 2)$$
$$UCL = 3.267 \times \overline{R}$$

Recent research has supported the idea that the moving range chart will necessarily identify the same process instability as the individuals chart: a large value in the observation on the individuals chart is likely to also create one (often two) moving range values that are out of control. It is certainly easier to present the individuals chart without the moving range chart; however, the analyst should be aware there are cases where the moving range chart has been useful in troubleshooting.

Table 9.4 contains 25 measurements. To facilitate comparison, the measurements are the first observations in each subgroup used in the previous average/range and average/standard deviation control chart examples.

The control limits for the moving range chart are calculated from this data as follows:

$$\overline{R} = \frac{\text{sum of ranges}}{\text{number of ranges}} = \frac{196}{24} = 8.17$$

$$LCL_R = D_3\overline{R} = 0 \times 8.17 = 0$$

$$UCL_R = D_4\overline{R} = 3.267 \times 8.17 = 26.69$$

Sample 1	Range
110	None
103	7
97	6
96	1
105	9
110	5
100	10
93	7
90	3
103	13
97	6
103	6
90	13
97	7
99	2
106	7
90	16
96	6
109	13
90	19
91	1
108	17
96	12
101	5
96	5

Table 9.4 Data for Individuals and Moving Ranges Control Charts

Since it is not possible to have a subgroup range less than zero, the LCL is not shown on the control chart for ranges.

Control Limit Equations for Individuals Charts

$$\bar{X} = \frac{\text{sum of measurements}}{\text{number of measurements}}$$

$$LCL = \bar{X} - E_2\bar{R} = \bar{X} - 2.66 \times \bar{R}$$

$$UCL = \bar{X} + E_2\bar{R} = \bar{X} + 2.66 \times \bar{R}$$

where $E_2 = 2.66$ is the constant used when individual measurements are plotted, and $\bar{R}$ is based on subgroups of $n = 2$.

Using the data in Table 9.4:

$$\bar{X} = \frac{\text{sum of measurements}}{\text{number of measurements}} = \frac{2475}{25} = 99.0$$

$$LCL_X = \bar{X} - E_2\bar{R} = 99.0 - 2.66 \times 8.17 = 77.27$$

$$UCL_X = \bar{X} + E_2\bar{R} = 99.0 + 2.66 \times 8.17 = 120.73$$

The completed individuals control chart is shown in Fig. 9.8.

In this case the conclusions are the same as with averages charts. However, averages charts always provide tighter control than X charts, due to the relative width of the distributions shown in Fig. 9.9. For a given set of data, the distribution of the raw observations is much wider (by a factor of the square root of *n*) than the distribution of the subgroup averages of size *n* formed from the data. As the process centerline (i.e., the center of each distribution in Fig. 9.9) moves away from the previous target, the tails of the distribution of averages is further from the target than the tails of the observational distribution. The average chart would detect the process shift more quickly than the individuals chart.

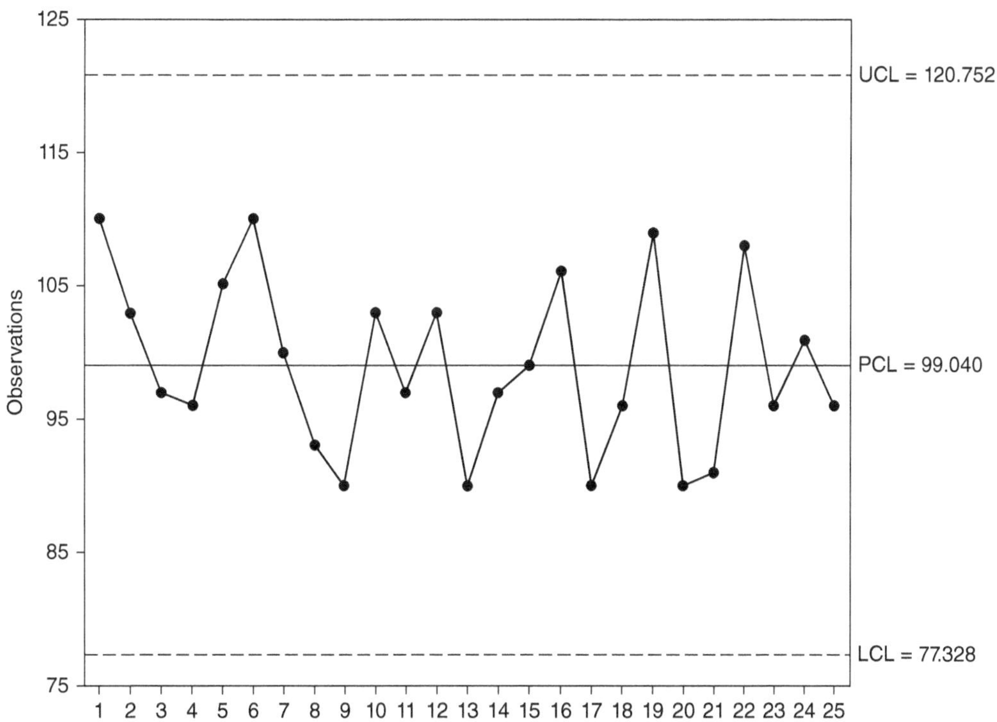

FIGURE 9.8 Completed individuals control chart.

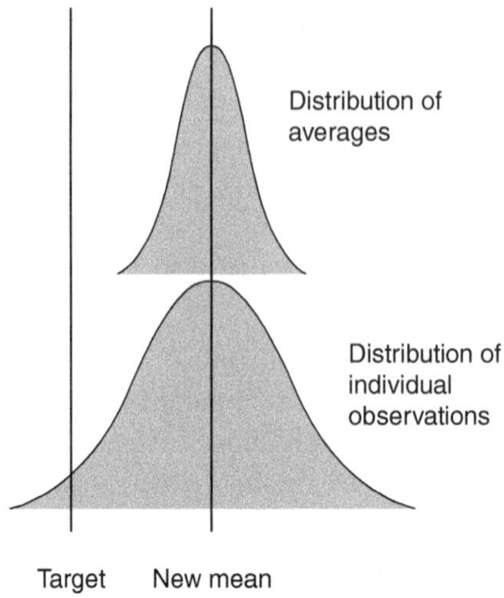

FIGURE 9.9 Comparison of distribution of individual observations with distribution of subgroup averages calculated from the observations.

The subgroup size is determined in consideration of the principles of the rational subgroup, the cost of obtaining the measurement, and the cost of failing to detect process shifts. Table 9.5 provides the expected number of subgroups (sometimes called the *average run length*) that will be plotted before the shift is detected as a special cause. For example, a subgroup

n/k	0.5	1	1.5	2	2.5	3
1	155	43	14	6	3	1
2	90	17	5	2	1	1
3	60	9	2	1	1	1
4	43	6	1	1	1	1
5	33	4	1	1	1	1
6	26	3	1	1	1	1
7	21	2	1	1	1	1
8	17	2	1	1	1	1
9	14	1	1	1	1	1
10	12	1	1	1	1	1

From Keller (2011b).

TABLE 9.5 Average Number of Subgroups Required to Detect Shift of *k* Sigma with Subgroup Size of *n*

of size 3 (leftmost column) would detect a process shift of 1 sigma (from top row) in 9 subgroups, on average. That is, sometimes it would detect it more quickly, and sometimes it would take more subgroups; but if you experienced that condition many times over, the average number of subgroups needed to detect the shift is 9. A subgroup of size 5 would detect the 1 sigma shift in 4 subgroups (on average). A subgroup of size 1 would need 43 subgroups (on average). Larger subgroups will provide better sensitivity to smaller shifts, but there is sometimes an unwarranted cost in obtaining the additional data. The cost implications of failing to detect that process shift as soon as possible must be weighed against the cost of the additional data. As a general rule, subgroups of size 3 to 5 are recommended, as they detect reasonable shifts of 1.5 sigma or larger fairly quickly. When a process has been in control for a period of time, and it is desirable to detect more subtle shift in the process (e.g., 0.5 sigma shifts), it is recommended to use EWMA charts, such as described in Keller (2011b), since larger subgroups are both costly and run the risk of having special causes occur in the subgroups collected over a longer period of time.

As a general rule, it's best to collect small subgroups more frequently (than larger subgroups less often). The more frequent subgroups provide more opportunity to detect process shifts more quickly. This is particularly useful when beginning to analyze a process and there is little information concerning the types or frequency of special causes.

It is recommended that a sufficient number of subgroups be collected to experience the process over a period of time (Keller, 2011b). If the control chart is limited to only a few days of data, it has hardly experienced the common cause variation that will predictably occur over longer periods of time. In some cases, it may be desirable to define the control limits over a short period of time, such as for a process capability study or prerelease study for your customer. In those situations, be aware that the control limits may be tighter than what the process will experience over longer periods, and you may find yourself chasing special causes for several weeks.

An additional consideration is that the constants in the table in Appendix 1 are really only constants for a "large" number of subgroups. Although many people quote 25 or 35 subgroups as the minimum number, this is an appropriate number for a subgroup of size five. Smaller subgroups require more subgroups before the constants approach constant value down to three decimals or so. For a subgroup of size three, 50 subgroups are recommended (Keller, 2011b). Subgroups of size one require 150 or more subgroups, which is also recommended so that the distribution can be verified.

The control limit calculations for the averages and individuals charts shown above are based on properties of the normal distribution. The use of three sigma limits provides adequate detection of special causes, without

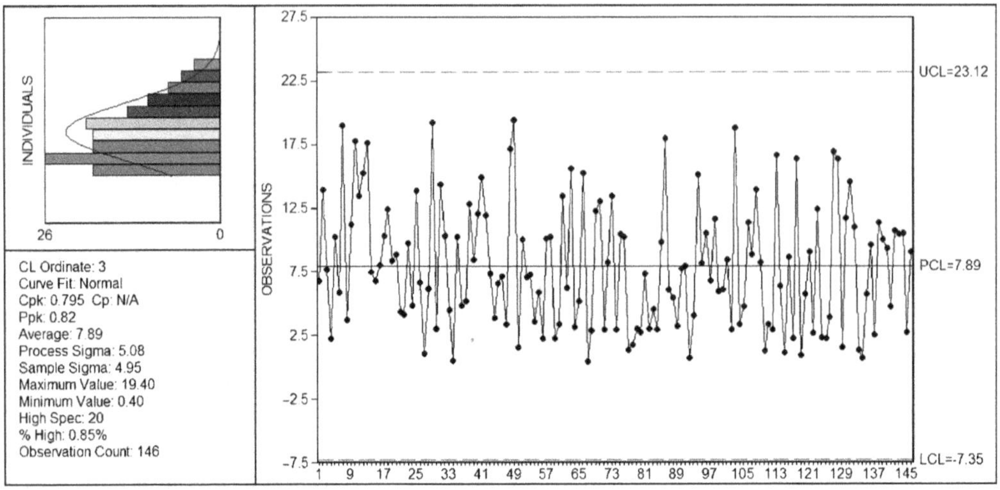

FIGURE 9.10 Example of misleading control limits using normal distribution assumptions (Keller 2011b).

an inordinate number of false alarms. The averages charts, in plotting the averages, are relatively insensitive to departures from normality of the data based on the central limit theorem. The central limit theorem, which has been extensively validated, holds that the distribution of the averages will approximate a normal distribution when the distribution of raw observations is non-normal, even for subgroups as small as three to five observations.

If it is suspected that the individuals' data is non-normal, or if it is not known, then there is a risk in using the standard individuals control limit calculations shown above. Consider the cycle time data shown in Fig. 9.10, plotted on an individual's chart. The calculated lower control limit is –7.35 minutes, which is obviously impossible. The histogram to the left of the control chart shows the data to be skewed and bounded close to zero. The negative control limit thus provides a wide area of insensitivity: if the process cycle time decreases due to a special cause (i.e., an improvement to the process), it would not be detected because of the inaccurate lower control limit. Consider the same data plotted in Fig. 9.11, where control limits are based on non-normal calculations, and it is clear this chart is more capable of detecting process shifts.

Control Charts for Attributes Data

Control Charts for Proportion Defective (*p* Charts)

p charts are statistical tools used to evaluate the proportion defective, or proportion non-conforming, produced by a process.

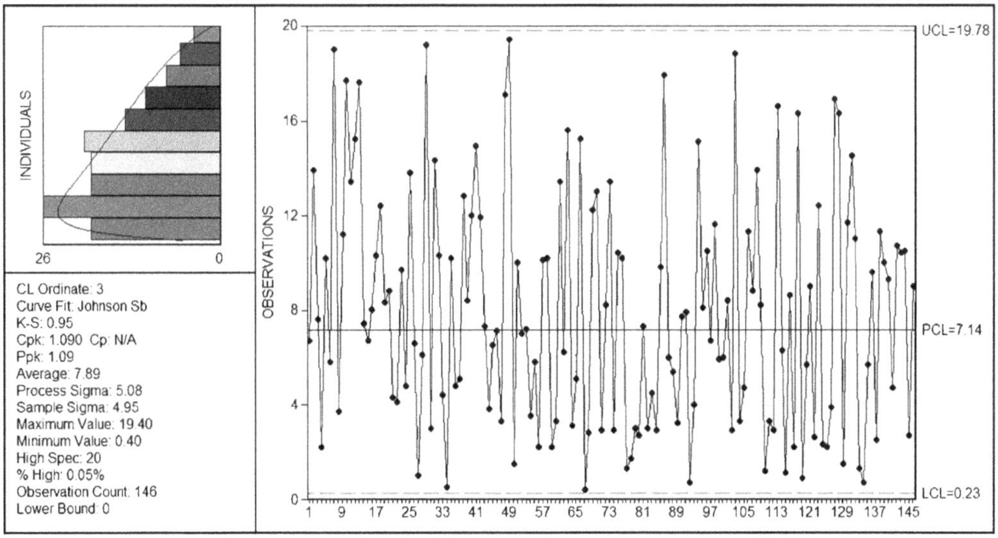

Figure 9.11 Non-normal control limits applied to data in Fig. 9.10 (Keller 2011b).

p charts can be applied to any variable where the appropriate performance measure is a unit count. *p* charts answer the question: "Has a special cause of variation caused the central tendency of this process to produce an abnormally large or small number of defective units over the time period observed?"

***p* Chart Control Limit Equations** Like all control charts, *p* charts consist of three guidelines: centerline, a lower control limit, and an upper control limit. The centerline is the average proportion defective and the two control limits are set at plus and minus three standard deviations. If the process is in statistical control, then virtually all proportions should be between the control limits and they should fluctuate randomly about the centerline.

$$p = \frac{\text{subgroup defective count}}{\text{subgroup size}}$$

$$\overline{p} = \frac{\text{sum of subgroup defective counts}}{\text{sum of subgroup sizes}}$$

$$LCL = \overline{p} - 3\sqrt{\frac{\overline{p}(1-\overline{p})}{n}}$$

$$UCL = \overline{p} + 3\sqrt{\frac{\overline{p}(1-\overline{p})}{n}}$$

In the above equations, n is the subgroup size. If the subgroup sizes varies, the control limits will also vary, becoming closer together as n increases.

As with all control charts, a special cause is probably present if there are any points beyond either the upper or the lower control limit. Analysis of p chart patterns between the control limits is extremely complicated if the sample size varies because the distribution of p varies with the sample size.

The data in Table 9.6 were obtained by opening randomly selected crates from each shipment and counting the number of bruised peaches. There are 250 peaches per crate. Normally, samples consist of one crate

Shipment Number	Crates	Peaches	Bruised	p
1	1	250	47	0.188
2	1	250	42	0.168
3	1	250	55	0.220
4	1	250	51	0.204
5	1	250	46	0.184
6	1	250	61	0.244
7	1	250	39	0.156
8	1	250	44	0.176
9	1	250	41	0.164
10	1	250	51	0.204
11	2	500	88	0.176
12	2	500	101	0.202
13	2	500	101	0.202
14	1	250	40	0.160
15	1	250	48	0.192
16	1	250	47	0.188
17	1	250	50	0.200
18	1	250	48	0.192
19	1	250	57	0.228
20	1	250	45	0.180
21	1	250	43	0.172
22	2	500	105	0.210
23	2	500	98	0.196
24	2	500	100	0.200
25	2	500	96	0.192
	TOTALS	8000	1544	

Table 9.6 Raw Data for p Chart

per shipment. However, when part-time help is available, samples of two crates are taken.

Using the above data, the centerline and control limits are found as follows:

$$p = \frac{\text{subgroup defective count}}{\text{subgroup size}}$$

these values are shown in the last column of Table 9.6.

$$\bar{p} = \frac{\text{sum of subgroup defective counts}}{\text{sum of subgroup size}} = \frac{1544}{8000} = 0.193$$

which is constant for all subgroups.

$n = 250$ (1 crate):

$$LCL = \bar{p} - 3\sqrt{\frac{\bar{p}(1-\bar{p})}{n}} = 0.193 - 3\sqrt{\frac{0.193 \times (1-0.193)}{250}} = 0.118$$

$$UCL = \bar{p} + 3\sqrt{\frac{\bar{p}(1-\bar{p})}{n}} = 0.193 + 3\sqrt{\frac{0.193 \times (1-0.193)}{250}} = 0.268$$

$n = 500$ (2 crates):

$$LCL = 0.193 - 3\sqrt{\frac{0.193 \times (1-0.193)}{500}} = 0.140$$

$$UCL = 0.193 + 3\sqrt{\frac{0.193 \times (1-0.193)}{500}} = 0.246$$

The control limits and the subgroup proportions are shown in Fig. 9.12.

Pointers for Using *p* Charts In some cases, the "moving control limits" may not be necessary, and the average sample size (total number inspected divided by the number of subgroups) may be used to calculate control limits. For instance, with our example the sample size doubled from 250 peaches to 500 but the control limits hardly changed at all. Table 9.7 illustrates the different control limits based on 250 peaches, 500 peaches, and the average sample size, which is $8000 \div 25 = 320$ peaches.

Notice that the conclusions regarding process performance are the same when using the average sample size as they are using the exact sample sizes. This is usually the case if the variation in sample size isn't too great. There are many rules of thumb, but most of them are extremely conservative. The best way to evaluate limits based on the average sample size is to check it out as shown above. SPC is all about improved decision making. In general, use the most simple method that leads to correct decisions.

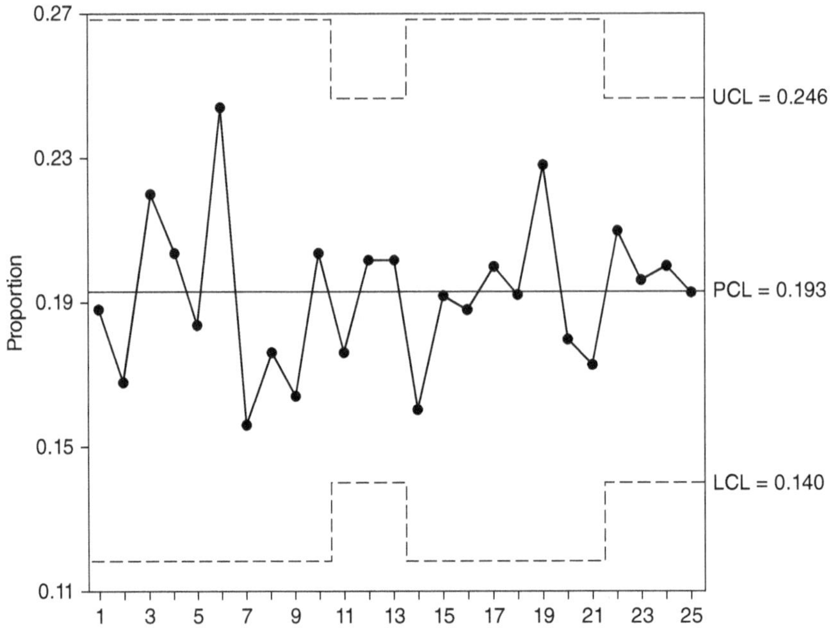

Figure 9.12 *p* chart example constructed using *Green Belt XL* software (*Courtesy of www.qualityamerica.com by permission*).

Sample Size	Lower Control Limit	Upper Control Limit
250	0.1181	0.2679
500	0.1400	0.2460
320	0.1268	0.2592

Table 9.7 Effect of Using Average Sample Size

Control Charts for Count of Defectives (*np* Charts)

np charts are statistical tools used to evaluate the count of defectives, or count of items non-conforming, produced by a process. *np* charts can be applied to any variable where the appropriate performance measure is a unit count and the subgroup size is held constant. Note that wherever an *np* chart can be used, a *p* chart can be used too.

Control Limit Equations for *np* Charts Like all control charts, *np* charts consist of three guidelines: centerline, a lower control limit, and an upper control limit. The centerline is the average count of defectives-per-subgroup and the two control limits are set at plus and minus three standard deviations. If the process is in statistical control, then virtually all subgroup counts will be between the control limits, and they will fluctuate randomly about the centerline.

$$np = \text{subgroup defective count}$$

$$n\bar{p} = \frac{\text{sum of subgroup defective counts}}{\text{number of subgroups}}$$

$$LCL = n\bar{p} - 3\sqrt{n\bar{p}(1-\bar{p})}$$

$$UCL = n\bar{p} + 3\sqrt{n\bar{p}(1-\bar{p})}$$

Note that

$$\bar{p} = \frac{n\bar{p}}{n}$$

The data in Table 9.8 were obtained by opening randomly selected crates from each shipment and counting the number of bruised peaches. There are 250 peaches per crate (constant n is required for np charts).

Using the above data the centerline and control limits are found as follows:

$$n\bar{p} = \frac{\text{sum of subgroup defective counts}}{\text{number of subgroups}} = \frac{838}{30} = 27.93$$

$$LCL = n\bar{p} - 3\sqrt{n\bar{p}\left(1-\bar{p}\right)} = 27.93 - 3\sqrt{27.93 \times \left(1 - \frac{27.93}{250}\right)} = 12.99$$

$$UCL = n\bar{p} + 3\sqrt{n\bar{p}\left(1-\bar{p}\right)} = 27.93 + 3\sqrt{27.93 \times \left(1 - \frac{27.93}{250}\right)} = 42.88$$

The control limits and the subgroup defective counts are shown in Fig. 9.13.

Control Charts for Average Occurrences-per-Unit (*u* Charts)

u charts are statistical tools used to evaluate the average number of occurrences-per-unit produced by a process. *u* charts can be applied to any variable where the appropriate performance measure is a count of how often a particular event occurs. *u* charts answer the question "Has a special cause of variation caused the central tendency of this process to produce an abnormally large or small number of occurrences over the time period observed?" Note that, unlike *p* or *np* charts, *u* charts do not necessarily involve counting physical items. Rather, they involve counting

Shipment Number	Bruised Peaches
1	20
2	28
3	24
4	21
5	32
6	33
7	31
8	29
9	30
10	34
11	32
12	24
13	29
14	27
15	37
16	23
17	27
18	28
19	31
20	27
21	30
22	23
23	23
24	27
25	35
26	29
27	23
28	23
29	30
30	28
TOTAL	838

TABLE 9.8 Raw Data for *np* Chart

of *events*. For example, when using a *p* chart one would count bruised peaches. When using a *u* chart one would count the *bruises*.

Control Limit Equations for *u* Charts Like all control charts, *u* charts consist of three guidelines: centerline, a lower control limit, and an upper control limit. The centerline is the average number of occurrences-per-unit and

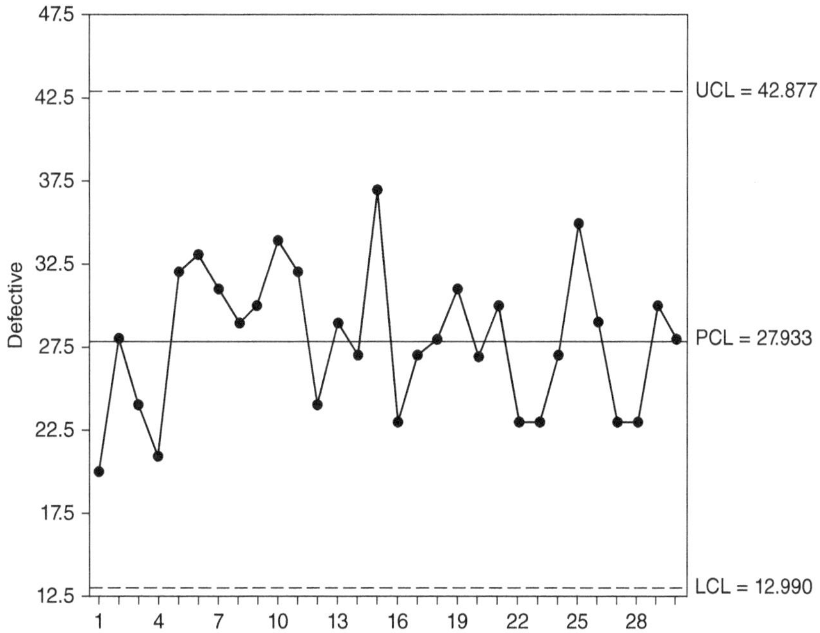

FIGURE 9.13 *np* chart example constructed using *Green Belt XL* software (*Courtesy of www.qualityamerica.com by permission*).

the two control limits are set at plus and minus three standard deviations. If the process is in statistical control then virtually all subgroup occurrences-per-unit should be between the control limits and they should fluctuate randomly about the centerline.

$$u = \frac{\text{subgroup count of occurrences}}{\text{subgroup size in units}}$$

$$\bar{u} = \frac{\text{sum of subgroup occurrences}}{\text{sum of subgroup sizes in units}}$$

$$LCL = \bar{u} - 3\sqrt{\frac{\bar{u}}{n}}$$

$$UCL = \bar{u} + 3\sqrt{\frac{\bar{u}}{n}}$$

In the above equations, n is the subgroup size in units. If the subgroup size varies, the control limits will also vary.

The data in Table 9.9 were obtained by opening randomly selected crates from each shipment and counting the number of bruises on peaches. There are 250 peaches per crate. Our unit size will be taken as one full crate;

Shipment Number	Units (Crates)	Flaws	Flaws-per-Unit
1	1	47	47
2	1	42	42
3	1	55	55
4	1	51	51
5	1	46	46
6	1	61	61
7	1	39	39
8	1	44	44
9	1	41	41
10	1	51	51
11	2	88	44
12	2	101	50.5
13	2	101	50.5
14	1	40	40
15	1	48	48
16	1	47	47
17	1	50	50
18	1	48	48
19	1	57	57
20	1	45	45
21	1	43	43
22	2	105	52.5
23	2	98	49
24	2	100	50
25	2	96	48
TOTALS	32	1544	

Table 9.9 Raw Data for *u* Chart

that is, we will be counting crates rather than the peaches themselves. Normally, samples consist of one crate per shipment. However, when part-time help is available, samples of two crates are taken.

Using the above data the centerline and control limits are found as follows:

$$u = \frac{\text{subgroup count of occurrences}}{\text{subgroup size in units}}$$

These values are shown in the last column of Table 9.9.

$$\bar{u} = \frac{\text{sum of subgroup count of occurrences}}{\text{sum of subgroup unit sizes}} = \frac{1544}{32} = 48.25$$

which is constant for all subgroups.

$n = 1$ unit:

$$LCL = \bar{u} - 3\sqrt{\frac{\bar{u}}{n}} = 48.25 - 3\sqrt{\frac{48.25}{1}} = 27.411$$

$$UCL = \bar{u} + 3\sqrt{\frac{\bar{u}}{n}} = 48.25 + 3\sqrt{\frac{48.25}{1}} = 69.089$$

$n = 2$ units:

$$LCL = 48.25 - 3\sqrt{\frac{48.25}{2}} = 33.514$$

$$UCL = 48.25 + 3\sqrt{\frac{48.25}{2}} = 62.986$$

The control limits and the subgroup occurrences-per-unit are shown in Fig. 9.14.

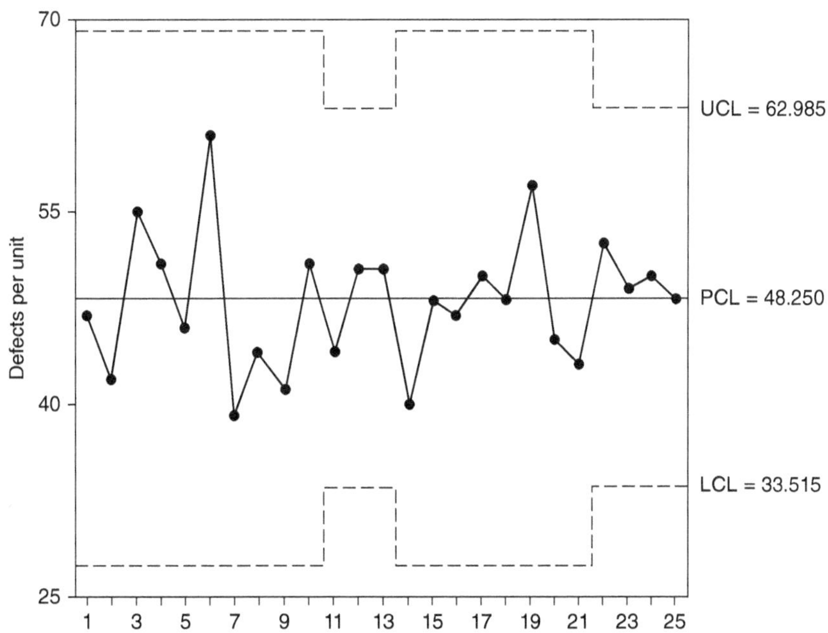

Figure 9.14 *u* chart example constructed using *Green Belt XL* software (*Courtesy of www.qualityamerica.com by permission*).

The reader may note that the data used to construct the u chart were the same as those used for the p chart, except that we considered the counts as being counts of occurrences (bruises) instead of counts of physical items (bruised peaches). The practical implications of using a u chart when a p chart should have been used, or vice versa, are usually not serious. The decisions based on the control charts will be quite similar in most cases regardless of whether a u or a p chart is used.

One way of helping determine whether or not a particular set of data is suitable for a u chart or a p chart is to examine the equation used to compute the centerline for the control chart. If the unit of measure is the same in both the numerator and the denominator, then a p chart is indicated; otherwise, a u chart is indicated. For example, if

$$\text{Centerline} = \frac{\text{bruises per crate}}{\text{number of crates}}$$

then the numerator is in terms of bruises while the denominator is in terms of crates, indicating a u chart.

The unit size is arbitrary but once determined it cannot be changed without recomputing all subgroup occurrences-per-unit and control limits. For example, if the occurrences were accidents and a unit was 100,000 hours worked, then a month with 250,000 hours worked would be 2.5 units and a month with 50,000 hours worked would be 0.5 units. If the unit size were 200,000 hours, then the two months would have 1.25 and 0.25 units respectively. The equations for the centerline and control limits would "automatically" take into account the unit size, so the control charts would give identical results regardless of which unit size is used.

As with all control charts, a special cause is probably present if there are any points beyond either the upper or lower control limit. Analysis of u chart patterns between the control limits is extremely complicated when the sample size varies and is usually not done.

Control Charts for Counts of Occurrences-per-Unit (*c* Charts)

c charts are statistical tools used to evaluate the number of occurrences-per-unit produced by a process. c charts can be applied to any variable where the appropriate performance measure is a count of how often a particular event occurs and samples of constant size are used. c charts answer the question: "Has a special cause of variation caused the central tendency of this process to produce an abnormally large or small number of occurrences over the time period observed?" Note that, unlike p or np charts, c charts do not involve counting physical items. Rather, they involve counting of *events*. For example, when using an np chart one would count bruised peaches. When using a c chart one would count the *bruises*.

Control Limit Equations for *c* Charts Like all control charts, *c* charts consist of three guidelines: centerline, a lower control limit, and an upper control limit. The centerline is the average number of occurrences-per-unit and the two control limits are set at plus and minus three standard deviations. If the process is in statistical control then virtually all subgroup occurrences-per-unit should be between the control limits and they should fluctuate randomly about the centerline.

$$\bar{c} = \frac{\text{sum of subgroup occurrences}}{\text{number of subgroups}}$$

$$LCL = \bar{c} - 3\sqrt{\bar{c}}$$

$$UCL = \bar{c} + 3\sqrt{\bar{c}}$$

The data in Table 9.10 were obtained by opening randomly selected crates from each shipment and counting the number of bruises. There are 250 peaches per crate. Our unit size will be taken as one full crate; that is, we will be counting crates rather than the peaches themselves. Every subgroup consists of one crate. If the subgroup size varied, a *u* chart would be used.

Using the above data the centerline and control limits are found as follows:

$$\bar{c} = \frac{\text{sum of subgroup occurrences}}{\text{number of subgroups}} = \frac{1006}{30} = 33.53$$

$$LCL = \bar{c} - 3\sqrt{\bar{c}} = 33.53 - 3\sqrt{33.53} = 16.158$$

$$UCL = \bar{c} - 3\sqrt{8\bar{c}} = 33.53 + 3\sqrt{33.53} = 50.902$$

The control limits and the occurrence counts are shown in Fig. 9.15.

One way of helping determine whether or not a particular set of data is suitable for a *c* chart or an *np* chart is to examine the equation used to compute the centerline for the control chart. If the unit of measure is the same in both the numerator and the denominator, then a *p* chart is indicated; otherwise, a *c* chart is indicated. For example, if

$$\text{Centerline} = \frac{\text{bruises}}{\text{number of crates}}$$

then the numerator is in terms of bruises while the denominator is in terms of crates, indicating a *c* chart.

The unit size is arbitrary but, once determined, it cannot be changed without recomputing all subgroup occurrences-per-unit and control limits.

Shipment Number	Flaws
1	27
2	32
3	24
4	31
5	42
6	38
7	33
8	35
9	35
10	39
11	41
12	29
13	34
14	34
15	43
16	29
17	33
18	33
19	38
20	32
21	37
22	30
23	31
24	32
25	42
26	40
27	21
28	23
29	39
30	29
TOTAL	1006

Table 9.10 Raw Data for c Chart

As with all control charts, a special cause is probably present if there are any points beyond either the upper or lower control limit. Analysis of *c* chart patterns between the control limits is shown later in this chapter.

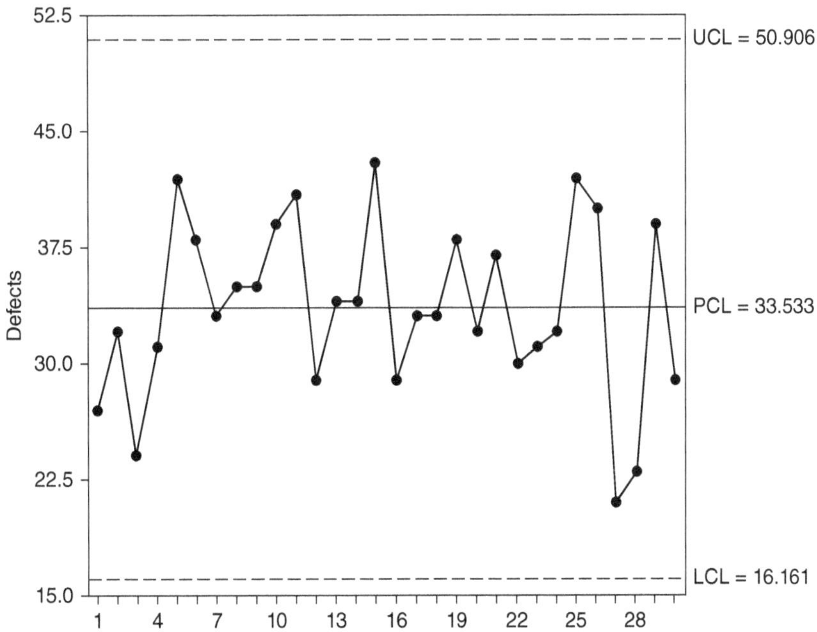

FIGURE 9.15 *c* chart example constructed using *Green Belt XL* software (*courtesy of www.qualityamerica.com by permission*).

Control Chart Selection

Selecting the proper control chart for a particular data set is a simple matter if approached properly as illustrated in Fig. 9.16.

To use the decision tree, begin at the left-most node and determine whether the data are measurements or counts. If measurements, then select the control chart based on the subgroup size. If the data are counts,

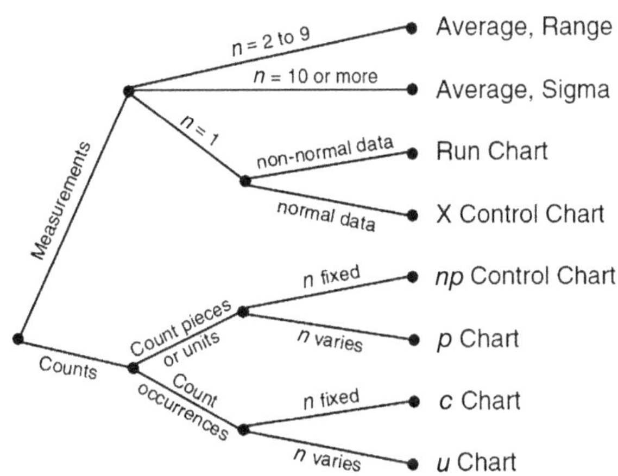

FIGURE 9.16 Decision tree for selecting control chart.

then determine whether the counts are of occurrences or pieces. An aid in making this determination is to examine the equation for the process average. If the numerator and denominator involve the same units, then a *p* or *np* chart is indicated. If different units of measure are involved, then a *u* or *c* chart is indicated. For example, if the average is in accidents-per-month, then a *c* or *u* chart is indicated because the numerator is in terms of accidents but the denominator is in terms of time.

Control Chart Interpretation

Control charts provide the operational definition of the term *special cause*. A special cause is simply anything that leads to an observation beyond a control limit. However, this simplistic use of control charts does not do justice to their power. Control charts are running records of the performance of the process and, as such, they contain a vast store of information on potential improvements. While some guidelines are presented here, control chart interpretation is an art that can only be developed by looking at many control charts and probing the patterns to identify the underlying system of causes at work.

Freak patterns are the classic special-cause situation (Fig. 9.17). Freaks result from causes that have a large effect but that occur infrequently. When investigating freak values, look at the cause-and-effect diagram for items that meet these criteria. The key to identifying freak causes is timeliness in collecting and recording the data. If you have difficulty, try sampling more frequently.

Drift is generally seen in processes where the current process value is partly determined by the previous process state. For example, if the process is a plating bath, the content of the tank cannot change instantaneously;

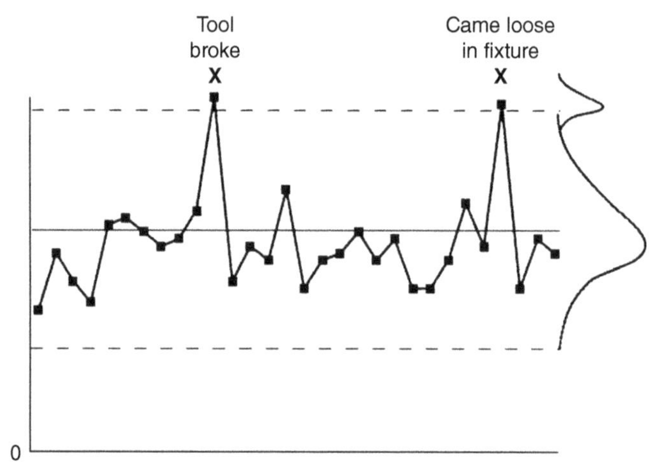

Figure 9.17 Control chart patterns: freaks.

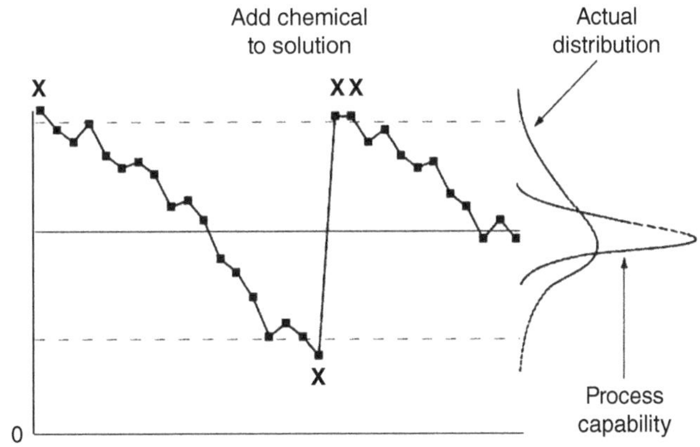

FIGURE 9.18 Control chart patterns: drift.

instead it will change gradually (Fig. 9.18). Another common example is tool wear: the size of the tool is related to its previous size. Once the cause of the drift has been determined, the appropriate action can be taken. Whenever economically feasible, the drift should be eliminated; for example, install an automatic chemical dispenser for the plating bath, or make automatic compensating adjustments to correct for tool wear. Note that the total process variability increases when drift is allowed, which adds cost. When drift elimination is not possible, the control chart can be modified in one of two ways:

1. Make the slope of the centerline and control limits match the natural process drift. The control chart will then detect departures from the natural drift.
2. Plot *deviations* from the natural or expected drift.

Cycles often occur due to the nature of the process. Common cycles include hour of the day, day of the week, month of the year, quarter of the year, week of the accounting cycle, etc. (Fig. 9.19). Cycles are caused by modifying the process inputs or methods according to a regular schedule. The existence of this schedule and its effect on the process may or may not be known in advance. Once the cycle has been discovered, action can be taken. The action might be to adjust the control chart by plotting the control measure against a variable base. For example, if a day-of-the-week cycle exists for shipping errors because of the workload, you might plot shipping errors per 100 orders shipped instead of shipping errors per day. Alternatively, it may be worthwhile to change the system to smooth out the cycle. Most processes operate more efficiently when the inputs are relatively stable and when methods are changed as little as possible.

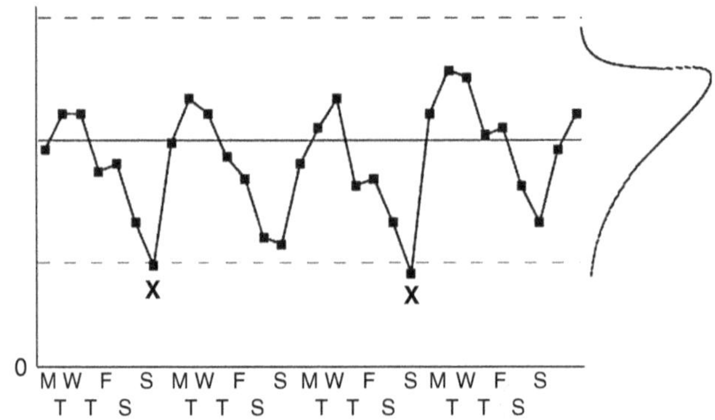

Figure 9.19 Control chart patterns: cycles.

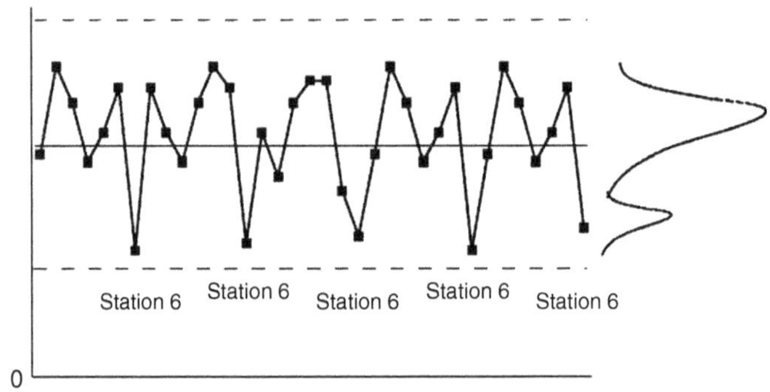

Figure 9.20 Control chart patterns: repeating patterns.

A controlled process will exhibit only "random looking" variation. A pattern where every *n*th item is different is, obviously, nonrandom (Fig. 9.20). These patterns are sometimes quite subtle and difficult to identify. It is sometimes helpful to see if the average fraction defective is close to some multiple of a known number of process streams. For example, if the machine is a filler with 40 stations, look for problems that occur 1/40, 2/40, 3/40, etc., of the time.

When plotting measurement data the assumption is that the numbers exist on a continuum; that is, there will be many different values in the data set. In the real world, the data are never completely continuous (Fig. 9.21). It usually doesn't matter much if there are, say, 10 or more different numbers. However, when there are only a few numbers that appear over and over it can cause problems with the analysis. A common problem is that the R chart will underestimate the average range, causing the control limits on both the average and range charts to be too close together. The result will be too many "false alarms" and a general loss of confidence in SPC.

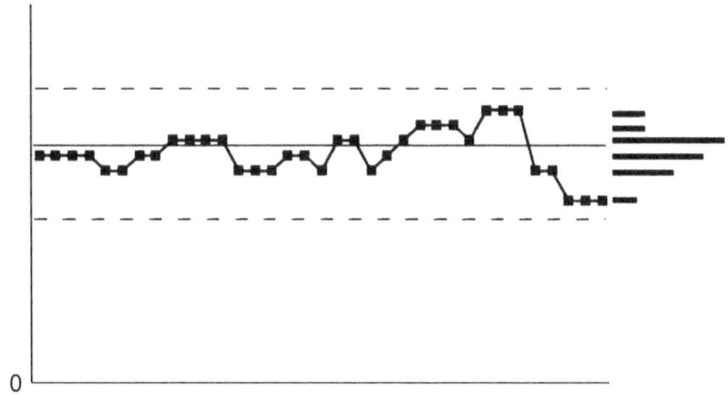

FIGURE 9.21 Control chart patterns: discrete data.

The usual cause of this situation is inadequate gage resolution. The ideal solution is to obtain a gage with greater resolution. Sometimes the problem occurs because operators, inspectors, or computers are rounding the numbers. The solution here is to record additional digits.

The reason SPC is done is to accelerate the learning process and to eventually produce an improvement. Control charts serve as historical records of the learning process and they can be used by others to improve other processes. When an improvement is realized the change should be written on the old control chart; its effect will show up as a less variable process. These charts are also useful in communicating the results to leaders, suppliers, customers, and others interested in quality improvement (Fig. 9.22).

Seemingly random patterns on a control chart are evidence of unknown causes of variation, which is not the same as *uncaused* variation. There should be an ongoing effort to reduce the variation from these so-called common causes. Doing so requires that the unknown causes of variation

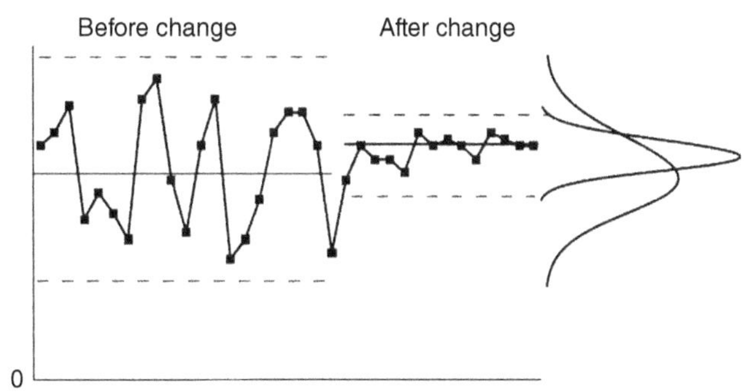

FIGURE 9.22 Control chart patterns: planned changes.

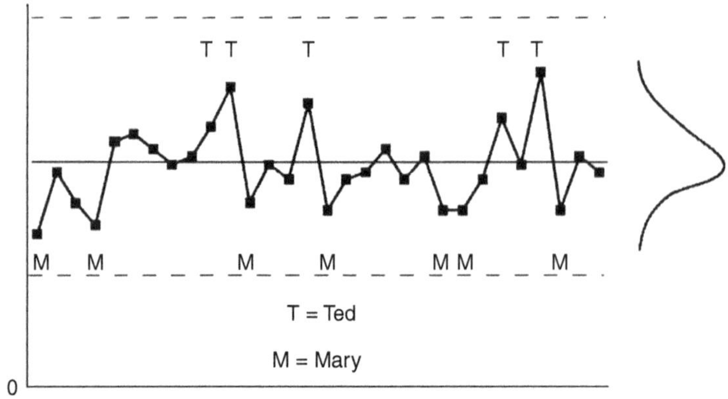

Figure 9.23 Control chart patterns: suspected differences.

be identified. One way of doing this is a retrospective evaluation of control charts. This involves brainstorming and preparing cause-and-effect diagrams, then relating the control chart patterns to the causes listed on the diagram. For example, if "operator" is a suspected cause of variation, place a label on the control chart points produced by each operator (Fig. 9.23). If the labels exhibit a pattern, there is evidence to suggest a problem. Conduct an investigation into the reasons and set up controlled experiments (prospective studies) to test any theories proposed. If the experiments indicate a true cause-and-effect relationship, make the appropriate process improvements. Keep in mind that a statistical *association* is not the same thing as a causal *correlation*. The observed association must be backed up with solid subject-matter expertise and experimental data.

Mixture exists when the data from two different cause systems are plotted on a single control chart (Fig. 9.24). It indicates a failure in creating

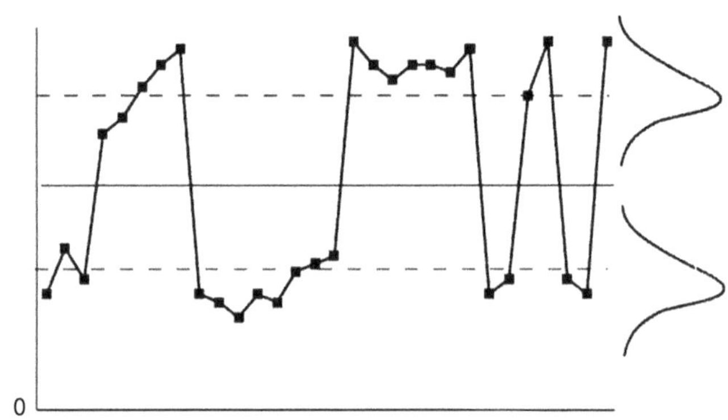

Figure 9.24 Control chart patterns: mixture.

rational subgroups. The underlying differences should be identified and corrective action taken. The nature of the corrective action will determine how the control chart should be modified.

For example, if the mixture represents two different operators who can be made more consistent, then a single control chart can be used to monitor the new, consistent process. Alternatively, if the mixture represents the difference in the number of emergency room cases received on Saturday evening, versus the number received during the week, then separate control charts should be used to monitor patient-load during the two different time periods.

Run Tests

If the process is stable, then the distribution of subgroup averages will be approximately normal. With this in mind, we can also analyze the *patterns* on the control charts to see if they might be attributed to a special cause of variation. To do this, we divide a normal distribution into zones, with each zone one standard deviation wide. Figure 9.25 shows the approximate percentage we expect to find in each zone from a stable process.

Zone C is the area from the mean to the mean plus or minus one sigma, zone B is from plus or minus one sigma to plus or minus two sigma, and zone A is from plus or minus two sigma to plus or minus three sigma.

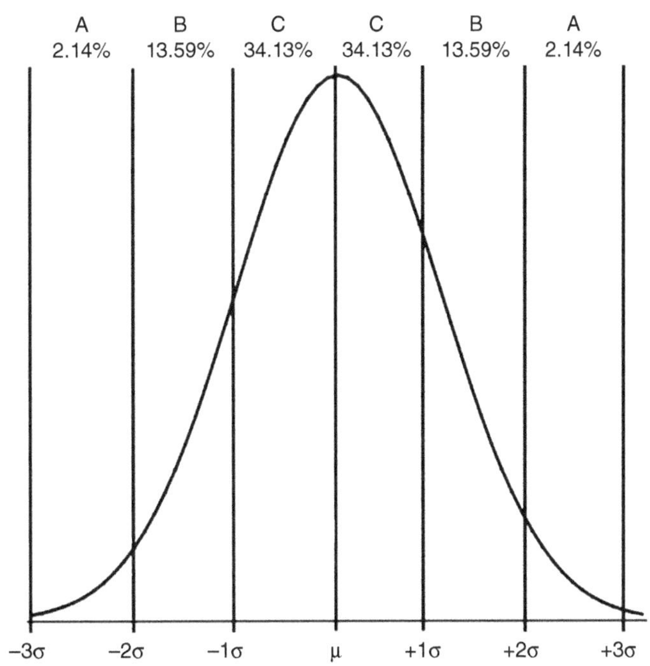

FIGURE 9.25 Percentiles of the normal distribution.

Figure 9.26 Normal distribution zone lines on a control chart.

Of course, any point beyond three sigma (i.e., outside of the control limit) is an indication of an out-of-control process.

Since the control limits are at plus and minus three standard deviations, finding the one and two sigma lines on a control chart is as simple as dividing the distance between the grand average and either control limit into thirds, which can be done using a ruler. This divides each half of the control chart into three zones. The three zones are labeled A, B, and C as shown on Fig. 9.26.

Based on the expected percentages in each zone, sensitive run tests can be developed for analyzing the patterns of variation in the various zones. Remember, the existence of a nonrandom pattern means that a special cause of variation was (or is) probably present. The averages, *np*, and *c* control chart run tests are shown in Fig. 9.27.

Using Specifications for Process Control

The control charts for differentiating between common and special causes of variation were invented nearly 100 years ago, yet there are many organizations that believe their processes are quite well controlled without them. Unfortunately, in most cases, they are mistaken.

The alternative offered by most practitioners is a standard trend chart, which looks much like the individual control chart, but uses specification limits in place of the calculated control limits. (Specification limits refer to customer or management requirements, defined as an upper bound, a lower bound, or both. Observations outside the specifications are referred to as defects.)

Consider an operator using this type of chart to monitor a production line. A sample is taken from the process and compared with the specification. If the sample is close to the specification, the operator may decide to adjust the process, usually by re-locating it by the same distance the latest

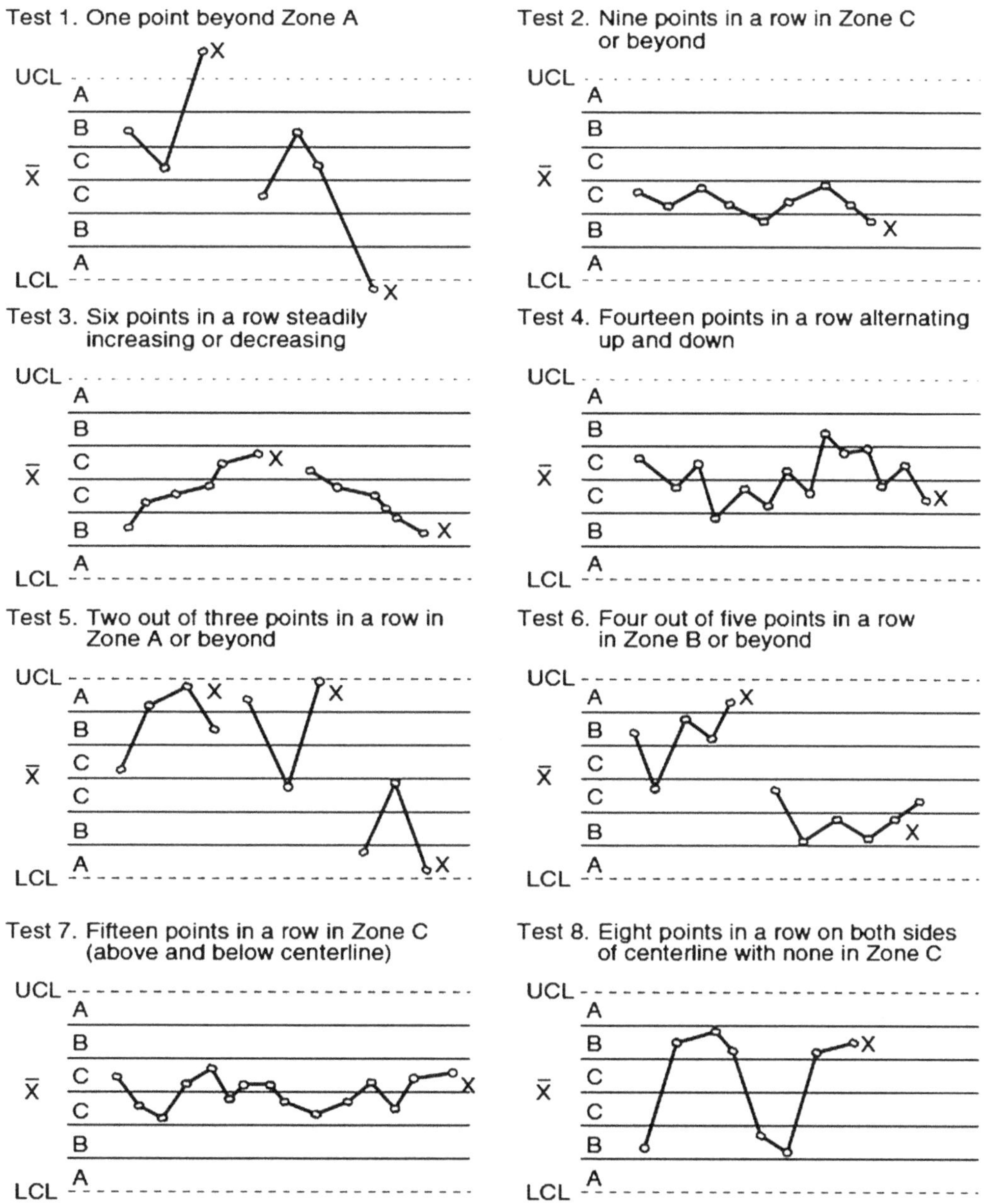

FIGURE 9.27 Tests for out-of-control patterns on control charts (Nelson, 1984).

observation was from the intended process target. For example, if the upper specification limit (usl) is 13 and the lower specification limit (lsl) is 7, the process target is 10 (calculated as the midpoint between the usl and lsl). If a process observation of 13 was observed, the operator would typically make

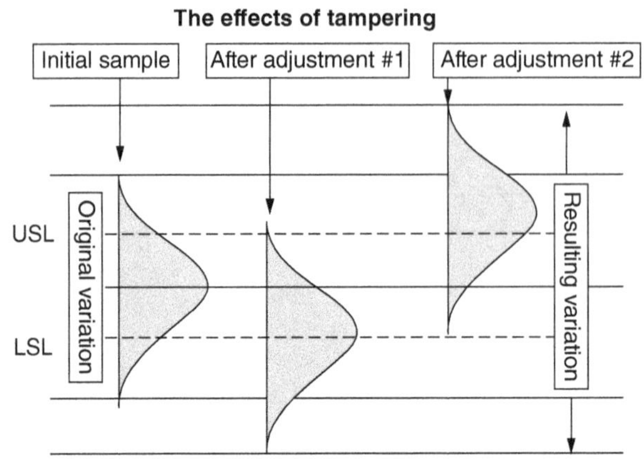

FIGURE 9.28 Effect of process tampering.

adjustments with the intent of moving the process centerline negative 3 units to the intended target of 10.

This situation is shown in the first two graphics (from the left) in Fig. 9.28. By adjusting the process, the operator has moved the entire process distribution minus 3 units. The leftmost distribution (i.e., before the adjustment) was actually (in this case) optimally located between the upper and lower specifications. Two problems masked that: (1) the operator had no idea where the process was located, since no control chart was used to define the process distribution, and (2) the process distribution is too wide for the specifications, so even though it may have been optimally located before the adjustment, there was still substantial likelihood the process would produce output beyond each of the specifications.

The operator could only appreciate these issues if a control chart were used to determine its common cause variation. When adjustments are made in the absence of a special cause, the adjustment causes additional variation, as shown in the figure. The minimum amount of variation results when the process is not adjusted. Since this process was not capable of reliably producing output within the specifications, 100 percent inspection of its output is required. In fact, any time a process is out of control, or its control status is unknown, only 100 percent sampling can be used.

When common cause variation is treated as special cause variation, and adjustments are made to the process, process tampering has occurred, and process variation has increased. This can happen in any type of process, including service/transactional processes.

The best means of diagnosing tampering is to conduct a process capability study and to use a control chart to provide guidelines for adjusting the process.

Perhaps the best analysis of the effects of tampering is from Deming (1986). Deming describes four common types of tampering by drawing the analogy of aiming a funnel to hit a desired target. These "funnel rules" are described by Deming (1986, p. 328):

1. "Leave the funnel fixed, aimed at the target, no adjustment."
2. "At drop k ($k = 1, 2, 3, \ldots$) the marble will come to rest at point z_k, measured from the target. (In other words, z_k is the error at drop k.) Move the funnel the distance $-z_k$ from the last position. Memory 1."
3. "Set the funnel at each drop right over the spot z_k, measured from the target. No memory."
4. "Set the funnel at each drop right over the spot (z_k) where it last came to rest. No memory."

Rule #1 is the best rule for stable processes. By following this rule, the process average will remain stable and the variance will be minimized. Rule #2 produces a stable output but one with twice the variance of rule #1. Rule #3 results in a system that "explodes"; that is, a symmetrical pattern will appear with a variance that increases without bound. Rule #4 creates a pattern that steadily moves away from the target, without limit (see Fig. 9.29).

At first glance, one might wonder about the relevance of such apparently abstract rules. However, upon more careful consideration, one finds many practical situations where these rules apply.

Rule #1 is the ideal situation and it can be approximated by using control charts to guide decision making. If process adjustments are made only when special causes are indicated and identified, a pattern similar to that produced by rule #1 will result.

Rule #2 has intuitive appeal for many people. It is commonly encountered in such activities as gage calibration (check the standard once and adjust the gage accordingly) or in some automated equipment (using an automatic gage, check the size of the last feature produced and make a

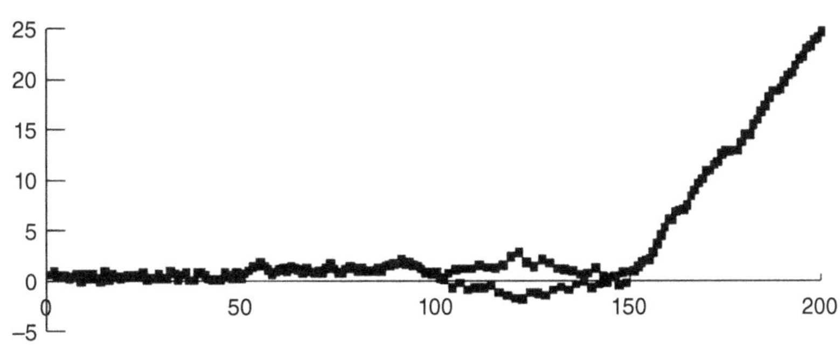

FIGURE 9.29 Funnel rule simulation results.

compensating adjustment). Since the system produces a stable result, this situation can go unnoticed indefinitely. However, as shown by Taguchi (1986), increased variance translates to poorer quality and higher cost.

The rationale that leads to rule #3 goes something like this: "A measurement was taken and it was found to be 10 units above the desired target. This happened because the process was set 10 units too high. I want the average to equal the target. To accomplish this I must try to get the next unit to be 10 units too low." This might be used, for example, in preparing a chemical solution. While reasonable on its face, the result of this approach is a wildly oscillating system.

A common example of rule #4 is the "train-the-trainer" method. A master spends a short time training a group of "experts," who then train others, who train others, etc. An example is on-the-job training. Another is creating a setup by using a piece from the last job. Yet another is a gage calibration system where standards are used to create other standards, which are used to create still others, and so on. Just how far the final result will be from the ideal depends on how many levels deep the scheme has progressed.

Process Capability Studies

Process capability analysis provides an indication of whether a controlled process is capable of reliably meeting the customer requirements. A capability analysis is a prediction, so it can only be obtained after it is verified the process is in statistical control. Process capability analysis is a two-stage process that involves:

1. Bringing a process into a state of statistical control for a reasonable period of time
2. Comparing the long-term process performance to management or engineering requirements

Process capability analysis can be done with either attribute data or continuous data if, and only if, the process is in statistical control, and has been for a reasonable period of time. Application of process capability methods to processes that are not in statistical control results in unreliable estimates of process capability and should never be done.

How to Perform a Process Capability Study

This section presents a step-by-step approach to process capability analysis (Pyzdek, 1985). The approach makes frequent reference to materials presented elsewhere in this book.

1. *Select a candidate for the study.* This step should be institutionalized. A goal of any organization should be ongoing process improvement.

However, because a company has only a limited resource base and can't solve all problems simultaneously, it must set priorities for its efforts. The tools for this include Pareto analysis and fishbone diagrams.

2. *Define the process.* It is all too easy to slip into the trap of solving the wrong problem. Once the candidate area has been selected in step 1, define the scope of the study. A process is a unique combination of machines, tools, methods, and personnel engaged in adding value by providing a product or service. Each element of the process should be identified at this stage. This is not a trivial exercise. The input of many people may be required. There are likely to be a number of conflicting opinions about what the process actually involves.
3. *Procure resources for the study.* Process capability studies disrupt normal operations and require significant expenditures of both material and human resources. Since it is a project of major importance, it should be managed as such. All of the usual project management techniques should be brought to bear. This includes planning, scheduling, and management status reporting.
4. *Evaluate the measurement system.* Using the techniques described in Chap. 14 evaluate the measurement system's ability to do the job. Again, be prepared to spend the time necessary to get a valid means of measuring the process before going ahead.
5. *Prepare a control plan.* The purpose of the control plan is twofold: (1) isolate and control as many important variables as possible, and (2) provide a mechanism for tracking variables that cannot be completely controlled. The object of the capability analysis is to determine what the process can do if it is operated the way it is designed to be operated. This means that such obvious sources of *potential* variation as operators and vendors will be controlled while the study is conducted. In other words, a single well-trained operator will be used and the material will be from a single vendor.

   There are usually some variables that are important, but that are not controllable. One example is the ambient environment, such as temperature, barometric pressure, or humidity. Certain process variables may degrade as part of the normal operation; for example, tools wear and chemicals are used. These variables should still be tracked using log sheets and similar tools. See Chap. 7 for information on designing data collection systems.
6. *Select a method for the analysis.* The SPC method will depend on the decisions made up to this point. If the performance measure is an attribute, one of the attribute charts will be used. Variables

charts will be used for process performance measures assessed on a continuous scale. Also considered will be the skill level of the personnel involved, need for sensitivity, and other resources required to collect, record, and analyze the data.

7. *Gather and analyze the data.* Use one of the control charts described in this chapter, plus common sense. It is usually advisable to have at least two people go over the data analysis to catch inadvertent errors in transcribing data or performing the analysis.
8. *Track down and remove special causes.* A special cause of variation may be obvious, or it may take months of investigation to find it. The effect of the special cause may be good or bad. Removing a special cause that has a bad effect usually involves eliminating the cause itself. For example, if poorly trained operators are causing variability, the special cause is the training system (not the operator), and it is eliminated by developing an improved training system or a process that requires less training. However, the removal of a beneficial special cause may actually involve incorporating the special cause into the normal operating procedure. For example, if it is discovered that materials with a particular chemistry produce better product, the special cause is the newly discovered material and it can be made a common cause simply by changing the specification to ensure that the new chemistry is always used.
9. *Estimate the process capability.* One point cannot be overemphasized: *the process capability cannot be estimated until a state of statistical control has been achieved!* After this stage has been reached, the methods described later in this section may be used. After the numerical estimate of process capability has been arrived at it must be compared to management's goals for the process, or it can be used as an input into economic models. Deming's all-or-none rules (see Acceptance Sampling) provide a simple model that can be used to determine if the output from a process should be sorted 100 percent or shipped as is.
10. *Establish a plan for continuous process improvement.* Once a stable process state has been attained, steps should be taken to maintain it and improve upon it. SPC is just one means of doing this. Far more important than the particular approach taken is a company environment that makes continuous improvement a normal part of the daily routine of everyone.

Statistical Analysis of Process Capability Data

This section presents several methods of analyzing the data obtained from a process capability study.

Control Chart Method: Attributes Data

1. Collect samples from 25 or more subgroups of consecutively produced units. Follow the guidelines presented in steps 1–10 above.

   Plot the results on the appropriate control chart (e.g., *c* chart). If all groups are in statistical control, go to step #3. Otherwise, identify the special cause of variation and take action to eliminate it. Note that a special cause might be beneficial. Beneficial activities can be "eliminated" as special causes by doing them all of the time. A special cause is "special" only because it comes and goes, not because its impact is either good or bad.

   Using the control limits from the preceding step (called operation control limits), put the control chart to use for a period of time. Once you are satisfied that sufficient time has passed for most special causes to have been identified and eliminated, as verified by the control charts, go to step #4.
2. The process capability is estimated as the control chart *centerline*. The centerline on attribute charts is the long-term expected quality level of the process, for example, the average proportion defective. This is the level created by the common causes of variation.

If the process capability doesn't meet management requirements, take immediate action to modify the process for the better. "Problem solving" (e.g., studying each defective) won't help, and it may result in tampering. Whether it meets requirements or not, always be on the lookout for possible process improvements. The control charts will provide verification of improvement.

Control Chart Method: Variables Data

1. Collect samples from 25 or more subgroups of consecutively produced units, following the 10-step plan described above.

   Plot the results on the appropriate control chart (e.g., $\overline{X}$ and R chart). If all groups are in statistical control, go to the step #3. Otherwise, identify the special cause of variation and take action to eliminate it.
2. Using the control limits from the preceding step (called operation control limits), put the control chart to use for a period of time. Once you are satisfied that sufficient time has passed for most special causes to have been identified and eliminated, as verified by the control charts, estimate process capability as described below.

The process capability is estimated from the process average and standard deviation, where the standard deviation is computed based on the average range or average standard deviation. When statistical control has been achieved, the capability is the level created by the common causes of process variation. The formulas for estimating the process standard deviation are:

Range chart method: $$\sigma_X = \frac{\bar{R}}{d_2}$$

Sigma chart method: $$\sigma_X = \frac{\bar{S}}{c_4}$$

The values d_2 and c_4 are constants from the table in Appendix 1.

Process Capability Indexes

Only now can the process be compared with engineering requirements. One way of doing this is by calculating capability indices. Table 9.11 shows the calculations for several capability indexes, when the distribution of the raw observations is approximated by a normal distribution. SPC software is usually used to calculate these values, especially when non-normality is suspected.

$C_P = \frac{\text{engineering tolerance}}{6\hat{\sigma}}$ where engineering tolerance = upper specification limit – lower specification limit
$C_R = 100 \times \frac{6\hat{\sigma}}{\text{engineering tolerance}}$ where engineering tolerance = upper specification limit – lower specification limit
$C_M = \frac{\text{engineering tolerance}}{8\hat{\sigma}}$ where engineering tolerance = upper specification limit – lower specification limit
$Z_{MIN} = \text{Minimum}\{Z_L, Z_U\}$
$C_{PK} = \frac{Z_{MIN}}{3}$
$C_{pm} = \frac{C_p}{\sqrt{1 + \frac{(\mu - T)^2}{\hat{\sigma}^2}}}$

Table 9.11 Process Capability Indexes

Interpreting Capability Indexes

Perhaps the biggest drawback of using process capability indexes is that they take the analysis a step away from the data. The danger is that the analyst will lose sight of the purpose of the capability analysis, which is to improve quality. To the extent that capability indexes help accomplish this goal, they are worthwhile. To the extent that they distract from the goal, they are harmful.

The quality engineer should continually refer to this principle when interpreting capability indexes.

> $\boldsymbol{C_P}$ Historically, this is one of the first capability indexes used. The "natural tolerance" of the process is computed as 6s. The index simply makes a direct comparison of the process natural tolerance with the engineering requirements. Assuming the process distribution is normal and the process average is exactly centered between the engineering requirements, a C_P index of 1 would give a "capable process." However, to allow a bit of room for process drift, the generally accepted minimum value for C_P is 1.33. In general, the larger C_P is, the better. The C_P index has two major shortcomings. First, it can't be used unless there are both upper and lower specifications. Second, it does not account for process centering. If the process average is not exactly centered relative to the engineering requirements, the C_P index will give misleading results. In recent years, the C_P index has largely been replaced by C_{PK} (see on the next page).
>
> $\boldsymbol{C_R}$ The C_R index is algebraically equivalent to the C_P index. The index simply makes a direct comparison of the process with the engineering requirements. Assuming the process distribution is normal and the process average is exactly centered between the engineering requirements, a C_R index of 100 percent would give a "capable process." However, to allow a bit of room for process drift, the generally accepted maximum value for C_R is 75 percent. In general, the smaller C_R is, the better. The C_R index suffers from the same shortcomings as the C_P index.
>
> $\boldsymbol{C_M}$ The C_M index is generally used to evaluate machine capability studies, rather than full-blown process capability studies. Since variation will increase when normal sources of process variation are added (e.g., tooling, fixtures, materials, etc.), C_M uses a four sigma spread rather than a three sigma spread.
>
> $\boldsymbol{Z_U}$ The Z_U index measures the process location (central tendency) relative to its standard deviation and the upper requirement. If the distribution is normal, the value of Z_U can be used to determine the percentage above the upper requirement by using the table in Appendix 2. In general, the bigger Z_U is, the better. A value of at least +3 is required to

assure that 0.1 percent or less defective will be produced. A value of +4 is generally desired to allow some room for process drift.

$\boldsymbol{Z_L}$ The Z_L index measures the process location relative to its standard deviation and the lower requirement. If the distribution is normal, the value of Z_L can be used to determine the percentage above the upper requirement by using the table in Appendix 2. In general, the bigger Z_L is, the better. A value of at least +3 is required to ensure that 0.1 percent or less defective will be produced. A value of +4 is generally desired to allow some room for process drift.

$\boldsymbol{Z_{\mathrm{MIN}}}$ The value of Z_{MIN} is simply the smaller of the Z_L or the Z_U values. It is used in computing C_{PK}.

$\boldsymbol{C_{PK}}$ The value of C_{PK} is simply Z_{MIN} divided by 3. Since the smallest value represents the nearest specification, the value of C_{PK} tells you if the process is truly capable of meeting requirements. A C_{PK} of at least +1 is required, and +1.33 is preferred. Note that C_{PK} is closely related to C_P, and that the difference between C_{PK} and C_P represents the potential gain to be had from centering the process.

$\boldsymbol{C_{PM}}$ A C_{PM} of at least 1 is required, and 1.33 is preferred. C_{PM} is closely related to C_P. The difference represents the potential gain to be obtained by moving the process mean closer to the target. Unlike C_{PK}, the target need not be the center of the specification range.

For example, assume that a process is in statistical control based on an $\overline{\mathrm{X}}$ and R chart with subgroups of 5. The grand average (or centerline of the $\overline{\mathrm{X}}$ chart) is calculated as 0.99832, and the average range (or centerline of the R chart) is calculated as 0.02205. From the table of d_2 values (Appendix 1), we find d_2 is 2.326 for subgroups of 5. Thus, using the equation above for calculating the process standard deviation using the Range chart method:

$$\hat{\sigma} = \frac{0.02205}{2.326} = 0.00948$$

If the process requirements are a lower specification of 0.980 and an upper specification of 1.020 (i.e., 1:000 ± 0:020), the Z values are calculated as:

$$Z_U = \frac{\text{upper specification} - \overline{\overline{\mathrm{X}}}}{\hat{\sigma}} = \frac{1.020 - 0.99832}{0.00948} = 2.3$$

$$Z_L = \frac{\overline{\overline{\mathrm{X}}} - \text{lower specification}}{\hat{\sigma}} = \frac{0.99832 - 0.980}{0.00948} = 1.9$$

$$C_P = \frac{\text{engineering tolerance}}{6\hat{\sigma}} = \frac{1.020 - 0.9800}{6 \times 0.00948} = 0.703$$

$$C_R = 100 \times \frac{6\hat{\sigma}}{\text{engineering tolerance}} = 100 \times \frac{6 \times 0.00948}{0.04} = 142.2\%$$

$$C_M = \frac{\text{engineering tolerance}}{8\hat{\sigma}} = \frac{0.04}{8 \times 0.00948} = 0.527$$

$$Z_{\text{MIN}} = \text{Minimum}\,\{1.9, 2.3\} = 1.9 \quad C_{PK} = \frac{Z_{\text{MIN}}}{3} = \frac{1.9}{3} = 0.63$$

Assuming that the target is precisely 1.000, we compute:

$$C_{PM} = \frac{C_p}{\sqrt{1 + \frac{\left(\overline{\overline{X}} - T\right)^2}{\hat{\sigma}^2}}} = \frac{0.703}{\sqrt{1 + \frac{(0.99832 - 1.000)^2}{0.00948^2}}} = 0.692$$

Since the minimum acceptable value for C_p is 1, the 0.703 result indicates that this process cannot meet the requirements. Furthermore, since the C_p index doesn't consider the centering process, we know that the process can't be made acceptable by merely adjusting the process closer to the center of the requirements. Thus, we would expect the Z_L, Z_U, and Z_{MIN} values to be unacceptable as well.

The C_R value always provides the same conclusions as the C_p index. The number implies that the "natural tolerance" of the process uses 142.2 percent of the engineering requirement, which is, of course, unacceptable.

The C_M index should be 1.33 or greater. Obviously it is not. If this were a machine capability study, the value of the C_M index would indicate that the machine was incapable of meeting the requirement.

The value of C_{PK} is only slightly smaller than that of C_P. This indicates that we will not gain much by centering the process. The actual amount we would gain can be calculated by assuming the process is exactly centered at 1.000 and recalculating Z_{MIN}. This gives a predicted total reject rate of 3.6 percent instead of 4.0 percent.

CHAPTER 10

Quality Audits

An audit is a comparison of observed activities and/or results with documented requirements. The evidence provided from audits forms the basis of improvement in either the element audited, or in the requirements. Effective quality auditing can prevent problems by uncovering situations that, while still acceptable, are trending toward an eventual problem. The attention of management brought on by an unfavorable audit report can often prevent future noncompliance.

ISO 19011, Guidelines for Auditing Quality Systems, defines a quality audit as a

> *systematic, independent and documented process for obtaining audit evidence (records, statements of fact or other information, which are relevant to the audit criteria and verifiable) and evaluating it objectively to determine the extent to which the audit criteria (set of policies, procedures or requirements) are fulfilled.*

Review activities must meet several criteria to be considered audits. Informal walk-throughs, while useful, do not qualify as systematic evaluation, so do not meet the criteria for an audit. Examinations by employees who report to the head of the function being examined are also important, but are not audits. The reference to "policies, procedures or requirements" implies the existence of written documentation.

Undocumented quality systems are not proper subject matter for quality audits. Implementation is audited by comparing the planned policies, procedures, or requirements with observed practices, with an eye toward whether or not (1) these are properly implemented, and (2) if so, do they accomplish the stated objectives?

It is no surprise to find that an activity as important and as common as quality audits is covered by a large number of different standards. It is in the best interest of all parties that the audit activity be standardized to the extent possible. One of the fundamental principles of effective auditing is "no surprises," something easier to accomplish if all parties involved use the same rule book. Audit standards are, in general, guidelines that are voluntarily adopted by auditor and auditee. Often the parties make compliance mandatory by specifying the audit standard as part of a contract.

When this occurs it is common practice to specify the revision of the applicable standard to prevent future changes from automatically becoming part of the contract.

Willborn (1993) reviews eight of the most popular quality audit standards and provides a comparative analysis of these standards in the following areas:

- General features of audit standards
- Auditor (responsibilities, qualifications, independence, performance)
- Auditing organizations (auditing teams, auditing departments/groups)
- Client and auditee
- Auditing (initiation, planning, implementation)
- Audit reports (drafting, form, content, review, distribution)
- Audit completion (follow-up, record retention)
- Quality assurance

Auditing standards exist to cover virtually every aspect of the audit. The reader is encouraged to consult these standards, or Willborn's summaries, to avoid reinventing the wheel.

Types of Quality Audits

There are three basic types of quality audits: systems, products, and processes. Systems audits are the broadest in terms of scope. The most commonly audited system is the quality system: the set of activities designed to ensure that the product or service delivered to the end user complies with all quality requirements. Product audits are performed to confirm that the system produced the desired result. Process audits are conducted to verify that the inputs, actions, and outputs of a given process match the requirements. All of these terms are formally defined in several audit standards.

Product Audits

Product audits are generally conducted from the customer's perspective. ISO 9000:2000 divides products into four generic categories: hardware, software, processed materials, and services.

The quality system requirements are essentially the same for all product categories. ISO 9000 defines four facets of product quality: quality due to defining the product to meet marketplace requirements, quality due to design, quality due to conformance with the design, and quality due to product support. Traditionally, product quality audits were conducted primarily to determine conformance with design. However, modern quality audit standards (e.g., the ISO 9000 family) are designed to determine all four facets of quality.

One purpose of product audit is to estimate the quality being delivered to customers; thus product audits usually take place after internal inspections have been completed.

Product audits differ from inspection in the following ways: (1) audits are broader in scope than inspections, (2) audits provide more depth than inspections, (3) audits provide information useful for product quality improvement, and (4) audits offer another level of assurance beyond routine inspection. Inspection normally focuses on a small number of important product characteristics. Inspection samples are selected at random in sizes large enough to produce statistically valid inferences regarding lot quality. Audits, on the other hand, are concerned with the quality being produced by the system. Thus, the unit of product is viewed as representing the common result of the system that produced it. Audit samples are sometimes seemingly quite small, but they serve the purpose of showing a system snapshot.

Audit samples are examined in greater depth than are product samples; that is, more information is gathered per unit of product. The sample results are examined from a systems perspective. The examination goes beyond mere conformance to requirements. Minor discrepancies are noted, even if they are not serious enough to warrant rejection. A common practice is to use a weighting scheme to assign "demerits" to each unit of product. Aesthetics can also be evaluated by the auditor (e.g., paint flaws, scratches, etc.). Table 10.1 presents an example of a publisher's audit of a sample of 1000 books of a given title.

These audit scores are presented on histograms and control charts to determine their distribution and to identify trends. Product audit results are compared with the marketing requirements (i.e., customer requirements) as well as the engineering requirements.

Product audits are often conducted in the marketplace itself. By obtaining the product as a customer would, the auditor can examine the impact of transportation, packaging, handling, storage, and so on. These audits also provide an opportunity to compare the condition of the product with that being offered by competitors.

Problem	**Seriousness**	**Weight**	**Frequency**	**Demerits**
Cover bent	Major	5	2	10
Page wrinkled	Minor	3	5	15
Light print	Incidental	1	15	15
Binding failure	Major	5	1	5
		TOTAL	23	45

TABLE 10.1 Book Audit Results

Process Audits

Process audits focus on specific activities or organizational units. Examples include engineering, marketing, calibration, inspection, discrepant materials control, corrective action, etc. Processes are organized, value-added manipulations of inputs that result in the creation of a product or service. Process audits compare the actual operations with the documented requirements of the operations. Process audits should begin with an understanding of how the process is supposed to operate. A process flowchart is a useful tool in helping to reach this understanding.

It has been said that a good reporter determines the answer to six questions: who? what? when? where? why? and how? This approach also works for the process auditor. For each important process task, ask:

- Who is supposed to do the job? (Are any credentials required?)
- What is supposed to be done?
- When is it supposed to be done?
- Where is it supposed to be done?
- Why is this task done?
- How is this task supposed to be done?

The documentation should contain the answers to every one of these questions. If it doesn't, the auditor should suspect that the process isn't properly documented. Of course, the actual process should be operated in conformance to documented requirements.

Systems Audits

Systems are arrangements of processes: a group of interacting, interrelated, or interdependent elements forming a complex whole. Systems audits differ from process audits primarily in their scope. Whereas process audits focus on an isolated aspect of the system, systems audits concentrate on the relationships between the various processes in the system. In the case of quality audits, the concern is with the quality system. The quality system is the set of all activities designed to ensure that all important quality requirements are determined, documented, and followed.

The level of requirements for quality systems varies with the type of organization being audited and, perhaps, with the size of the organization. Organizations that produce, distribute, and support a product have greater needs than organizations that sell a service. The changes made to the ISO 9000 series in 2000 delineated the requirements in broader terms that were more clearly applicable to most or all organizations.

Internal Audits

Considering the benefits that derive from quality auditing, it is not surprising that most quality audits are internal activities conducted by organizations interested in self-improvement. Of course, the same principles apply to internal audits as to external audits (e.g., auditor independence). Ishikawa (1985) describes four types of internal audits:

- Audit by the president
- Audit by the head of the unit (by division head, factory manager, branch office manager, etc.)
- Quality control (QC) audit by QC staff
- Mutual QC audit

President's audits are similar to what Tom Peters has called "management by wandering around" (MBWA). The president personally visits different areas of the organization to make firsthand observations of the effectiveness of the quality system. Audit by the head of the unit is equivalent to the president's audit, except the audit is limited to functional areas under the jurisdiction of the head person. Quality control audits are conducted by the quality department in various parts of the organization. Unlike presidents and unit heads who are auditing their own areas, quality department auditors must obtain authorization before conducting audits. In mutual QC audits, separate divisions of the company exchange their audit teams. This provides another perspective from a team with greater independence.

Two-Party Audits

Most audits are conducted between customers and suppliers. In this case suppliers usually provide a contact person to work with the customer auditor. In addition, suppliers usually authorize all personnel to provide whatever information the auditor needs, within reason, of course. Two-party audits are generally restricted to those parts of the quality system of direct concern to the parties involved. The customer will evaluate only those processes, products, and system elements that directly or indirectly impact upon their purchases.

Third-Party Audits

One problem with two-party audits is that a supplier will be subject to audits by many different customers, each with their own (sometimes conflicting) standards. Likewise, customers must audit many different suppliers, each with their own unique approach to quality systems design. Third-party audits are one way of overcoming these difficulties.

In a third-party audit the auditing organization is not affiliated with either the buyer or the seller. The audit is conducted to a standard that both the buyer and seller accept, such as the ISO 9000 series discussed in Chap. 2. As the use of ISO 9000 becomes more widespread, the incidence of third-party audits will continue to increase. However, ISO 9000 audits are conducted at a high system level. Product and process audits will continue to be needed to address specific issues between customers and suppliers.

Desk Audits

The emphasis of the discussion above has been on the on-site visit. However, a significant portion of the auditing activity takes place between the auditor and auditee, each working at their respective organizations. A great deal of the audit activity involves examination of documentation. The documentation reveals whether or not a quality system has been developed. It describes the system as the supplier wants it to be. From a documentation review, the auditor can determine if the quality system, as designed, meets the auditor's requirements. If not, a preliminary report can inform the auditee of any shortcomings. Corrective action can be taken either to modify the documentation or to develop new system elements. Once the documentation is in a form acceptable to the auditor, an on-site visit can be scheduled to determine whether the system has been properly implemented. Properly done, desk audits can save both auditor and auditee significant expense and bother.

Planning and Conducting the Audit

Most quality audits are pre-announced, which provides several advantages. A pre-announced audit is much less disruptive of operations. The auditee can arrange to have the right people available to the auditor. The auditor can provide the auditee with a list of the documentation he or she will want to review so the auditee can make it available. Much of the documentation can be reviewed prior to the on-site visit. The on-site visit is much easier to coordinate when the auditee is informed of the audit. Finally, pre-announced audits make it clear that the audit is a cooperative undertaking, not a punitive one.

Of course, when deliberate deception is suspected, surprise audits may be necessary. Surprise audits are usually very tightly focused and designed to document a specific problem. In most cases, quality professionals are not trained or qualified to conduct adversarial audits. Such audits are properly left to accounting and legal professionals trained in the handling of such matters.

Audits can be scheduled at various points in the buying cycle. The following timings of audits are all quite common:

- *Pre-award audit.* Conducted to determine whether the prospective supplier's quality system meets the customer's requirements.
- *Surveillance audit.* Conducted to ensure that an approved supplier's quality system continues to comply with established requirements.
- *Contract renewal.* Conducted to determine whether a previously approved supplier continues to meet the quality system requirements.
- *Problem resolution.* A tightly focused audit conducted to identify the root cause of a problem and to ensure that effective corrective action is taken to prevent future occurrences of the problem.
- *In-process observation.* On-site audits performed to ensure that processes are performed according to established requirements. These audits are often performed when it is difficult or impossible to determine whether or not requirements have been met by inspecting or testing the finished product.

At times periodic audits are automatically scheduled. For example, to maintain certification to the ISO 9000 series standards, an organization is periodically reassessed.

Auditor Qualifications

Willborn (1993, pp. 11–23) provides an extensive discussion of auditor qualifications. The first requirement for any auditor is absolute honesty and integrity. Auditors are often privy to information of a proprietary or sensitive nature. They sometimes audit several competing organizations. The information an auditor obtains must be used only for the purpose for which it was intended. It should be held in strict confidence. No amount of education, training, or skill can compensate for lack of ethics.

The auditor must be independent of the auditee. In addition, auditors must comply with professional standards, possess essential knowledge and skills, and maintain technical competence. Auditors must be fair in expressing opinions and should inspire the confidence of both the auditee and the auditor's parent organization.

The auditor acts as only an auditor and in no other capacity, such as management consultant or manager. Managers of audit organizations should have a working knowledge of the work they are supervising.

An auditor's qualifications must conform to any applicable standards, and they must be acceptable to all parties. The auditing organization should establish qualifications for auditors and provide training for technical specialists.

Some auditing activities, such as those of nuclear power plants, require special certification. Lead auditors require additional training in leadership skills and management. Third parties may also provide certification of auditors, such as by ASQ.

Auditors should be able to express themselves clearly and fluently both verbally and in writing. They should be well versed in the standards to which they are auditing. Where available, this knowledge should be verified by written examination and/or certification.

Auditors should master the auditing techniques of examining, questioning, evaluating and reporting, identifying methods, following up on corrective action items, and closing out audit findings. Industrial quality auditors should have knowledge of design, procurement, fabrication, handling, shipping, storage, cleaning, installation, inspection, testing, statistics, nondestructive examinations, maintenance, repair, operation, modification of facilities or associated components, and safety aspects of the facility/process. In a specific audit assignment, the knowledge of individual auditors might be complemented by other audit team members.

Internal Quality Surveys as Preparation

While quality audits are formal, structured evaluations involving independent auditors, quality surveys are internal, less formal reviews of quality systems, products, or processes often conducted at the request of internal management. The purpose of a quality survey is informational, so formal reports are generally not prepared. Rather, the survey results are presented in information-sharing sessions with concerned personnel. Quality surveys conducted prior to quality audits can assist the organization in preparing for the audit.

Steps in Conducting an Audit

Most quality systems audits involve similar activities. The checklist below is an adaptation of the basic audit plan described by Keeney (1995).

- Choose the audit team. Verify that no team member has a conflict of interest.
- Meet with the audit team and review internal audit procedures.
- Discuss forms to be used and procedures to be followed during the audit.
- Perform a desk audit of the quality manual and other documentation to verify the scope of the audit and provide an estimate of the duration of the audit.
- Assign audit subteams to their respective audit paths.
- Contact the auditee and schedule the audit.
- Perform the audit.
- Write corrective action requests (CARs) and the audit summary report, listing the CARs in the audit summary.
- Conduct a closing meeting (exit briefing).

- Issue the audit summary report.
- Present the complete audit findings, including all notes, reports, checklists, CARs, etc., to the quality manager.
- Prepare a final audit report.
- Follow up on CARs.

Audit Reporting Process

Audit results are reported while the audit is in progress and upon completion of the audit. The principle is simple: the auditor should keep the auditee up-to-date at all times. This is a corollary of the "no surprises" principle. In addition, it helps the auditor avoid making mistakes by misinterpreting observations; in general, the auditee is better informed about internal operations than the auditor.

Auditees should be informed before, during, and after the audit. Prior to the audit, the auditee is told the scope, purpose, and timing of the audit and allowed to play an active role in planning the audit. Upon arrival the auditor and auditee should meet to review plans and timetables for the audit. Verbal daily briefings should be made, presenting the interim results and tentative conclusions of the auditor. At these meetings the auditee is encouraged to present additional information and clarification to the auditor. Written minutes of these meetings should be maintained and published. Upon completion of the audit an exit briefing is recommended.

As a matter of courtesy, an interim report should be issued as soon as possible after completion of the audit. The interim report should state the main findings of the auditor and the auditor's preliminary recommendations.

Formal audit reports are the ultimate product of the audit effort. Formal audit reports usually include the following items:

- Audit purpose and scope
- Audit observations
- Conclusions and recommendations
- Objectives of the audit
- Auditor, auditee, and third-party identification
- Audit dates
- Audit standards used
- Audit team members
- Auditee personnel involved
- Statements of omission
- Qualified opinions
- Issues for future audits

- Auditee comments on the report (if applicable)
- Supplementary appendices

The audit report is the "official document" of the audit. Audit reports should be prepared in a timely fashion. Ideally, the deadline for issuing the report should be determined in the audit plan prepared beforehand. Audit reports should describe the purpose and scope of the audit; the entity audited; the membership of the audit team, including affiliation and potential conflicts of interest; the observations of the audit; and recommendations. Detailed evidence supporting the recommendations should be included in the report. Audit reports may include recommendations for improvement. They may also include acknowledgment of corrective action already accomplished. The formal report may draw upon the minutes of the meetings held with the auditee. It should also note the auditee's views of previously reported findings and conclusions.

Auditor opinions are allowed, but they should be clearly identified as opinions and supporting evidence should be provided.

Anyone who has been an auditor for any time knows that it is sometimes necessary to report unpleasant findings. These results are often the most beneficial to the auditee, providing information that in-house personnel may be unwilling to present. When presented properly by an outside auditor, the "bad news" may act as the catalyst to long-needed improvement. However, the auditor is advised to expect such reactions as denial, anger, and frustration when the findings are initially received. In general, unpleasant findings should be supported more extensively. Take care in wording the findings so that the fewest possible emotional triggers are involved. The sooner the auditee and auditor can begin work on correcting the problems, the better for both parties.

If a report is prepared by more than one person, one person will be designated as senior auditor. The senior auditor will be responsible for assembling and reviewing the completed report. If the audit was conducted by a second party, distribution of the report will be limited to designated personnel in the auditor and auditee's organizations, usually senior management. If a third-party audit was conducted, the audit report will also be sent to the client, who in turn is responsible for informing the auditee's senior management. In certain cases (e.g., government audits, financial audits), the results are also made available to the public. However, unless otherwise specified, audit reports are usually considered to be private, proprietary information that cannot be released without the express written permission of the auditee.

Post-Audit Activities (Corrective Action, Verification)

Audit reports often contain descriptions of problems and discrepancies encountered during the audit. However, not all problems are equal; some are more serious than others. A well-written audit report will classify the problems according to how serious they are. Product-defect-seriousness classification schemes are discussed below. Some organizations also apply seriousness classification to discrepancies found in planning, procedures, and other areas.

These seriousness classifications (e.g., "critical," "major," "minor") should be explicitly defined and understood by all parties. Generally, some sort of weighting scheme is used in conjunction with the classification scheme and an audit score is computed. Although the audit score contains information, it should not be the sole criterion for deciding whether or not the audit was passed or failed. Instead, consider the numerical data as additional input to assist in the decision-making process.

Most audits are not pass/fail propositions. Rather, they represent an effort on the part of buyers and sellers to work together for the long term. When viewed in this light, it is easy to see that identifying a problem is just the first step. Solving the problem requires locating the root cause of the problem, which is challenging work. Many times the problem is treated as if it were a cause; that is, action is taken to "manage the problem" rather than addressing its cause. Examples of this are inspection to remove defects or testing software to catch bugs. Wilson et al. (1993) define "root cause" as that most basic reason for an undesirable condition or problem, which, if eliminated or corrected, would have prevented it from existing or occurring. Root causes are usually expressed in terms of specific or systematic factors. A root cause usually is expressed in terms of the least common organizational, personal, or activity denominator.

In most cases the auditor is not capable of identifying the root cause of a problem. The auditee is expected to perform the necessary analysis and to specify the action taken to address the cause(s) of the problem. At this point the auditor can sometimes determine that the root cause has not been identified and can assist the auditee in pursuing the problem at a deeper level. At other times there is no choice but to validate the corrective action by additional audits, tests, or inspections. The final proof of the effectiveness of any corrective action must be in achieving the desired result.

Product, Process, and Materials Control

Work Instructions

Work instructions must establish quantitative or qualitative means for determining that each operation has been done satisfactorily. These criteria must also be suitable for use with related inspections or tests, because work instructions serve operating personnel, supervisors, inspectors, managers, and even customers. Compliance with instructions is subject to review and audit.

Work instructions include the documented procedures that define how production, installation, or servicing will take place. These instructions describe the operating environmental conditions as well as the activities necessary to ensure that the finished product meets all of the customer's requirements. Work instructions also includes "cheat sheets," "crib notes," and other tidbits that people keep to remind them of the way "it's really done." ISO 9000 makes these informal notes part of the official documentation of the process.

Just how far one should go in documenting a process is debatable. Clearly, if the documentation becomes so massive that no one has time to read it all, it no longer serves its purpose. Work instructions that include an overwhelming number of "tips" associated with rare problems over a period of years will make it more difficult to locate the truly useful information.

Consider, for example, your daily trip to work. Simple documentation might list the streets that you take under normal conditions. However, one day you find a traffic jam and take an alternate route. Should you write this down? Well, if the traffic jam is caused by a long-term construction project, perhaps. But if it's due to a rare water-main rupture, it's probably not necessary.

General George Patton famously said, "Don't tell people how to do things. Tell them what to do and let them surprise you with their results." Allowing flexibility in work instructions, when workers are properly trained in their cross-functional purpose in satisfying customers, can provide empowerment and lead to superior customer service. In this context, work instructions can provide the reasons for satisfying particular objectives of the function (the *why's* for the *what's*), rather than the specific *how's*, which may overly constrain discretion.

As technology improves, databases may be developed to quickly and effectively filter information relevant to the task at hand. This will effectively increase the amount of data that can be made available to the process operator. Until then, the documentation must be contained within human cognitive limits. The guiding principle should be minimum size subject to being reasonably complete and accessible to those who will use it.

Work instructions should cover the following items:

- The manner in which the work will be done
- The equipment needed to do the work
- The working environment
- Compliance with other procedures and documents
- Process parameters to be monitored and how they will be monitored (e.g., checklists, control charts)

- Product characteristics to be monitored and how they will be monitored
- Workmanship criteria
- Maintenance procedures
- Verification methods (process qualification)

Work instructions should be written in clear, simple terms, using the language that is easiest for the person doing the work to understand. The people doing the work should be intimately involved in preparing the work instructions. Pictures, diagrams, graphics, and illustrations should be used to make the documentation easy to understand and apply. If the instructions are voluminous, they should include such aids as indexes, tables of contents, tables of figures, tabs, etc. to assist in locating relevant information. Of course, to ensure that they are up-to-date, the documentation should be cross-indexed to the engineering drawings, purchase orders, or other documents that they implement. Work instructions should be part of the overall document control system of a firm.

Classification of Characteristics

All but the most simple products or services include large numbers of features or characteristics of interest to the customer. In theory, every feature of every unit produced, or every transaction conducted, could be inspected and judged against the requirements. This would add considerable cost to the product and, for most features, add little or no value to the customer. Instead, it is better to establish a hierarchy of importance for the various characteristics of the product or service. Which features are so important that they deserve a great deal of attention? Which need only a moderate amount of attention? Which need only a cursory inspection or review? The activity of arriving at this determination is known as classification of characteristics.

In practice, characteristics are usually classified into the categories *critical*, *major*, and *minor*. The terms can be defined in simple terms as follows:

Critical characteristic. Any feature whose failure can reasonably be expected to present a safety hazard either to the user of the product, or to anyone depending on the product functioning properly. For service, any characteristic that would lead to legal implications, or severely impact reputation.

Major characteristic. Any feature, other than critical, whose failure would likely result in a reduction of the usability of the product. For service, any characteristic that would lead to loss of goodwill or future business.

Minor characteristic. Any feature, other than major or critical, whose failure would likely be noticeable to the user.

Incidental characteristic. Any feature other than critical, major, or minor. While it is possible to develop classification schemes that are more detailed, the above definitions suffice for the vast majority of applications.

Identification of Materials and Status

Has this been inspected? If so, was it accepted? Rejected? Does it require rework? Re-inspection? Retest? Obtaining clear answers to these questions is a primary task in quality control. Virtually all quality systems standards and specifications require that systems that identify the status of purchased materials, customer-supplied materials, production materials, work-in-process, and finished goods be developed, well documented, and fully implemented.

Purchased Materials

Proper identification of purchased materials begins, of course, with the supplier. A key part of the supplier's quality system must include the identification of materials and status discussed below. Once received, the quality status of purchased materials should be identified in accordance with documented procedures. The procedures should cover how purchased material will be identified (e.g., with status tags), where the materials are to be stored until conformance to requirements has been established, how nonconforming material will be identified and stored, and how to process nonconforming purchased materials.

Customer-Supplied Materials

Procedures must be developed and documented for the control of verification, storage, and maintenance of customer-supplied product provided for incorporation into the supplies, or for related activities. The procedures must ensure that product that is lost, damaged, or otherwise unsuitable for use is recorded and reported to the customer.

Work-in-Process (WIP)

Procedures for the identification of the inspection and test status of all WIP should be developed and documented. The identification of inspection and test status should be part of the quality plan covering the entire cycle of production. The purpose of the procedures is to ensure that only product that has passed the necessary inspection and test operations is delivered. WIP procedures should also include any in-process observations, verifications, and tests that are required. For example, some products must undergo certain interim processing that cannot be verified except by direct observation as the processing is taking place.

Finished Goods

The quality plan should include procedures that document the tests or inspections required prior to the release of product for delivery to the customer. The procedures should specify how the inspection and test status of finished goods will be shown, where the goods will be stored while awaiting shipment, and proper methods of packaging, handling, and loading the goods for final delivery.

Lot Traceability

Documented procedures should be prepared to ensure that, when required, lot traceability is maintained. Traceability is largely a matter of record-keeping. The system should ensure that the units in the lot and the lot itself are identified, and the integrity of the lot is maintained (i.e., every unit that is part of the lot remains in the lot).

Lot traceability is generally required when there is reason to believe that the unit in question may need to be located at some time in the future. There are many reasons why this might be necessary, but the primary reason is that a safety defect might be discovered. The manufacturer should be able to quickly communicate with all those who are at risk from the defect. Items whose failure would cause an unsafe condition to exist are known as *critical components* or *critical items*.

Materials Segregation Practices

The previous sections describe various activities relating to the identification of various types of materials, for example, by type of defective, or by processing status. Once a "special" classification has been made (e.g., material to be scrapped or reworked), the procedure specifies how the affected material will be identified. Next, provision must often be made to physically remove the material from the normal processing stream. Formal, written procedures should be developed to describe the control, handling, and disposition of nonconforming materials to ensure that such materials are adequately identified and prevented from becoming mixed with acceptable materials.

The physical control of nonconforming materials varies widely from firm to firm and by type of problem. Some organizations require that discrepant-critical components be immediately removed to a locked storage area and require authorization from designated individuals for release.

Configuration Control

Configuration control is the systematic evaluation, coordination, approval or disapproval, and implementation of all approved changes in the configuration of an item after formal establishment of its configuration identification.

The intent of configuration management is to control the form, fit, and function of configuration items. Configuration control is primarily concerned with managing engineering changes.

Deviations and Waivers

While an engineering change involves a permanent change to the engineering design, a deviation is a *temporary* departure from an established requirement. Deviation requests should be formally evaluated and approved only when they result in significant benefit. Repeated deviations should be investigated; if the underlying requirements are too stringent, they should be modified.

CHAPTER 11

Supply Chain Management

The majority of the cost of most manufactured product is in purchased materials. In some cases the percentage is 80 percent or more; it is seldom less than 50 percent. The importance of consistently high levels of quality in purchased materials is clear.

It is important to remember that dealings between companies are really dealings between people. People work better together if certain ground rules are understood and followed. Above all, the behavior of both buyer and seller should reflect honesty and integrity. This is especially important in quality control, where many decisions are "judgment calls." There are certain guidelines that foster a relationship based on mutual trust:

- *Don't be too legalistic.* While it is true that nearly all buyer-seller arrangements involve a contract, it is also true that unforeseen conditions sometimes require that special actions be taken. If buyer and seller treat each other with respect, these situations will present no real problem.
- *Maintain open channels of communication.* This involves both formal and informal channels. Formal communication includes such things as joint review of contracts and purchase order requirements by both seller and buyer teams, on-site seller and buyer visits and surveys, corrective action request and follow-up procedures, record-keeping requirements, and so on. Informal communications involve direct contact between individuals in each company on an ongoing and routine basis. Informal communications to clarify important details, ask questions, gather background to aid in decision making, etc., will prevent many problems.
- *The buyer should furnish the seller with detailed product descriptions.* This includes drawings, workmanship standards, special processing instructions, or any other information the seller needs to provide product of acceptable quality. The buyer should ascertain that the seller understands the requirements.

- *The buyer should objectively evaluate the seller's quality performance.* This evaluation should be done in an open manner, with the full knowledge and consent of the seller. The buyer should also keep the seller informed of his *relative standing* with respect to other suppliers of the same product. However, this should be done in a manner that does not compromise the position of any other seller or reveal proprietary information.
- *The buyer should be prepared to offer technical assistance to the seller, or vice versa.* Such assistance may consist of on-site visits by buyer or seller teams, telephone assistance, or transfer of documents. Of course, both parties are obligated to protect the trade secrets and proprietary information they obtain from one another.
- *The seller should inform the buyer of any known departure from historic or required levels of quality.*
- *The buyer should inform the seller of any change in requirements in a timely fashion.*
- *The seller should be rewarded for exceptional performance.* Such rewards can range from plaques to increased levels of business.

The basic principles of ethical behavior have been very nicely summarized in the Code of Ethics for Members of the American Society for Quality, which is available on their Web site.

Scope of Vendor Quality Control

Most companies purchase several types of materials. Some of the materials are just supplies, not destined for use in the product to be delivered to the customer. Traditionally, vendor quality control does not apply to these supplies. Of those items destined for the product, some are simple items that have loose tolerances and an abundant history of acceptable quality. The quality of these items will usually be controlled informally, if at all. The third category of purchased goods involves items that are vital to the quality of the end product, complex, and with limited or no history. Purchase of these items may even involve purchase of the vendor's "expertise," for example, designs, application advice, etc. It is the quality of this category of items that will be the subject of subsequent discussions.

Vendor quality is aided by cooperation between Product Design, Purchasing and Quality functions, as shown in Table 11.1.

Evaluating Vendor Quality Capability

When making important purchases, companies often seek assurances they are making the right decision. The vendor quality survey has traditionally served as the "crystal ball" to provide this assurance. To some degree this

Activity	Participating Departments		
	Product Design	Purchasing	Quality Control
Establish a vendor quality policy	X	X	XX
Use multiple vendors for major procurements		XX	
Evaluate quality capability of potential vendors	X	X	XX
Specify requirements for vendors	XX		X
Conduct joint quality planning	X		X
Conduct vendor surveillance		X	XX
Evaluate delivered product	X		XX
Conduct improvement programs	X	X	XX
Use vendor quality ratings in selecting vendors		XX	X

(Juran and Gryna, 1980)

where X = shared responsibility XX = primary responsibility

Table 11.1 Responsibility Matrix for Vendor Relations

approach has been replaced by third-party audits such as ISO 9000. The vendor quality survey usually involves a visit to the vendor by a team from the buyer prior to the award of a contract, for this reason it is sometimes called a "pre-award survey." The team is usually composed of representatives from the buyer's design engineering, quality control, production, and purchasing departments. The quality control elements of the survey usually include, at a minimum:

- Quality management
- Design and configuration control
- Incoming material control
- Manufacturing and process control
- Inspection and test procedures
- Control of nonconforming material
- Gage calibration and control
- Quality information systems and records
- Corrective action systems

The evaluation typically employs a checklist with a numerical rating scheme. A simplified example of a supplier evaluation checklist is shown in Fig. 11.1. When conducted at the supplier's facility, it is known as a physical survey.

Criteria	Pass	Fail	See Note
The quality system has been developed	❑	❑	❑
The quality system has been implemented	❑	❑	❑
Personnel are able to identify problems	❑	❑	❑
Personnel recommend and initiate solutions	❑	❑	❑
Effective quality plans exist	❑	❑	❑
Inspection stations have been identified	❑	❑	❑
Management regularly reviews quality program status	❑	❑	❑
Contracts are reviewed for special quality requirements	❑	❑	❑
Processes are adequately documented	❑	❑	❑
Documentation is reviewed by quality	❑	❑	❑
Quality records are complete, accurate, and up-to-date	❑	❑	❑
Effective corrective action systems exist	❑	❑	❑
Nonconforming material is properly controlled	❑	❑	❑
Quality costs are properly reported	❑	❑	❑
Changes to requirements are properly controlled	❑	❑	❑
Adequate gage calibration control exists	❑	❑	❑

Figure 11.1 Vendor evaluation checklist.

The above checklist is a simplified version of those used in practice, which can consume 15 pages or more. Caution should be exercised in constructing overly cumbersome and difficult-to-use checklists. If you are not bound by some government or contract requirement, it is recommended that you prepare a brief checklist similar to the one above and supplement the checklist with a report that documents your personal observations. Properly used, the checklist can help guide you without tying your hands.

While a checklist is a useful aid, it can never substitute for the knowledge of a skilled and experienced evaluator. Numerical scores should be supplemented by the observations and interpretations of the evaluator. The input of vendor personnel should also be included. If there is disagreement between the evaluator and the vendor, the position of both sides should be clearly described.

In spite of their tremendous popularity, physical vendor surveys are only one means of evaluating the potential performance of a supplier, and studies suggest they are limited in their usefulness. One such study by Brainard (1966) showed that 74 of 151 vendor surveys resulted in incorrect predictions; that is, either a good supplier was predicted to be bad or vice versa; a coin flip would've been as good a predictor, and a lot cheaper! Desk surveys (discussed later in this chapter) may provide a satisfactory alternative. Perhaps more important than surveys is the actual performance of the vendor. Vendors should submit "correlation samples" with the first shipments, which should be numbered and accompanied by results of the vendor's quality inspection and test results. Verify that the vendor has correctly checked each important characteristic and that your results agree, or "correlate" with his. Finally, keep a running history of quality performance. The best predictor of future good performance seems to be a record of good performance in the past. If you are a subscriber to GIDEP, the Government Industry Data Exchange Program, you have access to a wealth of data on many suppliers. (GIDEP subscribers must also contribute to the data bank.) The Coordinated Aerospace Supplier Evaluation (CASE) database for the aviation industry is another useful resource. If relevant to your application, these compilations of the experience and audits of a large number of companies can be a real money- and time-saver.

Vendor Quality Planning

Vendor quality planning involves efforts directed toward preventing quality problems, appraisal of product at the vendor's plant as well as at the buyer's place of business, corrective action, disposition of nonconforming merchandise, and quality improvement. The process usually begins in earnest after a particular source has been selected. Most preaward evaluation is general in nature; after the vendor has been selected it is time to get down to the detailed level.

A first step in the process is the transmission of the buyer's requirements to the vendor. Even if the preliminary appraisal of the vendor's capability indicated that the vendor could meet your requirements, it is important that the requirements be studied in detail again before actual work begins. Close contact is required between the buyer and the vendor to ensure that the requirements are clearly understood. The vendor's input should be solicited; could a change in requirements help them produce better-quality parts?

Next it is necessary to work with the vendor to establish procedures for inspection, test, and acceptance of the product. How will the product be inspected? What workmanship standards are to be applied? What in-process testing and inspection are required? What level of sampling will be employed? These and similar questions must be answered at this stage. It is good practice to have the first few parts completely inspected by both the vendor and the buyer to ensure that the vendor knows which features must be checked as well as how to check them. The buyer may want to be at the vendor's facility when production first begins.

Corrective action systems must also be developed. Many companies have their own forms, procedures, etc., for corrective action. If you want your system to be used in lieu of the vendor's corrective action system, the vendor must be notified. Bear in mind that the vendor may need additional training to use your system. Also, the vendor may want some type of compensation for changing their established way of doing things. If at all possible, let the vendor use their own systems.

Statistical process control should be used by the vendor to ensure lot-to-lot stability. When statistical control is established, there is little need for end-process final inspection at the vendor location or incoming receiving inspection at the buyer's location.

If special record keeping will be required, this needs to be clearly defined. For example, most major defense items have traceability and configuration control requirements. Government agencies, such as the FDA, often have special requirements. Automotive companies have record-keeping requirements designed to facilitate possible future recalls. The vendor may not be aware of some of these requirements, and it is to your mutual benefit to discuss your specific needs early in the process.

Post-Award Surveillance

Our focus up to this point has been to develop a process that will minimize the probability of the vendor's producing items that don't meet your requirements. This effort must continue after the vendor begins production. However, after production has begun the emphasis can shift from an evaluation of systems to an evaluation of *actual* program, process, and product performance.

Program evaluation is the study of a supplier's facilities, personnel, and quality systems. While this is the major thrust during the pre-award phase of an evaluation, program evaluation doesn't end when the contract is awarded. Change is inevitable, and the buyer should be kept informed of changes to the vendor's program. This may be accomplished by providing the buyer with a registered copy of the vendor's quality manual, which is updated routinely. Periodic follow-up audits may also be required, especially if product quality indicates a failure of the quality program.

A second type of surveillance involves surveillance of the vendor's process. Process evaluations involve a study of methods used to produce an end result. Process performance can usually be best evaluated by statistical methods, and it is becoming common to require that statistical process control (SPC) be applied to critical-process characteristics. Many large companies require that their suppliers perform statistical process control studies, called process capability studies, as part of the pre-award evaluation. (See Chap. 9 for further details on process capability studies.)

The final evaluation, product evaluation, is also the most important. Product evaluation consists of evaluating conformance to requirements. This may involve inspection at the vendor's site, submission of objective evidence of conformance by the vendor, inspection at the buyer's receiving dock, or actual use of the product by the buyer or the end user. This *must* be the final proof of performance. If the end product falls short of requirements, it matters little that the vendor's program looks good, or that all of the in-process testing meets established requirements.

The surveillance activity is a communications tool. To be effective it must be conducted in an ethical manner, with the full knowledge and cooperation of the vendor. The usual business communications techniques, such as advance notification of visits, management presentations, exit briefings, and follow-up reports, should be utilized to ensure complete understanding.

Vendor Rating Schemes

Evaluating vendors involves comparing a large number of factors, some quantitative and some qualitative. Vendor rating schemes attempt to simplify this task by condensing the most important factors into a single number, the vendor rating, that can be used to evaluate the performance of a single vendor over time, or to compare multiple sources of the same item.

Most vendor rating systems involve assigning weights to different important measures of performance, such as quality, cost, and delivery. The weights are selected to reflect the relative importance of each measure. Once the weights are determined, the performance measure is multiplied by the weight and the results are totaled to get the rating.

For example, consider the rating scheme shown in Table 11.2:

The performance for each of three vendors (A, B, and C) for a common length of time are shown in Table 11.3. This performance data is then used to attain the ratings calculations for the three hypothetical suppliers as shown in Table 11.4.

As you can see in the example shown above, even simple rating schemes combine reject rates, delivery performance, and dollars into a single composite number. Using this value we would conclude that vendors B and C are approximately the same. What vendor B lacks in the pricing category, they make up for in quality and delivery. Vendor A has a rating much lower than B or C.

Performance	Measure	Weight
Quality	% of lots accepted	5
Cost	lowest cost/cost	300
Delivery	% on-time shipments	2

Table 11.2 Vendor Rating Scheme

Vendor	% Lots Accepted	Price ($)	% On-Time Deliveries
A	90	60	80
B	100	70	100
C	85	50	95

TABLE 11.3 Example Vendor Performance Data

Vendor	Quality Rating	Price Rating	Delivery Rating	Overall Rating
A	450	250	160	860
B	500	214	200	914
C	425	300	190	915

TABLE 11.4 Calculated Vendor Ratings

Generally, attempts to condense multifaceted evaluations into a single index are subject to fundamental error. As such, it's important to consider them in proper context, and allow room for some level of subjective interpretation. Characteristics of useful rating schemes include:

1. The scheme is clearly defined and understood by both the buyer and the seller.
2. Only relevant information is included.
3. The plan should be easy to use and update.
4. Rating schemes should be applied only where they are needed.
5. The ratings should "make sense" when viewed in light of other known facts.

Special Processes

A special process is one that has an effect that can't be readily determined by inspection or testing subsequent to processing. The difficulty may be due to some physical constraint, such as the difficulty in verifying grain size in a heat-treated metal, or the problem may simply be economics, such as the cost of 100 percent X-ray of every weld in an assembly. In these cases, special precautions are required to ensure that processing is carried out in accordance with requirements.

The two most common approaches to control of special processes are certification and process audit. Certification can be applied to the skills of key personnel, such as welder certification, or to the processes themselves.

With processes the certification is usually based on some demonstrated capability of the process to perform a specified task. For example, a lathe may machine a special test part designed to simulate product characteristics otherwise difficult or impossible to measure. The vendor is usually responsible for certification. Process audit involves establishing a procedure for the special process, then reviewing actual process performance for compliance to the procedure. A number of books exist to help with the evaluation of special processes. In addition, there are inspection service companies that allow you to hire experts to verify that special processes meet established guidelines. These companies employ retired quality control professionals, as well as full-time personnel. In addition to reducing your costs, these companies can provide a level of expertise you may not otherwise have.

Partnership and Alliances

Research suggests that purchased items account for 60 percent of sales, 50 percent of all quality problems, and 75 percent of all warranty claims. Yet, even these impressive figures understate the importance of suppliers to a firm's success. The emphasis on just-in-time (JIT) inventory management systems has created a situation where any slippage in quality or delivery commitments causes an immediate detrimental impact on production schedules and the firm's ability to ship finished product. The interdependence of the supplier and the purchaser is now painfully obvious. This has led many firms to reduce the number of suppliers in an effort to better manage supplier relationships. The new approach is to treat suppliers as partners rather than as adversaries. Suppliers are given larger and longer contracts and, in turn, they are expected to work closely with the purchaser to ensure that the purchaser's customers are satisfied.

The conventional wisdom in American quality control was, for decades, that multiple vendors would keep all suppliers "on their toes" through competition. Multiple vendors provided a hedge against unforeseen problems like fire, flood, or labor disputes, and became the de facto standard for most firms (and required by major government agencies, including the Department of Defense).

In the 1980s, the consensus on multiple sources of supply began to erode, as Japan's enormous success with manufacturing in general and quality in particular inspired American businesspeople to study the Japanese methods. Japanese businesses *discourage* multiple-source purchases whenever possible, in keeping with the philosophy of W. Edwards Deming (see points 2 and 4 in Deming's 14 Points, Chap. 3). The advocates of single-source procurement argue that it encourages the supplier to take long-term actions on your behalf and makes suppliers more loyal and committed to your success. Statistically, minimum variability in product can be obtained if the sources of variation are minimized and multiple suppliers are an obvious source of variation.

Recently, floods caused by tsunamis in Japan have caused a devastating effect on the supplier chain, and the use of single long-term local suppliers has hampered recovery of the nation as a whole. This would seem to be a strong argument for maintaining multiple sources of key components, especially from a geographical perspective.

The decision regarding single-source versus multiple-source must be made on a case-by-case basis. In most cases, the many benefits of sole sourcing are only outweighed when the risks are fairly substantial.

Traditional interactions between supplier and purchaser focused on various forms of product inspection, which was often adversarial. Improved levels of interaction might focus on prevention of quality-related issues, based on the premise that quality must be built by the supplier with the purchaser's help. The relationship is not adversarial, but it is still arm's-length.

When customer-supplier partnerships are developed, purchasers and suppliers work closely on such issues as joint training, planning, and sharing confidential information (e.g., sales forecasts). Communications channels are wide open and include such new forms as designer to designer or quality engineer to quality engineer. This approach results in parallel communication, a dramatic change from the purchasing agent to sales representative approach used in the past (serial communication). A research study showed that, for serial communications channels, quality levels stayed flat or declined while, for parallel communication, quality levels improved dramatically.

Process improvement teams may also be formed and chartered as described in Part IV, where the process to be improved is the supplier-purchaser communication process. These teams meet at the supplier's and purchaser's facilities to set goals for the relationship and to develop plans for achieving their goals. Contact personnel and methods of contact are established, including a schedule of meetings. These meetings are used to update each other on progress and new plans, not merely for discussion of problems. Purchasers show suppliers how the product is designed, where the supplier's materials are used, and the manufacturing processes involved.

In some cases, joint technological plans (JTP) are developed cooperatively by suppliers and purchasers. The plans include specific performance requirements, including quality, reliability, and maintainability requirements, and the supplier's role in meeting those requirements is defined.

JTP also encompasses the processes to be used by suppliers and purchasers. Process control plans are prepared, including the identification of key process variables and how they are to be controlled and monitored. Special tasks to be done by the supplier (e.g., in-process inspections, SPC) are described. Classification of characteristics and defects is performed.

Purchasers work with suppliers to prepare aesthetic and sensory standards. Test and measurement methods are standardized.

Finally, finished product inspection and audit requirements are established. Sampling requirements are specified as well as acceptable quality levels. Lot identification and traceability systems are developed. Corrective action and follow-up systems acceptable to supplier and purchaser are instituted.

Contract administration is a cooperative effort. Both supplier and purchaser participate in evaluation of initial samples of product. Systems are implemented to ensure that design information and key information about changes in requirements are rapidly and accurately communicated.

Provisions are made for routine surveillance of supplier processes. All parties are involved in evaluating delivered product, and the data collected are used for improvement and learning, not for judgment. Action on nonconforming product is rational and based on what is best for the end user. Supplier quality personnel are part of the team.

In Japan, "Keiretsu" is the name for the close coordination of companies, suppliers, banks, and many other companies that work together for the good of the whole. Control is based on equity holdings of their suppliers. The Japanese system, commonly called a cartel, is precluded by the current U.S. antitrust laws. However, as described by Burt and Doyle (1993), American firms have created Keiretsu-type relationships with suppliers through strategic supply management. The focus is on value-added outsourcing relationships. Significant efficiencies can be gained through these approaches to strategic supply management. The basis for the American Keiretsu-type relationships includes shared long-term objectives and commitments. The key ingredients that must be present are strategic fit and mutual trust.

American Keiretsu-type relationships are compatible and interdependent with other corporate strategies, such as total quality management, strategic cost management, just-in-time manufacturing, simultaneous engineering, flexible manufacturing, core competencies, and value-chain management. The approach can be applied to procurement of hardware or nonproduction and service requirements, such as legal services, consulting services, and maintenance, repair, and operating supplies.

In these relationships, it is important that companies ensure protection of a company's core competencies through early agreements and during subsequent discussions and actions. Disciplines to control interim contacts at all levels are needed to ensure that the relationships stay focused within pre-established boundaries and do not flow over into product or technical activities that are beyond the scope of the agreed cooperation. The focus is on cooperation, trust, and synergism; however, this does not mean that executives can be careless in their business practices. They should not contribute to the development of a new, stronger competitor.

PART IV

Continuous Improvement

There are two acceptable methods for implementing quality improvements: improve performance given the current system, or improve the system itself. Performance improvements within a current system can often be accomplished by individuals working alone. For example: an operator might make certain adjustments to one or more machine settings, then inform the local supervision of the superior results; an order processing clerk may notice that orders can be more quickly completed under certain conditions, and she works with the supervisor to standardize on this method. Studies indicate that this sort of action will be responsible for about 5 to 15 percent of the improvements. The remaining 85 to 95 percent of all improvements will require changing the system itself. This is seldom accomplished by individuals working alone. It requires group action. Thus, the vast majority of quality improvement activity will take place in a group setting.

While continuous improvement should be a part of everyone's routine, conditions do not always encourage this behavior. Operating precedents and formal procedures are designed to maintain the status quo. Systems are established to detect negative departures from the status quo and react to them. Continuous improvement implies that we constantly attempt to change the status quo for the better. Doing this wisely requires an understanding of the nature of cause systems. Systems will always exhibit variable levels of performance, but the nature of the variation provides the key to what type of action is appropriate. If a system is "in control" in a statistical sense, then all of the observed variability is from common causes of variation that are inherent in the system itself. Improving performance of this stable process calls for fundamental changes to the system. Other times, systems will exhibit nonrandom variability, detected as "special causes" of variation on a statistical control chart (see Chap. 9). When special causes of variation are present, the special cause should be identified and addressed. It is unwise to take action on the system itself in this case. Likewise, looking for "the problem" when the variability is due to common causes is also counterproductive, as it increases process variation. Determining whether variability is from special causes or common causes requires an understanding of the statistical methods of control charts. This simple fact is cause for concern with regard to continuous improvement by individuals working without the benefit of meaningful systems-level thinking. In these environments, it is critical for local management to be experienced in these statistical tools, to ensure they are monitoring for adverse effects from small-scale improvement efforts. A cross-functional team with the right blend of experience, trained in the proper methods, is often a safer alternative.

Process improvement teams focus on improving one or more important characteristics of a process, for instance, quality, cost, cycle time, etc. The focus is on an entire process, rather than on a particular aspect of the process. Process improvement teams work on both incremental improvement (KAIZEN) and radical change (re-engineering or breakthrough projects). The team is composed of members who work with the process on a routine basis. Team members typically report to different bosses to provide a cross-functional perspective of the process, and their positions can be on different levels of the organization's hierarchy.

CHAPTER 12

Effective Change Management

Effective improvement requires changing the fundamentals of the process or product. Experts agree: change is difficult, disruptive, expensive, and a major cause of error. Yet, there are some common reasons organizations choose to face the difficulties involved with change:

- *Leadership.* Some organizations choose to maintain product or service leadership as a matter of policy. Change is a routine.
- *Competition.* When competitors improve their products or services such that their offering provides greater value than yours, you are forced to change. Refusal to do so will result in the loss of customers and revenues and can even lead to complete failure.
- *Technological advances.* Effectively and quickly integrating new technology into an organization can improve quality and efficiency and provide a competitive advantage. Of course, doing so involves changing management systems.
- *Training requirements.* Many companies adopt training programs without realizing that many such programs implicitly involve change. For example, a company that provides employees with SPC training should be prepared to implement a process control system. Failure to do so leads to morale problems and wastes training dollars.
- *Rules and regulations.* Change can be forced on an organization from internal regulators via policy changes and changes in operating procedures. Government and other external regulators and rule-makers (e.g., ISO for manufacturing, JCAHO for hospitals) can also mandate change.
- *Customer demands.* Customers, large and small, are not bound by your policies. While some may request, or even demand that you change your policy and procedures, others will say nothing at all, and simply take their business elsewhere.

Johnson (1993b, p. 233) gives the following summary of change management:

1. Change will meet resistance for many different reasons.
2. Change is a balance between the stable environment and the need to implement TQM. Change can be painful while it provides many improvements.
3. There are four times change can most readily be made by the leader: when the leader is new on the job, receives new training, or has new technology, or when outside pressures demand change.
4. Leaders must learn to implement change they deem necessary, change suggested from above their level, and change demanded from above their level.
5. There are all kinds of reaction to change. Some individuals will resist, some will accept, and others will have mixed reactions.
6. There is a standard process that supports the implementation of change. Some of the key requirements for change are leadership, empathy, and solid communications.
7. It is important that each leader become a change leader. This requires self-analysis and the will to change those things requiring change.

Roles

Change requires new behaviors from everyone involved. However, four specific roles commonly appear during most successful change processes (Hutton 1994, pp. 2–4):

- *Official change agent.* An officially designated person who has primary responsibility for helping management plan and manage the change process.
- *Sponsors.* Senior leaders with the formal authority to legitimize the change. The sponsor makes the change a goal for the organization and ensures that resources are assigned to accomplish it. No major change is possible without committed and suitably placed sponsors.
- *Advocate.* Someone who sees a need for change and sets out to initiate the process by convincing suitable sponsors. This is a selling role. Advocates often provide the sponsor with guidance and advice. Advocates may or may not hold powerful positions in the organization.

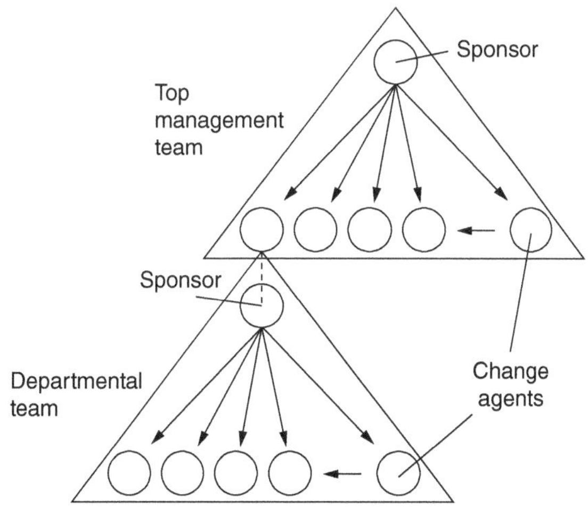

FIGURE 12.1 Cascading of sponsorship (Hutton, D.W., 1994).

- *Informal change agent.* Persons other than the official change agent who voluntarily help plan and manage the change process. While the contribution of these people is extremely important, it is generally not sufficient to cause truly significant, organization-wide change.

The position of these roles within a typical organizational hierarchy is illustrated graphically in Fig. 12.1.

Goals

There are three goals of change:

1. Change the way people think or act in the organization. All change begins with the individual, at a personal level. Unless the individual is willing to change his or her behavior, no real change is possible. Changing behavior requires a change in thinking.
2. Change the norms. Norms consist of standards, models, or patterns that guide behavior in a group. All organizations have norms or expectations of their members. Change cannot occur until the organization's norms change.
3. Changing the organization's systems or processes. This is the "meat" of the change. Ultimately, all work is a process, and quality improvement requires change at the process and system level. However, this cannot occur on a sustained basis until individuals change their behavior and organizational norms are changed.

Mechanisms Used by Change Agents

The change agents help accomplish the above goals in a variety of ways. Education and training are important means of changing individual perceptions and behaviors. In this discussion, a distinction is made between training and education. Training refers to instruction and practice designed to teach a person how to perform some task. Training focuses on concrete tasks that need to be done. Training will be an integral aspect of instituting any process-level change.

Education refers to instruction in how to think. Education focuses on integrating abstract concepts into one's knowledge of the world. Educated people will view the world differently after being educated than they did before. This is an essential part of the process of change.

As part of the change initiative, an effective change agent will organize an assessment of the organization to identify its strengths and weaknesses. Change is usually undertaken to either reduce areas of weakness, or exploit areas of strength. The assessment guides the training and education. Knowing one's specific strengths and weaknesses is useful in mapping the process for change.

Building Buy-in

Most organizations still have a hierarchical, command-and-control organizational structure, sometimes called "smoke stacks" or "silos." The functional specialists in charge of each smoke stack tend to focus on optimizing their own functional area, often to the detriment of the organization as a whole. In addition, the hierarchy gives these managers a monopoly on the authority to act on matters related to their functional specialty. The combined effect is both a desire to resist change and the authority to resist change, which often creates insurmountable roadblocks to quality improvement projects.

It is important to realize that organizational rules are, by their nature, a barrier to change. The formal rules take the form of written standard operating procedures (SOPs). The very purpose of SOPs is to standardize behavior. The quality profession has historically overemphasized formal documentation, and it continues to do so by advocating such approaches as ISO 9000 and ISO 14000. Formal rules are often responses to past problems, and they often continue to exist long after the reason for their existence has passed. In an organization that is serious about its written rules, even senior leaders find themselves helpless to act without submitting to a burdensome rule-changing process. The true power in such an organization is the bureaucracy that controls the procedures. If the organization falls into the trap of creating written rules for too many things, it can find itself moribund in a fast-changing external environment. This is a recipe for disaster.

Restrictive rules need not take the form of management limitations on staff, nor procedures that define hourly work in burdensome detail (e.g., union work rules). Projects almost always require that work be done differently and such procedures prohibit such change. Organizations that tend to be excessive in SOPs also tend to be heavy on work rules. The combination is often deadly to quality improvement efforts.

Organization structures preserve the status quo in ways beyond formal, written restrictions in the form of procedures and rules. Another effective method of limiting change is to require permission from various departments, committees, councils, boards, experts, etc. Even though the organization may not have a formal requirement that "permission" be obtained, the effect may be the same, for instance, "You should run that past accounting" or "Ms. Reimer and Mr. Evans should be informed about this project." When permission for vehicles for change (e.g., project budgets, plan approvals) is required from a group that meets infrequently it creates problems for project planners. Plans may be rushed so they can be presented at the next meeting, lest the project be delayed for months. Plans that need modifications may be put on hold until the next meeting, months away. Or projects may miss the deadline and be put off indefinitely.

External Roadblocks

Modern organizations do not exist as islands. Powerful external forces take an active interest in what happens within the organization. Government bodies have created a labyrinth of rules and regulations that the organization must negotiate to utilize its human resources without incurring penalties or sanctions. The restrictions placed on modern businesses by outside regulators is challenging to say the least. When research involves people, ethical and legal concerns sometimes require that external approvals be obtained. The approvals are contingent on such issues as informed consent, safety, cost and so on.

Many industries have "dedicated" agencies to deal with, such as the Food and Drug Administration (FDA) for the pharmaceutical industry. These agencies must often be consulted before undertaking projects. For example, a new treatment protocol for treatment of pregnant women may involve using a drug in a new way (e.g., administered on an outpatient basis instead of on an inpatient basis).

Many professionals face liability risks that are part of every decision. Often these fears create a "play it safe" mentality that acts as a barrier to change. The fear is even greater when the project involves new and untried practices and technology.

Individual Barriers to Change

Individuals will likely experience a range of emotions when change occurs in an organization.

- *People may be apprehensive or even fearful of change.* Organizational change requires that we adopt new policies or procedures, which can be unsettling to some members of the organization. There is security in the status quo, and fear of the unknown is a barrier to change.
- *It is natural, perhaps even productive, to be skeptical of change.* Change is not always for the better, at least in everyone's eyes, and even marginal improvements in some areas may require accommodation in others. Even when there is general acknowledgment that change could result in improvement, there may be skepticism that the improvement will be attained.

Forsha (1992) provides the process for personal change shown in Fig. 12.2.

The adjustment path results in preservation of the status quo. The action path results in change. The well-known PDCA cycle can be used once a commitment to action has been made by the individual. The goal of such change is continuous self-improvement.

Within an organizational context, the individual's reference group plays a part in personal resistance to change. A reference group is the aggregation of people a person thinks of when he or she uses the word "we." If "we" refers to the company, then the company is the individual's reference group and he or she feels connected to the company's success or failure. However, "we" might refer to the individual's profession or trade group, for example, "We doctors," "We engineers," "We union members." In this case the leaders shown on the formal organization chart will have little influence on the individual's attitude toward the success or failure of the project. When a project involves external reference groups with competing agendas, the task of building buy-in and consensus is daunting indeed.

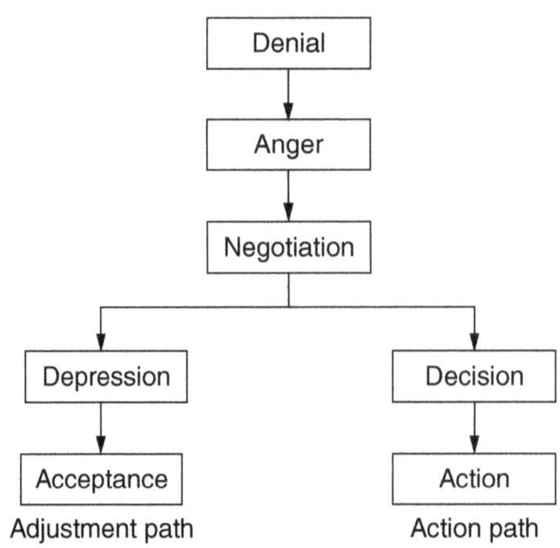

FIGURE 12.2 The process of personal change.

Ineffective Management Support Strategies

Strategy #1: Command people to act as you wish. With this approach the senior leadership simply commands people to act as the leaders wish. The implication is that those who do not comply will be subjected to disciplinary action. People in less senior levels of an organization often have an inflated view of the value of raw power. The truth is that even senior leaders have limited power to rule by decree. Human beings by their nature tend to act according to their own best judgment. Thankfully, commanding that they do otherwise usually has little effect. The result of invoking authority is that the decision-maker must constantly try to divine what the leader wants him or her to do in a particular situation. This leads to stagnation and confusion as everyone waits on the leader. Another problem with commanding as a form of "leadership" is the simple communication problem. Under the best of circumstances people will often simply misinterpret the leadership's commands.

Strategy #2: Change the rules by decree. When rules are changed by decree the result is again confusion. What are the rules today? What will they be tomorrow? This leads again to stagnation because people don't have the ability to plan for the future. Although rules make it difficult to change, they also provide stability and structure that may serve some useful purpose. Arbitrarily changing the rules based on force (which is what "authority" comes down to) instead of a set of guiding principles does more harm than good.

Strategy #3: Authorize circumventing of the rules. Here the rules are allowed to stand, but exceptions are made for the leader's "pet projects." The result is general disrespect for and disregard of the rules, and resentment of the people who are allowed to violate rules that bind everyone else. An improvement is to develop a formal method for circumventing the rules, for instance, deviation request procedures. While this is less arbitrary, it adds another layer of complexity and still doesn't change the rules that are making change difficult in the first place.

Strategy #4: Redirect resources to the project. Leaders may also use their command authority to redirect resources to the project. Rather than use logical prioritization schemes to prioritize projects, political clout is the basis of the allocation.

Effective Management Support Strategies

Strategy #1: Transform the formal organization and the organization's culture. By far the best solution to the problems posed by organizational roadblock is to transform the organization into one where these roadblocks no longer exist. As discussed earlier, this process can't be implemented by decree. As the leader helps project teams succeed, he or she will learn about the need for transformation. Using persuasive powers, the leader-champion can undertake the exciting challenge of creating a culture that embraces change instead of fighting it.

Strategy #2: Mentoring. In Greek mythology, Mentor was an elderly man, the trusted counselor of Odysseus, and the guardian and teacher of his son Telemachus. Today the term "mentor" is still used to describe a wise and trusted counselor or teacher. When this person occupies an important position in the organization's hierarchy, he or she can be a powerful force for eliminating roadblocks. Modern organizations are complex and confusing. It is often difficult to determine just where one must go to solve a problem or obtain a needed resource. The mentor can help guide the project manager through this maze by clarifying lines of authority. At the same time, the mentor's senior position enables him or her to see the implications of complexity and to work to eliminate unnecessary rules and procedures.

Strategy #3: Identify informal leaders and enlist their support. Because of their experience, mentors often know that the person whose support the project really needs is not the one occupying the relevant box on the organization chart. The mentor can direct the project leader to the person whose opinion really has influence. For example, a project may need the approval of, say, the vice president of engineering. The engineering VP may be balking because his senior metallurgist hasn't endorsed the project.

Strategy #4: Find legitimate ways around people, procedures, resource constraints, and other roadblocks. It may be possible to get approvals or resources through means not known to the project manager. Perhaps a minor change in the project plan can bypass a cumbersome procedure entirely. For example, adding an engineer to the team might automatically place the authority to approve process experiments within the team rather than in the hands of the engineering department.

These concerns must be addressed. Change cannot occur without buy-in from the key stakeholders responsible for change.

The following steps can be used to achieve buy-in within the organization (Keller, 2011a):

1. ***Define*** *key stakeholders.* These are the individuals or groups who can make or break the change initiative.
2. Measure baseline level of buy-in for each key stakeholder. How committed are they to change?
3. ***Analyze*** *buy-in reducers and boosters for each key stakeholder or stakeholder group.* Understand the concerns of the stakeholders, which may vary from one stakeholder to another.
4. Improve by addressing issues.
5. ***Control*** *with a plan to maintain buy-in.*

Notice the DMAIC approach, which will be discussed shortly as the model of process improvement, applied here to the problem of buy-in.

These steps for achieving buy-in may be used several times for different issues within a given project, and can even be used informally within a team meeting.

There are five levels that may be used to rate the buy-in (or lack of buy-in) from stakeholder or stakeholder groups (Forum Corporation, 1996):

- Hostility
- Dissent
- Acceptance
- Support
- Buy-in

The lowest level, *hostility*, is easy to recognize. *Dissent* may go unnoticed until stakeholders are questioned about the change initiative, such as through discussion or survey. *Acceptance* is the lowest level of buy-in that should be considered for proceeding with the change initiative, but is insufficient. *Support* must be achieved from a majority of the critical stakeholder groups. True *Buy-in* is the desired level, when stakeholders are enthusiastic in their commitment for change.

There are a number of issues that serve to reduce stakeholder buy-in (Forum Corporation, 1996).

- *Unclear goals.* Goals need to be clearly communicated throughout the stakeholder groups.
- *No personal benefit.* Goals should be stated in terms that provide a clear link to personal benefits for stakeholders, such as decreased hassles or improved working conditions.
- *Predetermined solutions.* When teams are given the solution without chance for analysis of alternatives, they will likely be skeptical of the result.
- *Lack of communication.* Analyses and results should be communicated throughout the stakeholder groups.
- *Too many priorities.* Teams need to be focused on achievable results.
- *Short-term focus.* Goals should provide clear benefits over short and longer terms.
- *No accountability.* Clearly defined project sponsors, stakeholders, and team members provide accountability.
- *Disagreement on who the customer is.* Clearly defined stakeholder groups are needed for project success.
- *Low probability of implementation.* Formal project sponsorship and approvals provide a clear implementation channel.

- *Insufficient resources.* Stakeholder groups need to understand that the project is sufficiently funded.
- *Midstream change in direction or scope.* Changes in project scope or direction provide a potential for a loss of buy-in. Changes must be properly communicated to stakeholder groups to prevent this reduction in buy-in.

The following chapters will show how these issues are addressed through the use of project charters, sponsored by management and executed by cross-functional stakeholder teams.

Project Deployment

Design and improvement projects will address one or more key areas: cost, schedule, or quality. Projects may be developed by senior leaders for deployment at the business level (a top-down approach), or developed with process owners at an operational level (bottoms-up approach). In either case, projects should be directly linked to the strategic goals of the organization. GE CEO Jack Welch considered the best projects those that solved customers' problems.

Projects are effectively owned by their sponsor. The sponsor, being a leader in the organization, works with the team leader to set the scope, objective, and deliverables of the project. The sponsor ensures that resources are available for the project members, and builds buy-in for the project at upper levels of management as needed. Each of these issues is documented in the project charter, which serves as a contract between the sponsor and the project team. The structure of the project and its charter keep the project focused. The project has a planned conclusion date with known deliverables, as well as buy-in from top management. Together, these requirements ensure project success.

Selecting Projects

Projects designed to improve processes should be limited to processes that are important. Important processes impact such things as product cost, delivery schedules, and product features, things that customers notice. Customers cannot help you identify these processes because they aren't familiar with your internal operations. However, customers can help you identify what's important to them; you must then relate this to your processes. Furthermore, projects should be undertaken only when success is feasible. Feasibility is determined by considering the scope and cost of a project and the support it is likely to receive from the process owner.

The well-known *Pareto principle* refers to the observation that a small percentage of processes cause a large percentage of the problems. The Pareto principle is useful in narrowing a list of choices to those few projects that offer the greatest potential.

Initially problems create "pain signals," such as schedule disruptions and customer complaints. Often these symptoms are treated rather than their underlying causes. For example, if quality problems cause schedule slippages that lead to customer complaints, the "solution" might be to keep a large inventory and sort the good from the bad. The result is that the schedule is met and customers stop complaining, but at huge cost. These opportunities are often greater than those causing more visible problems, but they are built into the process and difficult to see. One solution to the hidden problem phenomenon is reengineering, which is focused on processes rather than symptoms. Some guidelines for identifying dysfunctional processes for potential improvement are shown in Table 12.1 (Hammer and Champy, 1993).

The "symptom" column is useful in identifying problems and setting priorities. The "disease" column focuses attention on the underlying causes of the problem, and the "cure" column is helpful in chartering quality improvement project teams and preparing mission statements.

Pareto Prioritization Index

After a serious search for improvement opportunities, the organization's leaders will probably find themselves with more projects to pursue than they have resources. The Pareto priority index (PPI) is a simple way of

Symptom	Disease	Cure
Extensive information exchange, data redundancy, rekeying	Arbitrary fragmentation of a natural process	Discover why people need to communicate with each other so often
Inventory, buffers, and other assets stockpiled	System slack to cope with uncertainty	Remove the uncertainty
High ratio of checking and control to value-added work (internal controls, audits, etc.)	Fragmentation	Eliminate the fragmentation, integrate processes
Rework and iteration	Inadequate feedback in a long work process	Process control
Complexity, exceptions, and special causes	Accretion onto a simple base	Uncover original "clean" process and create new process(es) for special situations; eliminate excessive standardization of processes

Table 12.1 Dysfunctional Process Symptoms and Diseases

prioritizing these opportunities. The PPI is calculated as follows (Juran and Gryna, 1993):

$$\text{PPI} = \frac{\text{savings} \times \text{probability of success}}{\text{cost} \times \text{completion time}}$$

The inputs are, of course, estimates and the result is totally dependent on the accuracy of the inputs. The resulting number is an index value for a given project. The PPI values allow comparison of various projects; they have no intrinsic meaning in and of themselves. If there are clear standouts, the PPI can make it easier to select a project. Table 12.2 shows the PPIs for several hypothetical projects.

In this example, the PPI would indicate that resources should be allocated first to reducing wave solder defects, then to improving NC machine capability, and so on. The PPI may not always give such a clear ordering of priorities. When two or more projects have similar PPIs, a judgment must be made on other criteria.

Prioritization Matrix Approach to Project Selection

Prioritization matrices are designed to rationally narrow the focus of the team to those key issues and options that are most important to the organization. Brassard (1989, pp. 102–103) presents three methods for developing prioritization matrices: the full analytical criteria method, the combination interrelationship digraph (ID)/matrix method, and the consensus criteria method.

An example is provided in Figs. 12.3 to 12.5 (Keller, 2011a), based on an aerospace company's project selection criteria, which were established based on detailed feedback from a high-profile client. A review of the

Project	Savings $1,000s	Probability	Cost, $1,000s	Time, Years	PPI
Reduce wave solder defects 50%	$70	0.7	$25	0.75	2.61
NC machine capability improvement	$50	0.9	$20	1.00	2.25
ISO 9001 certification	$150	0.9	$75	2.00	0.90
Eliminate customer delivery complaints	$250	0.5	$75	1.50	1.11
Reduce assembly defects 50%	$90	0.7	$30	1.50	1.40

Table 12.2 Illustration of the Pareto Priority Index (PPI)

	Qualification of new or revised processes	Design reviews	Incorporation / control of Engineering Changes	Reality-based scheduling	Work procedures / training	Benefit/Cost ratio	Time to Implement	Probability of Success	Criteria Total	Criteria Weight
Qualification of new or revised processes	-	5	1	1	10	1	10	1	29	0.17
Design reviews	1/5	-	1	1/10	10	5	5	1	22 3/10	0.13
Incorporation / control of Engineering Changes	1	1	-	1/10	1	1/5	1/5	1/5	3 7/10	0.02
Reality-based scheduling	1	10	10	-	10	5	5	5	46	0.27
Work procedures / training	1/10	1/10	1	1/10	-	1/10	1/10	1/10	1 3/5	0.01
Benefit/Cost ratio	1	1/5	5	1/5	10	-	5	5	26 2/5	0.16
Time to Implement	1/10	1/5	5	1/5	10	1/5	-	5	20 7/10	0.12
Probability of Success	1	1	5	1/5	10	1/5	1/5	-	17 3/5	0.11

Figure 12.3 Criteria weighting matrix (using Quality America *Green Belt XL* software).

Criteria: Benefit/Cost ratio	Cell 12 Scrap Reduction	Proposal Cycle Time Reduction	ECO Cycle Time Reduction	Supplier A Reject Reduction	Option Total	Option Weight
Cell 12 Scrap Reduction	-	10	5	1/5	15 1/5	0.33
Proposal Cycle Time Reduction	1/10	-	1/5	1/10	2/5	0.01
ECO Cycle Time Reduction	1/5	5	-	1/10	5 3/10	0.12
Supplier A Reject Reduction	5	10	10	-	25	0.54

Figure 12.4 Options rating matrix for benefits/cost ratio criteria (using Quality America *Green Belt XL* software).

Summary Matrix	*Qualification of new or revised processes*	*Design reviews*	*Incorporation / control of Engineering Changes*	*Reality-based scheduling*	*Work procedures / training*	*Benefit/Cost ratio*	*Time to Implement*	*Probability of Success*	Option Rating	Option Rank
Cell 12 Scrap Reduction	0.043	0.005	0.001	0.003	0.001	0.052	0.059	0.051	0.215	3
Proposal Cycle Time Reduction	0.043	0.024	0.006	0.161	0.002	0.001	0.021	0.018	0.277	2
ECO Cycle Time Reduction	0.043	0.096	0.015	0.072	0.007	0.018	0.040	0.034	0.326	1
Supplier A Reject Reduction	0.043	0.008	0.001	0.039	0.000	0.086	0.002	0.002	0.182	4
Criteria Met Rating	0.17	0.13	0.02	0.27	0.01	0.16	0.12	0.11	1	
Criteria Met Ranking	2	4	7	1	8	3	5	6		

Figure 12.5 Summary matrix (using Quality America *Green Belt XL* software).

feedback prompted an internal review, which validated the prevalence and importance of the issues cited, which were summarized as follows:

- Qualifications of new or revised processes
- Design reviews
- Incorporation/control of engineering changes
- Reality-based scheduling
- Work procedures/training

The team added three additional selection criteria: benefit/cost ratio, time to implement, and probability of success. Using the analytical hierarchy process developed by Saaty (1988), the team compared each criterion with each of the other criteria using scores of 1/10, 1/5, 1, 5, and 10 to indicate significantly less desirable, less desirable, equally desirable, more desirable, or significantly more desirable, respectively. The results are shown in Fig. 12.3.

A number of potential project were defined, and these project options were compared against each criterion, one at a time, to build consensus within the team on how well each project option met the criteria. The options rating matrix for the benefit/cost ratio criteria is shown in Fig. 12.4. Note that there will be an options matrix for each criterion, so eight option matrices in this case.

A summary matrix can then be constructed by applying the criteria weights (shown in Fig. 12.3) to each of the eight option matrices. The summary matrix for the example, shown in Fig. 12.5, indicates that the best overall benefit relative to the weighted criteria is provided by the ECO Cycle Time Reduction project.

When criteria have equal weight, a simpler matrix diagram can be used to directly compare each option with each criteria. The analytical hierarchy process, also known as the full analytical method, is more time-consuming but allows a team to develop consensus on the criteria importance and the relative benefits of each project as it moves through the generation of each score in each cell of each matrix.

Project Selection Using Constraint Theory

Another approach for project selection uses the theory of constraints (TOC), discussed in Chap. 5. Pyzdek and Keller (2010) provide the following approach (used by permission), based on Goldratt's methods for constraint management (1990):

1. Identify the system's constraint(s). Consider a fictitious company that produces only two products, P and Q (Fig. 12.6). The market demand for P is 100 units per week and P sells for $90 per unit.

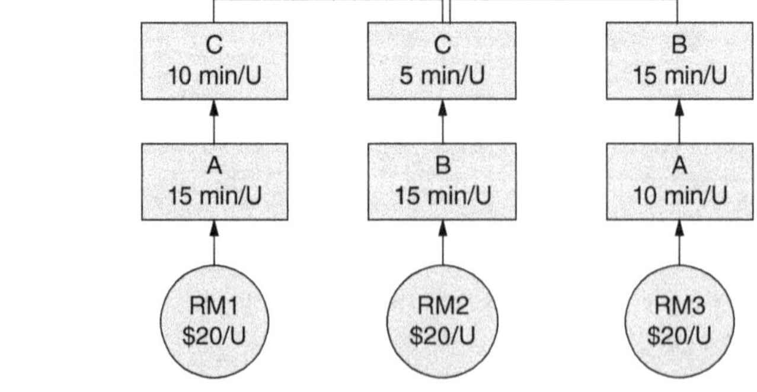

FIGURE 12.6 A simple process with a constraint.

The market demand for Q is 50 units per week and Q sells for $100 per unit. Assume that A, B, C, and D are workers who have different non-interchangeable skills and that each worker is available for only 2,400 minutes per week (8 hours per day, 5 days per week). For simplicity, assume that there is no variation, waste, etc. in the process. Assume this process has a constraint, Worker B. This fact has profound implications for selecting Six Sigma projects.

2. Decide how to exploit the system's constraint(s). Look for projects that minimize waste of the constraint. For example, if the constraint is (feeding) the market demand (i.e., a capacity constraint), then we look for projects that provide 100 percent on-time delivery. Let's not waste anything! If the constraint is a machine, or process step, as in this example, focus on reducing set-up time, eliminating errors or scrap, and keeping the process step running as much as possible.
3. Subordinate everything else to the above decision. Choose projects to maximize throughput of the constraint. After completing step 2, choose projects to eliminate waste from downstream processes; once the constraint has been utilized to create something we don't want to lose it due to some downstream blunder. Then choose projects to ensure that the constraint is always supplied with adequate non-defective resources from upstream processes. We pursue upstream processes last because by definition they have slack resources, so small amounts of waste upstream that are detected before reaching the constraint are less damaging to throughput.
4. Elevate the system's constraint(s). Elevate means "Lift the restriction." This is step #4, not step #2! Often the projects pursued in steps 2 and 3 will eliminate the constraint. If the constraint continues to exist after performing steps 2 and 3, look for projects that provide additional resources to the constraint. These might involve, for example, purchasing additional equipment or hiring additional workers with a particular skill.
5. If, in the previous steps, a constraint has been broken, go back to step 1. There is a tendency for thinking to become conditioned to the existence of the constraint. A kind of mental inertia sets in. If the constraint has been lifted, then you must rethink the entire process from scratch. Returning to step 1 takes you back to the beginning of the cycle.

Comparison of TOC with Traditional Approaches

It can be shown that the throughput-based approach is superior to the traditional approaches to project selection. For example, consider the data in Table 12.3. If you were to apply Pareto analysis to scrap rates you would

Process	A	B	C	D
Scrap rate	8%	3%	5%	7%

(Pyzdek and Keller, 2010)

Table 12.3 Example Process Scrap Rates

begin with projects that reduced the scrap produced by Worker A. In fact, assuming the optimum product mix, Worker A has about 25 percent slack time, so the scrap loss can be made up without shutting down Worker B, who is the constraint. The TOC would suggest that the scrap loss of Worker B and the downstream processes C and D be addressed first, the precise opposite of what Pareto analysis recommends.

Of course, before making a decision as to which projects to finance, cost/benefit analyses are still necessary, and the probability of the project succeeding must be estimated. But by using the TOC you will at least know where to look first for opportunities.

Using Constraint Information to Focus Six Sigma Projects

Applying the TOC strategy described earlier tells us *where* in the process to focus. Adding CTX information (see Table 12.4) can help tell us which

Project Type	Discussion
CTQ	Any unit produced by the constraint is especially valuable because if it is lost as scrap additional constraint time must be used to replace it or rework it. Since constraint time determines throughput (net profit of the entire system), the loss far exceeds what appears on scrap and rework reports. CTQ projects at the constraint are the highest priority.
CTS	CTS projects can reduce the time it takes the constraint to produce a unit, which means that the constraint can produce more units. This directly impacts throughput. CTS projects at the constraint are the highest priority.
CTC	Since the constraint determines throughput, the unavailability of the constraint causes lost throughput *of the entire system*. This makes the cost of constraint downtime extremely high. The cost of *operating* the constraint is usually minuscule by comparison. Also, CTC projects can have an adverse impact on quality or schedule. Thus, CTC projects at the constraint are low priority.

(Pyzdek and Keller, 2010)

Table 12.4 Throughput Priority of CTX Projects That Affect the Constraint

type of project to focus on; that is, should we focus on quality, cost, or schedule projects? Assume that you have three Six Sigma candidate projects, all focusing on process step B, the constraint. The area addressed is correct, but which project should you pursue first? Let's assume that we learn that one project will primarily improve quality, another cost, and another schedule. Does this new information help? Definitely! Take a look at Table 12.4 to see how this information can be used. Projects in the same priority group are ranked according to their impact on throughput.

The same thought process can be applied to process steps before and after the constraint. The results are shown in Table 12.5.

Note that Table 12.5 assumes that projects *before* the constraint do not result in problems *at* the constraint. Remember, impact should always be measured in terms of throughput. If a process upstream from the constraint has an adverse impact on throughput, then it can be considered a constraint. If an upstream process *average* yield is enough to feed the constraint on the average, it may still present a problem. For example, an upstream process producing 20 units per day with an average yield of 90 percent will produce, on average, 18 good units. If the constraint requires 18 units, things will be okay about 50 percent of the time, but the other 50 percent of the time things won't be okay. One solution to this problem is to place a work-in-process (WIP) inventory between the process and the constraint as a safety buffer. Then on those days when the process yield is below 18 units, the inventory can be used to keep the constraint running. However, there is a cost associated with carrying a WIP inventory. A Six Sigma project that can improve the yield will reduce or eliminate the need for the inventory and should be considered even if it doesn't impact the constraint directly, assuming the benefit-cost analysis justifies the project. On the other hand, if an upstream process can easily make up any deficit before the constraint needs it, then a project for the process will have a low priority.

Focus of Six Sigma Project				
CTX:		**Before the Constraint**	**At the Constraint**	**After the Constraint**
Characteristic addressed is critical to ...	Quality (CTQ)	△	⊙	⊙
	Cost (CTC)	○	△	○
	Schedule (CTS)	△	⊙	○

△Low throughput priority.
○Moderate throughput priority.
⊙High throughput priority.

(Pyzdek and Keller, 2010.)

Table 12.5 Project Throughput Priority versus Project Focus

Knowing the project's throughput priority will help you make better project selection decisions. Of course, the throughput priority is just one input into the project selection process; other factors may lead to a different decision, for example, impact on other projects, a regulatory requirement, a better payoff in the long term, etc.

DMAIC/DMADV Methodology

A somewhat standard five-stage methodology has been developed for improvement projects: Define, Measure, Analyze, Improve, Control (DMAIC). DMAIC is an extension of Shewhart's PDCA (Plan Do Check Act) and Deming's PDSA (Plan Do Study Act) cycles for improvement. Once you reach the final step, you may repeat the process for an additional cycle of improvement.

Motorola used the MAIC (Measure, Analyze, Improve, Control) acronym. GE and Allied Signal used DMAIC, which has become more the standard. Some consultants brand the methodology by adding even more steps. Harry and Schroeder (2000) added recognize to the front, and standardize and integrate to the end, referring to the product as their breakthrough strategy. Juran first developed the concept of breakthrough years earlier to describe methods for achieving orders of magnitude improvements in quality. A review of Harry and Schroeder's definitions of these additional terms shows similarity to others' descriptions of DMAIC. A casual review of Six Sigma practitioners found the DMAIC to be the most commonly used for improvement projects. When applied to product, service, or process design, the acronym DMADV is often used, where the Design stage is substituted for Improve, and Verify is substituted for Control. As seen in the following chapters, there is much commonality between DMAIC and DMADV, and this text will often refer to DMAIC for brevity, even when the comments apply to both methods.

The importance of DMAIC is in its structured approach, which ensures that projects are clearly defined and implemented, and that results are standardized into the daily operations.

The DMAIC methodology should be applied from leadership levels of the organization down to the process level. The methodology is the same, with variation on the scope or application. The upper levels of the organization apply these methods to larger business problems, such as market penetration, while the process-level projects improve a given aspect of a particular process, such as reducing cycle time for order processing.

Business-level projects are championed at the top level of the organization. They concentrate on vital aspects of the business success, such as market share, viability, profitability, employee retention, etc. They may involve purchasing or selling of business units, or ways to attract or maintain customer base. Because of the scope of the project, the time scale is measured in years, rather than months. Some business-level projects may take 3 to 5 years to cycle through DMAIC (Harry and Schroeder, 2000), while others are completed in less than 1 year.

Business-level projects may be defined at the top of a particular business unit within a larger corporation, as well as at the executive level of the corporation. GE, for example, set stretch goals for itself as a corporation, and extended them to each of the particular business units.

Operations-level projects concentrate on metrics particular to the functional areas within the organization, although the projects are typically cross-functional, and may even involve customers and/or suppliers. Operations-level projects may seek to improve yield, reduce material or labor costs, and remove the system-wide Hidden Factories responsible for rework in the organization. Operations-level projects may be defined to achieve goals within a stated fiscal year.

Process-level projects are much smaller in scope, and are designed for a much shorter duration. A given black belt will typically work three to four process-level projects a year, although smaller projects are not unheard of. In fact, it's recommended that process-level projects be defined to allow conclusion within a 3- or 4-month period (sometimes less). This prevents "world peace" projects that process-level project teams will have difficulty seeing to conclusion. A typical goal is for each process-level project to save $100,000 or more on an annualized basis. Process-level projects deal with issues such as cycle time reduction, defect reduction, process capability improvement, etc.

At their best, business-level, operations-level, and process-level projects are intertwined. Results from the business level provide projects defined at the operations level, which can in turn create projects at the process level. This top-down approach is generally preferred to a bottom-up approach, where projects are proposed by team members with a vested interest in the outcome. The top-down approach ensures that process-level projects are aligned with strategic business objectives and customer needs. Top-down developed projects also offer greater exposure of the project team to upper management, which can make those projects appealing to up-and-coming project team leaders. That said, projects developed at the process level can also offer great rewards: commitment from team members who understand the process; local recognition from affected co-workers; and improved processes affecting (at least) short-term goals, budget concerns, and customer orders. With that in mind, it's useful to find a happy medium between top-down and bottom-up project definition.

CHAPTER 13

Define Stage

The Define stage includes the following objectives (Keller, 2011a):

- *Project definition.* Define the project's scope, goals, and objectives; its team members and sponsors; and its schedule and deliverables.
- *Top-level process definition.* Define the stakeholders, inputs and outputs, and broad functions.
- *Team formation.* Assemble highly capable team from the key stakeholder groups; create common understanding of issues and benefits for project.

Project Definition

When applied at the business level, the project scope pertains to key business practices and customer interactions. Thus, the definition requires an understanding of the business, as well as the contribution of the business to its shareholders and customers.

GE used its concept of "stretch goals" in defining projects, particularly business-level projects. Stretch goals were those that went beyond what was foreseeable with the current corporate structure, resources, and/or technology. The idea was to expand beyond incremental improvements, to rethink the business, operation, or process to a point where orders of magnitude improvements could be achieved. Bear in mind that an organization going from 4 sigma to 6 sigma needs to reduce defects per million opportunities (DPMO) from 6210 to 3.4 (a 99.95 percent reduction).

A general guideline adopted by GE involved setting the project goal relative to the existing level of performance:

- If the process currently operates at or below a 3 sigma level of performance (an error rate of approximately 6.7 percent or larger), then the project should seek a 10 times reduction in errors. For example, reduce an 10 percent error rate to a 1 percent error rate.
- If the process currently operates better than 3 sigma, reduce the error rate by 50 percent. For example, reduce the error rate from 4 percent to 2 percent.

However, particularly at the process level, you'll need to balance stretch goals with a reasonable project completion time. In some cases, stretch goals can be divided into several projects assigned to several project teams. A work breakdown structure is often effective at identifying useful scope limiting statements.

Work Breakdown Structure

The work breakdown structure is a special-purpose tree diagram used to break down problems or projects into their components. An example is shown in Fig. 13.1. It reduces "big and complex" down to "tiny and manageable." By breaking the process into its components, subprocesses are exposed that might serve as logical break-points for separate improvement efforts. Limiting the project to one or only a few closely related categories will lead to a better chance of project success. The potential deliverables (in financial terms) for each of these subprocesses is the preferred means of justifying a given project proposal.

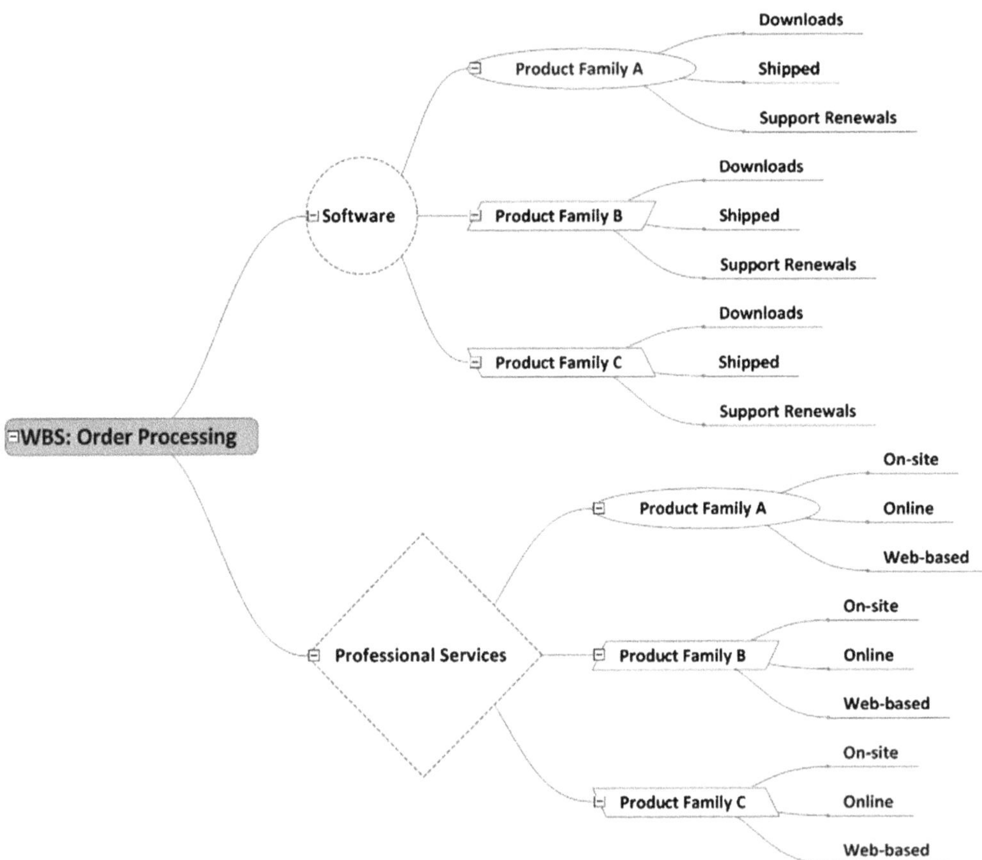

FIGURE 13.1 Example work breakdown structure (Pyzdek and Keller, 2010).

Pareto Diagrams

A Pareto diagram is another useful tool for focusing the project scope, particularly as applied to the unique categories obtained using the work breakdown structures.

Pareto analysis is the process of ranking opportunities to determine which of many potential opportunities should be pursued first. It is also known as "separating the vital few from the trivial many." Pareto analysis should be used at various stages in quality improvement to determine the next step. Pareto analysis is used to answer such questions as "What department should have the next project improvement team?" or "On what type of defect should we concentrate our efforts?"

The following steps are recommended to perform a Pareto analysis:

1. Determine the classifications (Pareto categories) for the graph. If the desired information does not exist, obtain it by designing check sheets and log sheets.
2. Select a time interval for analysis. The interval should be long enough to be representative of typical performance.
3. Determine the total occurrences (i.e., cost, defect counts, etc.) for each category. Also determine the grand total. If there are several categories that account for only a small part of the total, group these into a category called "other."
4. Compute the percentage for each category by dividing the category total by the grand total and multiplying by 100.
5. Rank order the categories from the largest total occurrences to the smallest.
6. Compute the "cumulative percentage" by adding the percentage for each category to that of any preceding categories.
7. Construct a chart with the left vertical axis scaled from 0 to at least the grand total. Put an appropriate label on the axis. Scale the right vertical axis from 0 to 100 percent, with 100 percent on the right side being the same height as the grand total on the left side.
8. Label the horizontal axis with the category names. The leftmost category should be the largest, the next category the second largest, and so on.
9. Draw in bars representing the amount of each category. The height of the bar is determined by the left vertical axis.
10. Draw a line that shows the cumulative percentage column of the Pareto analysis table. The cumulative percentage line is determined by the right vertical axis.

Category	Peaches Lost
Bruised	100
Undersized	87
Rotten	235
Under-ripe	9
Wrong variety	7
Wormy	3

TABLE 13.1 Raw Data for Pareto Analysis

For example, the data in Table 13.1 have been recorded for peaches arriving at Super Duper Market during August.

The completed Pareto diagram is shown in Fig. 13.2.

Project Charters

The project goals, objectives, deliverables, and so forth are documented on a project charter, which serves as a contract between the project team and its sponsors. Project charters typically have several key elements, answering the *what*, *who*, *why*, and *when* of the team's planned activities. (It's important that *how* be left to the team, as discussed in Chap. 10). A sample charter is shown in Fig. 13.3.

One or more project sponsors in mid- to upper-level managerial positions will sponsor a given project. Sponsors fund the project, allocate resources, and develop the initial charter (which is then managed by the assigned project team leader, who is usually either a black belt and/or

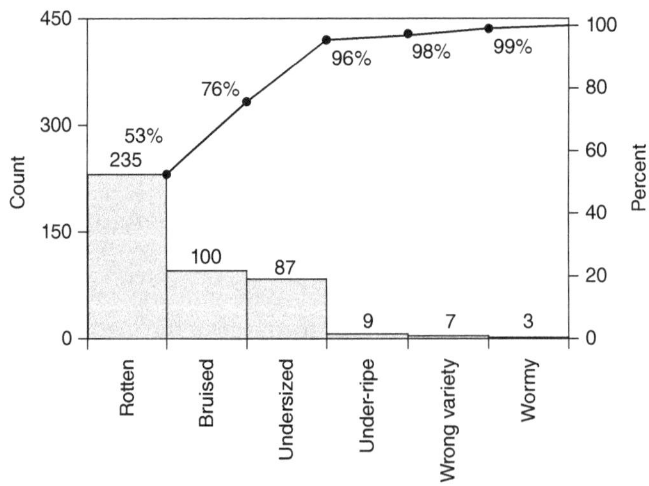

FIGURE 13.2 Example Pareto analysis constructed using *Green Belt XL* software (www.qualityamerica.com).

Project Charter

Project Name/Title: Order Processing Efficiency **Start Date: 9/17/07**

Problem/Project Description:

Current capacity in Sales/Customer Support area is constrained, while there are untapped opportunities for increased sales. We should limit, wherever possible, Sales involvement in order processing to free up resource for active lead follow-up and sales generation. In addition, errors and/or gaps in information acquired during Order Processing procedure have a negative impact on time required to generate, and/or receipt rate of, email marketing and software renewals to existing clients. This has an especially large potential impact, since it requires correction by senior sales staff, who might otherwise have more time to engage with clients, develop marketing efforts, or work with product development staff.

Project Scope (Process, Product, functional areas):

Limited to software products.

Project Objectives & Goals:	Metric	Baseline	Goal
To decrease cycle time & costs of specific Sale Department activities: ➢ Order Processing by 50%+ ➢ Marketing to existing clients by 80+% ➢ Software renewals by 80+%	Cost/Order Time/campaign Time/update	$32 download $40 shipped 2-4 hours 2-4 hours	$16 download $20 shipped 20 minutes 20 minutes

Business Need

Customer Impact:

Improved notification rate for renewals & upgrades; reduction in total cycle time as procedure more streamlined.

Shareholder Impact:

Increased sales potential, immediately on upgrades, but also for future sales with availability of sales staff; Reduced cost for order processing. Reduced costs for marketing & renewal campaigns.

Employee Impact:

Clearer responsibilities; Less interruption in process flow.

Project Sponsor:	*Stakeholder Group:*	*Signature / Date*
Peter Keene, VP	Sales & Operations	
Team Black Belt: Patrick Killihan		
Team Members: Don Debuski	Customer Support	
Helen Winkleham	Shipping & Packaging	
Anne Sheppard	Accounting	

DEFINE	MEASURE	ANALYZE	IMPROVE	CONTROL
Objective DateComplete	*Objective DateComplete*	*Objective DateComplete*	*Objective DateComplete*	*Objective Date Complete*
➢ Project Def. 9/17/07 ➢ Top level Process Def. 9/19/07 ➢ Team Formation 9/19/07	➢ Process Definition ______ ➢ Metric Def. ______ ➢ Estimate Baseline	➢ Value Stream Analysis____ ➢ Analyze Variation ______ ➢ Determine Drivers____	➢ Implement Process____ ➢ Assess Benefits___ ➢ Evaluate Failure Mode______	➢ Standardize Methods____ ➢ Control Plan______ ➢ Lessons Learned____

FIGURE 13.3 Sample project charter (Keller, 2011a).

green belt). As a member of management, the sponsor builds support for the project in the managerial ranks of the organization. The sponsor's managerial position in the functional area that is the subject of the improvement project helps to build awareness and support for the project in the operational ranks, as well as to clear roadblocks that might inhibit the timely progress of the project. When stakeholders are from different functional areas, the sponsor may be the level above the functional area management, so that resource allocation and departmental commitment are achieved. To avoid having top levels of the organization sponsor too many projects, co-sponsors from the top ranks of the affected functional areas may also be used.

The team leader is responsible for regular updates to the sponsor and all other stakeholder groups. This helps prevent undesirable surprises during the later stages of the project.

Project Scheduling

There is a wide variety of tools and techniques available to help the project manager develop a realistic project timetable to allocate resources, and to track progress during the implementation of the project plan. These systems are used to:

- Aid in planning and control of projects.
- Determine the feasibility of meeting specified deadlines.
- Identify the most likely bottlenecks in a project.
- Evaluate the effects of changes in the project requirements or schedule.
- Evaluate the effects of deviating from schedule.
- Evaluate the effect of diverting resources from the project, or redirecting additional resources to the project.

A modern version of a Gantt chart for a 4-month DMAIC improvement project, developed using MS Project, is shown in Fig. 13.4. Traditional Gantt charts were developed to show the relationships among the project tasks, along with time constraints. The horizontal axis of a Gantt

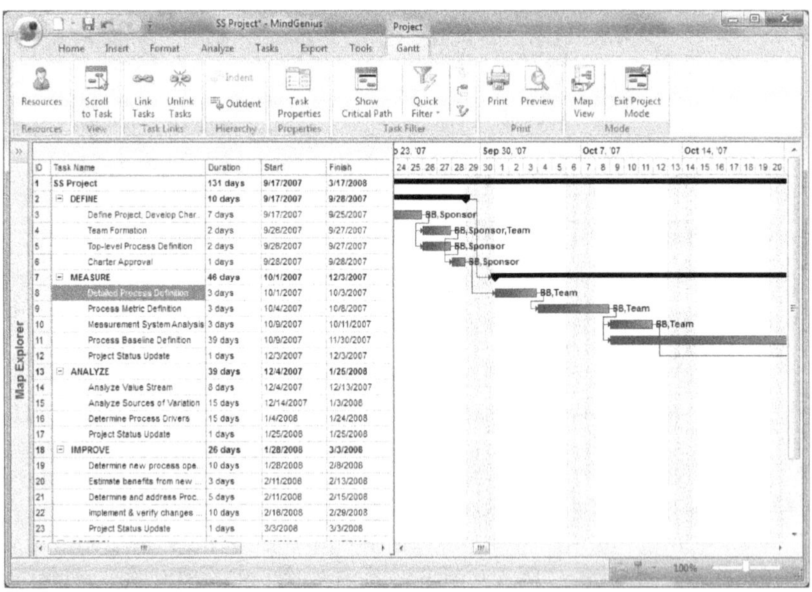

Figure 13.4 Example Gantt chart for a DMAIC improvement project (Keller, 2011a).

chart shows the units of time (days, weeks, months, etc.). The vertical axis shows the activities to be completed. Bars show the estimated start time and duration of the various activities.

Modern Gantt charts usually include designation of milestones (events that take zero time), as well as the individual responsible for each task. The completed chart clearly shows the task dependencies (i.e., which activities must be completed before any given activity may be started), and is often labeled with the critical path. (Historically, CPM [Critical Path Method] was considered an alternative to Gantt. Similarly, PERT [Program Evaluation and Review Technique] was developed to evaluate scheduling based on probabilistic activity times. Today, PERT, CPM, and Gantt actually comprise one technique.)

Project scheduling consists of four basic phases: planning, scheduling, improvement, and controlling. The planning phase involves breaking the project into distinct activities. The time estimates for these activities are then determined and a network (or arrow) diagram is constructed, with each activity being represented by an arrow.

The ultimate objective of the scheduling phase is to construct a time chart showing the start and finish times for each activity as well as its relationship to other activities in the project. The schedule must identify activities that are "critical" in the sense that they must be completed on time to keep the project on schedule.

It is vital not to merely accept the schedule as a given. The information obtained in preparing the schedule can be used to improve the project schedule. Activities that the analysis indicates to be critical are candidates for improvement. Pareto analysis can be used to identify those critical elements that are most likely to lead to significant improvement in overall project completion time. Cost data can be used to supplement the time data, and the combined time/cost information can be analyzed using Pareto analysis.

The final phase in project management is project control. This includes the use of the network diagram and time chart for making periodic progress assessments.

Constructing Network Charts

A common means of evaluating a project schedule is to graphically portray the interrelationships among the elements of a project. This network representation of the project plan shows all the precedence relationships, that is, the order in which the tasks must be completed. Arrows in the network chart represent activities, while boxes or circles represent events; in preparing and understanding this technique, it is very important to keep these two terms distinct. An arrow goes from one event to another only if the first event is the immediate predecessor of the second. If more than one activity must be completed before an event can occur, then there will be several arrows entering the box corresponding to that event. Sometimes one event must wait for another event, but no activity intervenes

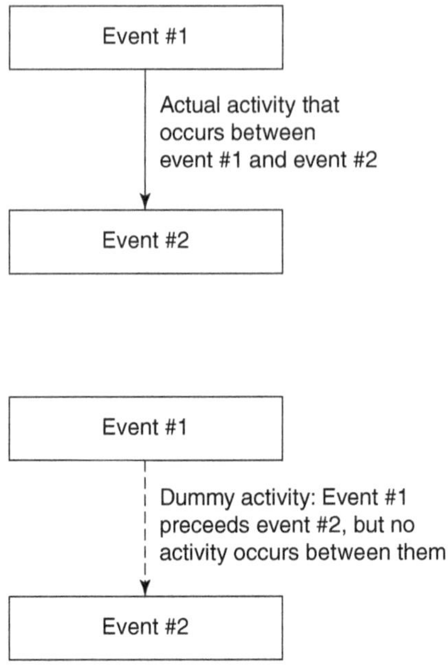

Figure 13.5 Network diagram terms and drawing conventions.

between the two events. In this case, the two events are joined with a dotted arrow, representing a dummy activity. Dummy activities take no time to complete; they merely show precedence relationships.

These drawing conventions are illustrated in Fig. 13.5.

The node toward which all activities lead, the final completion of the project, is called the sink of the network. Taha (1976) offers the following rules for constructing the arrow diagram:

Rule 1: *Each activity is represented by one and only one arrow in the network.* No single activity can be represented twice in the network. This does not mean that one activity cannot be broken down into segments.

Rule 2: *No two activities can be identified by the same head-and-tail events.* This situation may arise when two activities can be performed concurrently. The proper way to deal with this situation is to introduce dummy events and activities, as shown in Fig. 13.6. This rule facilitates the analysis of network diagrams with computer programs for project analysis.

Rule 3: *In order to ensure the correct precedence relationship in the arrow diagram, the following questions must be answered as each activity is added to the network:*

a. What activities must be completed immediately before this activity can start?
b. What activities immediately follow this activity?
c. What activities must occur concurrently with this activity?

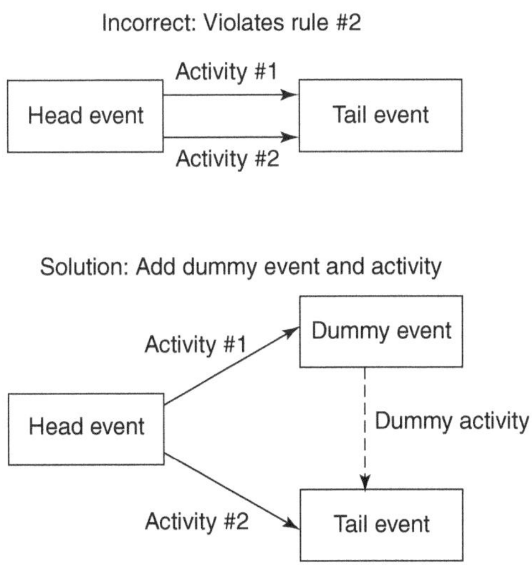

FIGURE 13.6 Parallel activities: network representation.

The data shown in Table 13.2 consist of the activities and their estimated completion times for constructing a house.

Now, it is important that certain of these activities be done in a particular order. For example, one cannot put on the roof until the walls are built. This is called a precedence relationship; that is, the walls must

Activity	Time to Complete (days)
Excavate	2
Foundation	4
Rough wall	10
Rough electrical work	7
Rough exterior plumbing	4
Rough interior plumbing	5
Wall board	5
Flooring	4
Interior painting	5
Interior fixtures	6
Roof	6
Exterior siding	7
Exterior painting	9
Exterior fixtures	2

From Hillier and Lieberman (1980)

TABLE 13.2 Activities Involved in Constructing a House

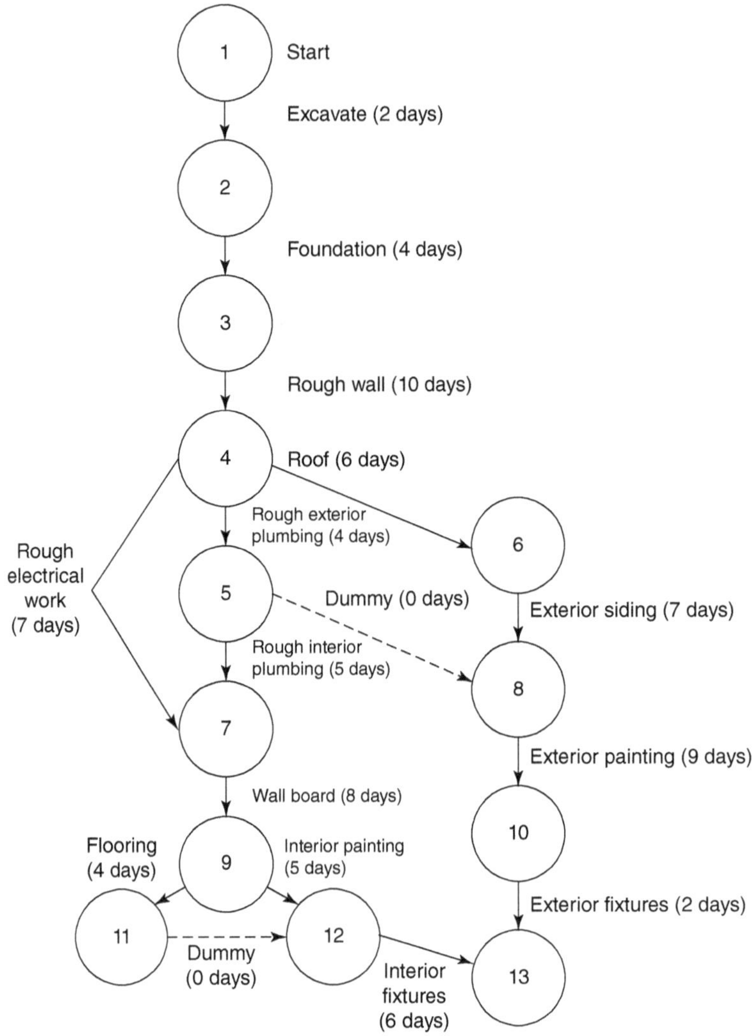

FIGURE 13.7 Network diagram for constructing a house.

precede the roof. The network diagram graphically displays the precedence relationships involved in constructing a house. A PERT network for constructing a house is shown in Fig. 13.7. (Incidentally, the figure is also an arrow diagram.)

Finding the Critical Path

There are two time-values of interest for each event: its *earliest time of completion* and its *latest time of completion*. The earliest time for a given event is the estimated time at which the event will occur if the preceding activities are started as early as possible. The latest time for an event is the

Event	Immediately Preceding Event	Earliest Time + Activity Time	Maximum = Earliest Completion Time
1	—	—	0
2	1	0 + 2	2
3	2	2 + 4	6
4	3	6 + 10	16
5	4	16 + 4	20
6	4	16 + 6	22
7	4	16 + 7	*
	5	20 + 5	25
8	5	20 + 0	*
	6	22 + 7	29
9	7	25 + 8	33
10	8	29 + 9	38
11	9	33 + 4	37
12	9	33 + 5	38
	11	37 + 0	*
13	10	38 + 2	*
	12	38 + 6	44

TABLE 13.3 Calculation of Earliest Completion Times for House Construction Example

estimated latest time the event can occur without delaying the completion of the project beyond its earliest time. Earliest times of events are found by starting at the initial event and working forward, successively calculating the time at which each event will occur if each immediately preceding event occurs at its earliest time and each intervening activity uses only its estimated time. Table 13.3 shows the process of finding the earliest completion time for the house construction example. (Event numbers refer to the network diagram in Fig. 13.7.) The reader is advised to work through the results in Table 13.3, line-by-line, using Fig. 13.7.

Thus, for example, the earliest time event #8 can be completed is 29 days. (Note that the asterisks in Table 13.3 denote calculations that resulted in the non-maximum condition. For example, event #8 occurs when both events #5 and #6 have been completed, so the maximum time calculation is used).

Latest times are found by starting at the final event and working backward, calculating the latest time an event will occur if each immediately following event occurs at its latest time. Table 13.4 displays the calculated latest completion times for the house construction example.

Event	Immediately Following Event	Latest Time – Activity Time	Minimum = Latest Time
13	—	—	44
12	13	44 – 6	38
11	12	38 – 0	38
10	13	44 – 2	42
9	12	38 – 5	33
	11	38 – 4	*
8	10	42 – 9	33
7	9	33 – 8	25
6	8	33 – 7	26
5	8	33 – 0	
	7	25 – 5	20
4	7	25 – 7	
	6	26 – 6	
	5	20 – 4	16
3	4	16 – 10	6
2	3	6 – 4	2
1	2	2 – 2	0

TABLE 13.4 Calculation of Latest Completion Times for House Construction Example

Slack time for an event is the difference between the latest and earliest times for a given event. Thus, assuming everything else remains on schedule, the slack for an event indicates how much delay in reaching the event can be tolerated without delaying the project completion. Slack times for the events in the house construction project are shown in Table 13.5.

Event	Slack	Event	Slack
1	0 – 0 = 0	7	25 – 25 = 0
2	2 – 2 = 0	8	33 – 29 = 4
3	6 – 6 = 0	9	33 – 33 = 0
4	16 – 16 = 0	10	42 – 38 = 4
5	20 – 20 = 0	11	38 – 37 = 1
6	26 – 22 = 4	12	38 – 38 = 0
Continued ...	*Continued ...*	13	44 – 44 = 0

TABLE 13.5 Calculation of Slack Times for House Construction Events

The slack time for an activity *x,y* is the difference between

1. The latest time of event *y*
2. The earliest time of event *x* plus the estimated activity time

Slack time for an activity is the difference between the latest and earliest times for a given activity. Thus, assuming everything else remains on schedule, the slack for an activity indicates how much delay in reaching the activity can be tolerated without delaying the project completion. Slack times for the activities in the house construction project are shown in Table 13.6.

Events and activities with slack times of zero are said to lie on the *critical path* for the project. Conversely, a critical path for a project is defined as a path through the network such that the activities on this path have *zero slack*. All activities and events having zero slack must lie on a critical path, but no others can. Figure 13.8 shows the activities on the critical path for the housing construction project as thick lines.

Control and Prevention of Schedule Slippage

Project managers can use the network and the information obtained from the network analysis in a variety of ways to help them manage their projects. One way is, of course, to pay close attention to the activities that lie on the critical path. Any delay in these activities will result in a delay for

Activity	Slack
Excavate (1,2)	2 – (0 + 2) = 0
Foundation (2,3)	6 – (2 + 4) = 0
Rough wall (3,4)	16 – (6 + 10) = 0
Rough exterior plumbing (4,5)	20 – (16 + 4) = 0
Roof (4,6)	26 – (16 + 6) = 4
Rough electrical work (4,7)	25 – (16 + 7) = 2
Rough interior plumbing (5,7)	25 – (20 + 5) = 0
Exterior siding (6,8)	33 – (22 + 7) = 4
Wall board (7,9)	33 – (25 + 8) = 0
Exterior painting (8,10)	42 – (29 + 9) = 4
Flooring (9,11)	38 – (33 + 4) = 1
Interior painting (9,12)	38 – (33 + 5) = 0
Exterior fixtures (10,13)	44 – (38 + 2) = 4
Interior fixtures (12,13)	44 – (38 + 6) = 0

Table 13.6 Calculation of Slack times for House Construction Activities

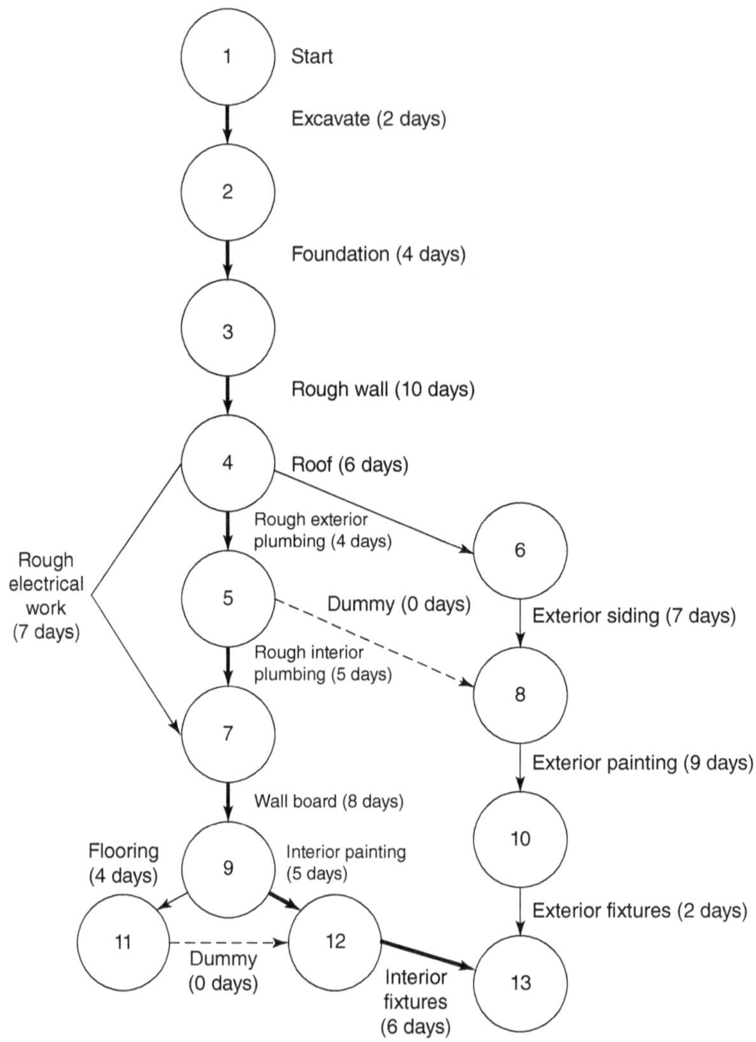

Figure 13.8 Critical path for house construction example.

the project. However, the manager should also consider assembling a team to review the network with an eye toward modifying the project plan to reduce the total time needed to complete the project. The manager should also be aware that the network times are based on estimates. In fact, it is likely that the completion times will vary. When this occurs it often happens that a new critical path appears. Thus, the network should be viewed as a dynamic entity that should be revised as conditions change.

Primary causes of slippage include poor planning and poor management of the project. Outside forces beyond the control of the project manager will often play a role. However, it isn't enough to be able to simply identify "outside forces" as the cause and beg forgiveness. Astute project managers will anticipate as many such possibilities as possible and prepare contingency plans to deal with them. The PDPC technique is useful in this endeavor (see Chap. 16). Schedule slippage should also be addressed rigorously via reviews conducted at intervals frequent enough to ensure that any unanticipated problems are identified before schedule slippage becomes a problem.

Resources

Resources are those assets of the firm, including employees' time, that are used to accomplish the objectives of the project. The project manager should define, negotiate, and secure resource commitments for the personnel, equipment, facilities, and services needed for the project. Resource commitments should be as specific as possible.

The following items should be defined and negotiated:

- What will be furnished?
- By whom?
- When?
- How will it be delivered?
- How much will it cost?
- Who will pay?
- When will payment be made?

Of course, there are always other opportunities for utilizing resources. On large projects, conflicts over resource allocation are inevitable. It is best if resource conflicts can be resolved between those managers directly involved. However, in some cases, resource conflicts must be addressed by higher levels of management. Senior managers should view resource conflicts as potential indications that the management system for allocating resources must be modified or redesigned. Often, such conflicts create ill will among managers and lead to lack of support, or even active resistance to the project. Too many such conflicts can lead to resentment toward quality improvement efforts in general.

Cost Considerations in Project Scheduling

Most project schedules can be compressed, if one is willing to pay the additional costs. For the analysis here, costs are defined to include direct elements only. Indirect costs (administration, overhead, etc.) will be considered in the final analysis. Assume that a straight line relationship exists between the cost of performing an activity on a normal schedule, and the cost of performing the activity on a crash schedule. Also assume that there is a crash time beyond which no further time savings are possible, regardless of cost. Figure 13.9 illustrates these concepts.

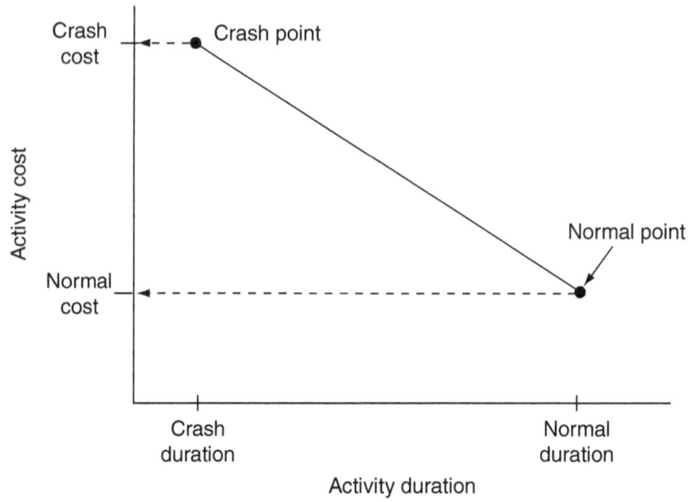

Figure 13.9 Cost-time relationship for an activity.

For a given activity the cost per unit of time saved is found as

(crash cost – normal cost)/(normal time – crash time)

When deciding which activity on the critical path to improve, one should begin with the activity that has the smallest cost per unit of time saved. The project manager should be aware that once an activity time has been reduced there may be a new critical path. If so, the analysis should proceed using the updated information; that is, activities on the new critical path should be analyzed.

The data for the house construction example is shown below, with additional data for costs and crash schedule times for each activity.

Activities shown in bold are on the critical path; only critical path activities are being considered since only they can produce an improvement in overall project duration. Thus, the first activity to consider improving would be foundation work, which costs $800 per day saved on the schedule (identified with an asterisk [*] in Table 13.7). Directing additional resources toward this activity would produce the best "bang for the buck" in terms of reducing the total time of the project. The next activities for consideration, assuming the critical path doesn't change, would be excavation, then exterior painting, etc.

As activities are addressed one-by-one, the time it takes to complete the project will decline, while the direct costs of completing the project will increase. Figure 13.10 illustrates the cost-duration relationship graphically.

Conversely, indirect costs such as overhead, etc., are expected to increase as projects take longer to complete. When the indirect costs are

Activity	Normal Schedule		Crash Schedule		Slope
	Time (days)	Cost	Time (days)	Cost	
Excavate	**2**	**1000**	**1**	**2000**	**1000**
Foundation	**4**	**1600**	**3**	**2400**	**800***
Rough wall	**10**	**7500**	**6**	**14000**	**1625**
Rough electrical work	7	7000	4	14000	2333
Rough exterior plumbing	**4**	**4400**	**3**	**6000**	**1600**
Rough interior plumbing	**5**	**3750**	**3**	**7500**	**1875**
Wall board	**5**	**3500**	**3**	**7000**	**1750**
Flooring	4	3200	2	5600	1200
Interior painting	**5**	**3000**	**3**	**5500**	**1250**
Interior fixtures	**6**	**4800**	**2**	**11000**	**1550**
Roof	6	4900	2	12000	1775
Exterior siding	7	5600	3	12000	1600
Exterior painting	9	4500	5	9000	1125
Exterior fixtures	2	1800	1	3200	1400

Table 13.7 Schedule Costs for Activities Involved in Constructing a House

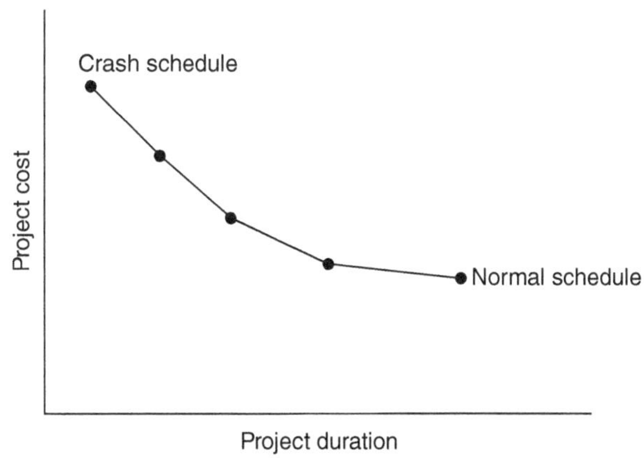

Figure 13.10 Direct costs as a function of project duration.

added to the direct costs, total costs will generally follow a pattern similar to that shown in Fig. 13.11.

To optimize resource utilization, the project manager will seek to develop a project plan that produces the minimum cost schedule. Of course, the organization will likely have multiple projects being conducted simultaneously, which places additional constraints on resource allocation.

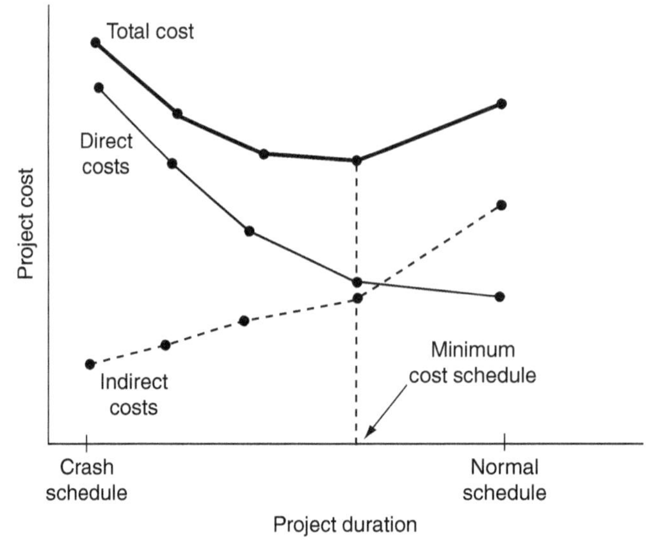

FIGURE 13.11 Total costs as a function of project duration.

Other Performance Measurement Methodology

Project information should be collected on an ongoing basis as the project progresses. Information obtained should be communicated in a timely fashion to interested parties and decision makers. The people who receive the information can often help the project manager to maintain or recover the schedule. There are two types of communication involved: feedback and feed-forward. Feedback is historical in nature and includes such things as performance to schedule, cost variances (relative to the project budget), and quality variances (relative to the quality plan). The reader will recall that initial project planning called for special control plans in each of these three areas. Feed-forward is oriented toward the future and is primarily concerned with heading off future variances in these three areas. Information reporting formats commonly fall into one of the following categories:

- Formal, written reports
- Informal reports and correspondence
- Presentations
- Meetings
- Guided tours of the project, when feasible
- Conversations

The principles of effective communication discussed in Chap. 20 should be kept constantly in mind. The choice of format for the communication should consider the nature of the audience and their needs and the time and resources available.

Top-Level Process Definition

A SIPOC (Supplier-Inputs-Process-Outputs-Customer) analysis is a preferred tool for defining the top-level view of the process. The SIPOC will ensure the key stakeholders are identified, which is needed at this stage to construct a relevant project team. An example SIPOC is shown in Fig. 13.12.

Team Formation

Effective team formation is critical to build stakeholder buy-in. Credible team members are selected from each of the key stakeholder groups to represent their functional areas in the design/improvement project. It is helpful if the candidates are enthusiastic about the change, but as noted earlier healthy skepticism is often productive as well. They will need to commit some time to group activities, away from their functional area, so local management support is necessary, as is their willingness to serve on the team.

Effective teams are generally limited to five to seven participants. Larger teams are more difficult to manage, and members may lose a sense of responsibility to the team. Additional team members may be ad hoc members from non–key stakeholder groups, who participate only as needed, such as for process expertise.

The team leader must clearly communicate personal responsibilities to team members in an initial meeting and fairly enforce these responsibilities in subsequent meetings. Typical responsibilities include:

- Take responsibility for success
- Follow through on commitments
- Contribute to discussions
- Actively listen
- Communicate clearly
- Provide constructive feedback, especially to team leader
- Accept feedback

The team leader is generally responsible for keeping the team focused. The project charter often serves as an effective focusing tool to avoid scope creep. The charter includes a project schedule (usually via a Gantt chart), which provides a time constraint. The DMAIC/DMADV methodology will also enforce considerable focus, when properly followed.

Effective team leaders will ensure that conflicts are resolved in a positive manner. Enforcing ground rules is necessary, and using the various tools within the DMAIC structure will allow the team to work through issues constructively. A critical aspect of DMAIC is data-driven

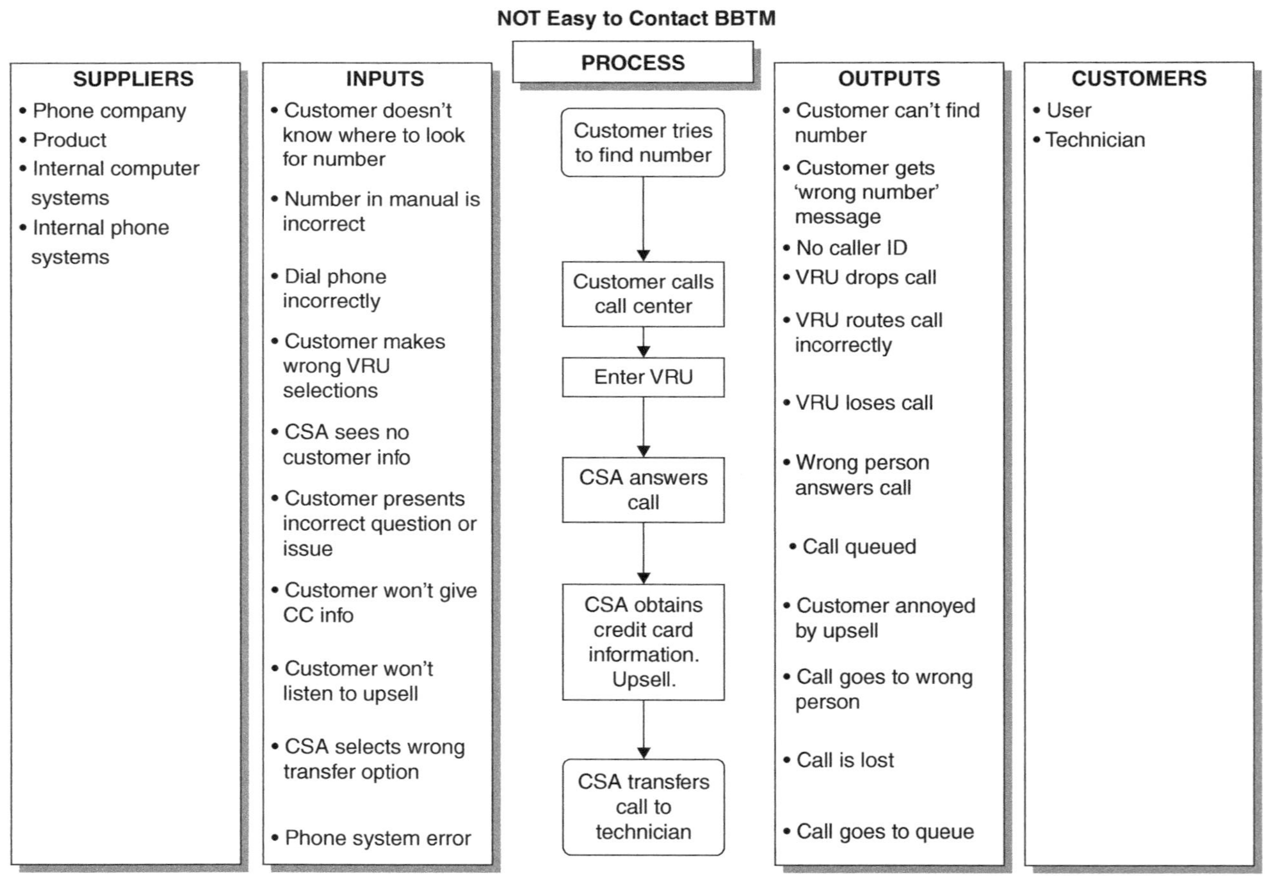

FIGURE 13.12 Example SIPOC analysis for identifying key stakeholder groups (Pyzdek and Keller, 2010).

decision making. While subjective insight is valued in brainstorming throughout each stage of DMAIC, the overall DMAIC process will move the team toward collecting and analyzing data to achieve more objective conclusions. The process will provide ample opportunity for the team leader to develop the cooperative problem-solving and communicative skills of the team members, which is another important role of the team leader.

Consensus is the preferred approach to team decision making. Consensus does not mean that everyone is in absolute agreement, nor that everyone prefers the proposal. Rather, consensus implies that the parties are willing to accept the proposal in spite of their differences. A good question to ask stakeholders to gauge their consensus level is: *Can you live with it?* We accept that there are differences in viewpoint, and strive to reconcile these with our analytical tools. Achieving consensus allows us to move forward, so the merits of the proposal can be proven through data analysis. Alternatives to consensus, such as majority voting, arbitrary flipping of a coin, or exchanging of votes for reciprocal votes (bartering), undermine the team's eventual results and must be avoided.

Perhaps the most obvious ground rule is respectful, inviting communication, to allow all members to participate. Toward this end, members should "leave their badge at the door," meaning that there is no seniority in a team meeting. The team leader is not senior to anyone else on the team, nor is it his or her personal project. Rather, the project is owned by the sponsor, and all team members are serving at the sponsor's request.

Finally, team members need to accept responsibility for action items and be prompt in following up on these items. Team leaders should ensure team members' time is used responsibly. Generally, it's best to meet only when necessary, although default times can be established for meetings to help members allocate time. The team is essentially providing their process expertise, and the team should only be convened when those skills are necessary.

Team Dynamics Management, Including Conflict Resolution

Conflict is a natural part of the creative process, and the team leader must ensure that creative conflict is not repressed, but encouraged. The effective team leader will explore the underlying reasons for the conflict. If personality disputes are involved that threaten to disrupt the team meeting, arrange one-on-one meetings between the parties and attend the meetings to help mediate.

The first step in establishing an effective group is to create a consensus decision rule for the group. For example:

> *No judgment may be incorporated into the group decision until it meets at least tacit approval of every member of the group.*

This minimum condition for group movement can be facilitated when the team leader adopts the following behaviors:

- *Avoid arguing for your own position.* Present it as lucidly and logically as possible, but be sensitive to and consider seriously the reactions of the group in any subsequent presentations of the same point.
- *Avoid "win-lose" stalemates in the discussion of opinions.* Discard the notion that someone must win and someone must lose in the discussion; when impasses occur, look for the next most acceptable alternative for all the parties involved.
- *Avoid changing your mind only to avoid conflict and to reach agreement and harmony.* Withstand pressures to yield that have no objective or logically sound foundation. Strive for enlightened flexibility, but avoid outright capitulation.
- *Avoid conflict-reducing techniques such as the majority vote, averaging, bargaining, coin-flipping, trading out, and the like.* Treat differences of opinion as indicative of an incomplete sharing of relevant information on someone's part, either about task issues, emotional data, or gut-level intuitions.
- *View differences of opinion as both natural and helpful rather than as a hindrance in decision making.* Generally, the more ideas expressed, the greater the likelihood of conflict will be, but the richer the array of resources will be as well.
- *View initial agreement as suspect.* Explore the reasons underlying apparent agreements; make sure people have arrived at the same conclusions for either the same basic reasons or for complementary reasons before incorporating such opinions into the group decision.
- *Avoid subtle forms of influence and decision modification.* For example, when a dissenting member finally agrees, don't feel that he must be rewarded by having his own way on some subsequent point.
- *Be willing to entertain the possibility that your group can achieve all the foregoing and actually excel at its task.* Avoid doomsaying and negative predictions for group potential.

Collectively, the above steps are sometimes known as the "consensus technique." In tests it was found that 75 percent of the groups that were instructed in this approach significantly outperformed their best individual resources.

Stages in Group Development

Groups of many different types tend to evolve in similar ways. It often helps to know that the process of building an effective group is proceeding normally. Bruce W. Tuckman identified four stages in the development of a group: forming, storming, norming, and performing.

During the *forming* stage a group tends to emphasize procedural matters. Group interaction is very tentative and polite. The leader dominates the decision-making process and plays a very important role in moving the group forward.

The *storming* stage follows forming. Conflict among members, and between members and the leader, are characteristic of this stage. Members question authority as it relates to the group objectives, structure, or procedures. It is common for the group to resist the attempts of their leader to move them toward independence. Members are trying to define their role in the group.

It is important that the leader deal with the conflict constructively. There are several ways in which this may be done:

- Do not tighten control or try to force members to conform to the procedures or rules established during the forming stage. If disputes over procedures arise, guide the group toward new procedures based on a group consensus.
- Probe for the true reasons behind the conflict and negotiate a more acceptable solution.
- Serve as a mediator between group members.
- Directly confront counterproductive behavior.
- Continue moving the group toward independence from its leader.

During the *norming* stage the group begins taking responsibility, or ownership, of its goals, procedures, and behavior. The focus is on working together efficiently. Group norms are enforced on the group by the group itself.

The final stage is *performing*. Members have developed a sense of pride in the group, its accomplishments, and their role in the group. Members are confident in their ability to contribute to the group and feel free to ask for or give assistance.

Common Team Problems

Table 13.8 lists some common problems with teams, along with recommended remedial action (Scholtes, 1988).

Productive Group Roles

There are two basic types of roles assumed by members of a group: task roles and group maintenance roles. Group task roles are those functions concerned with facilitating and coordinating the group's efforts to select, define, and solve a particular problem. The group task roles shown in Table 13.9 are generally recognized.

Problem	Action
Floundering	• Review the plan • Develop a plan for movement
The expert	• Talk to offending party in private • Let the data do the talking • Insist on consensus decisions
Dominating participants	• Structure participation • Balance participation • Act as gatekeeper
Reluctant participants	• Structure participation • Balance participation • Act as gatekeeper
Using opinions instead of facts	• Insist on data • Use scientific method
Rushing things	• Provide constructive feedback • Insist on data • Use scientific method
Attribution (i.e., attributing motives to people with whom we disagree)	• Don't guess at motives • Use scientific method • Provide constructive feedback
Ignoring some comments	• Listen actively • Train team in listening techniques • Speak to offending party in private
Wanderlust	• Follow a written agenda • Restate the topic being discussed
Feuds	• Talk to offending parties in private • Develop or restate ground rules

Table 13.8 Common Team Problems and Remedial Action

Another type of role played in small groups is the group maintenance roles. Group maintenance roles are aimed at building group cohesiveness and group-centered behavior. They include those behaviors shown in Table 13.10.

The development of task and maintenance roles is a vital part of the team-building process. Team building is defined as the process by which a group learns to function as a unit, rather than as a collection of individuals.

Counterproductive Group Roles

In addition to developing productive group-oriented behavior, it is also important to recognize and deal with individual roles that may block the building of a cohesive and effective team. These roles are shown in Table 13.11.

Role	Description
Initiator	Proposes new ideas, tasks, or goals; suggests procedures or ideas for solving a problem or for organizing the group
Information seeker	Asks for relevant facts related to the problem being discussed
Opinion seeker	Seeks clarification of values related to problem or suggestion
Information giver	Provides useful information about subject under discussion
Opinion giver	Offers his/her opinion of suggestions made; emphasis is on values rather than facts
Elaborator	Gives examples
Coordinator	Shows relationship among suggestions; points out issues and alternatives
Orientor	Relates direction of group to agreed-upon goals
Evaluator	Questions logic behind ideas, usefulness of ideas, or suggestions
Energizer	Attempts to keep the group moving toward an action
Procedure technician	Keeps group from becoming distracted by performing such tasks as distributing materials, checking seating, etc.
Recorder	Serves as the group memory

Table 13.9 Group Task Roles

Role	Description
Encourager	Offers praise to other members; accepts the contributions of others
Harmonizer	Reduces tension by providing humor or by promoting reconciliation; gets people to explore their differences in a manner that benefits the entire group
Compromiser	This role may be assumed when a group member's idea is challenged; admits errors, offers to modify his/her position
Gatekeeper	Encourages participation, suggests procedures for keeping communication channels open
Standard setter	Expresses standards for group to achieve; evaluates group progress in terms of these standards
Observer/commentator	Records aspects of group process; helps group evaluate its functioning
Follower	Passively accepts ideas of others; serves as audience in group discussions

Table 13.10 Group Maintenance Roles

Role	Description
Aggressor	Expresses disapproval by attacking the values, ideas, or feelings of others; shows jealousy or envy
Blocker	Prevents progress by persisting on issues that have been resolved; resists attempts at consensus; opposes without reason
Recognition-seeker	Calls attention to himself/herself by boasting, relating personal achievements, etc.
Confessor	Uses group setting as a forum to air personal ideologies that have little to do with group values or goals
Playboy	Displays lack of commitment to group's work by cynicism, horseplay, etc.
Dominator	Asserts authority by interrupting others, using flattery to manipulate, claiming superior status
Help-seeker	Attempts to evoke sympathy and/or assistance from other members through "poor me" attitude
Special-interest pleader	Asserts the interests of a particular group; this group's interest matches his/her self-interest

TABLE 13.11 Counterproductive Group Roles

The leader's role includes that of process observer. In this capacity, the leader monitors the atmosphere during group meetings and the behavior of individuals. The purpose is to identify counterproductive behavior. Of course, once counterproductive behavior has been identified, the leader must tactfully and diplomatically provide feedback to the group and its members.

Management's Role

As discussed in Chap. 12 and earlier in this chapter, management plays a key role in successful change efforts. Within improvement projects, they provide a critical role as project sponsor, ensuring teams have the necessary authority to investigate and implement changes, and resources are allocated on a timely basis.

In addition to these critical responsibilities, perhaps the most important thing management can do for a group is to give it time to become effective. This requires, among other things, that management work to maintain consistent group membership. Group members must not be moved out of the group without very good reason. Nor should there be a constant stream of new people temporarily assigned to the group. If a group is to progress through the four stages described earlier in this chapter, to the crucial performing stage, it will require a great deal of discipline from both the group and management.

CHAPTER 14

Measure Stage

Measure stage objectives include (Keller, 2011a):

- *Process definition.* Define the process at a detailed level including decision points and functions.
- *Metric definition.* Define metric to reliably establish process estimates.
- *Process baseline.* Use the defined metrics to establish the current state of the process, which should verify the assumptions of the Define stage. Determine whether the process is in statistical control.
- *Measurement systems analysis.* Quantify errors associated with the metric.

Process Definition

A process flowchart is simply a tool that graphically shows the inputs, actions, and outputs of a given system. These terms are defined as follows:

Inputs. The factors of production or service: land, materials, labor, equipment, and management.

Actions. The way in which the inputs are combined and manipulated in order to add value. Actions include procedures, handling, storage, transportation, and processing.

Outputs. The products or services created by acting on the inputs. Outputs are delivered to the customer or other user. Outputs also include *unplanned* and *undesirable* results, such as scrap, rework, pollution, etc. Flowcharts should contain these outputs as well.

Flowcharting is such a useful activity that the symbols have been standardized by various ANSI standards. There are special symbols for special processes, such as electronics or information systems. However, in most cases activities are contained within simple rectangles; decision points within diamonds, with one input and only two potential outputs (a yes/no path).

Flowcharts can be made either more complex or less complex. As a rule of thumb, to paraphrase Albert Einstein, "Flowcharts should be as

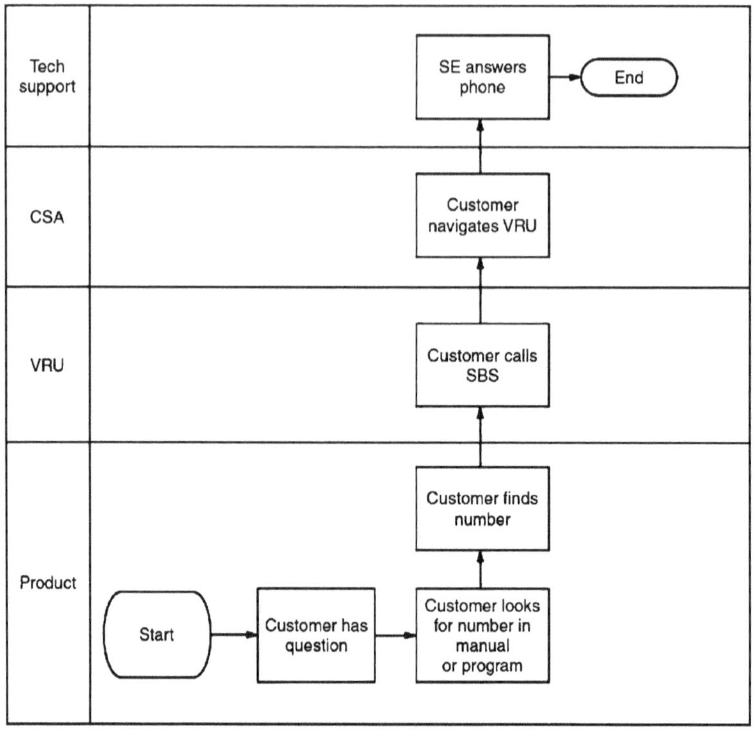

FIGURE 14.1 Example process map (Pyzdek and Keller, 2010).

simple as possible, but not simpler." The purpose of the flowchart is to help people understand the process, and this is not accomplished with flowcharts that are either too simple or too complex.

When flowcharts indicate a larger number of decision points, it is often a sign of an overly complicated process that has potential for error. Fortunately, the decision points are also a potential focus for improvement.

Process maps are flowcharts that also show, via swim lanes, how the process items move between functional areas, as shown in Fig. 14.1.

Metric Definition

The project definition includes a statement of the metric(s) that will be used to evaluate the process. It's important to choose metrics that will ensure you are actually improving quality, business performance, and customer satisfaction. Measuring profit and growth is not recommended for this purpose because they are typically slow to respond to customer dissatisfaction and can indicate short-term improvements at the cost of long-term viability. Properly chosen metrics will provide input for data-based decision making, and will become the language used to communicate the status and well-being of the business or process.

Once established, these metrics can be used to determine the relative importance of various factors affecting processes and business units, as well as comparing the various processes', business units', and competitors' contributions to the overall success of the business. They also provide a baseline to gauge the results of Six Sigma improvement efforts.

In short, metrics provide us with a statement of what is critical to quality, cost, and/or scheduling; how these will be measured and reported; and how these critical to quality (CTQ), critical to cost (CTC), and critical to schedule (CTS) metrics correlate with key process variables and controls to achieve system-wide improvements.

At the process level, you may establish a baseline by measuring the current process capability. Other common metrics in Six Sigma are the Throughput Yield, the Normalized Yield, and the Sigma Level for individual process steps, and the Rolled Throughput Yield for complete processes.

Establishing Process Baselines

The performance of the existing process must be established via a baseline to ascertain how well the current process meets the customer requirements and validate the project justification noted in the Define stage. When the process is in a state of statistical control, then its process capability index may be calculated, as discussed in Chap. 9. The capability index compares the calculated process variation with the stated customer requirements. Process variation may be estimated only after process stability is established using a control chart. Only stable processes can be predicted; unstable processes are the combination of multiple processes, each occurring at separate intervals in time.

When the process is unstable, or not in statistical control, a process performance index may be used as rough estimate to compare observed variation within a given sample with the customer requirements.

A control chart is critical at this stage so results of improvement achieved later in the project can be reliably estimated. When improvement efforts are carried out without the use of a control chart, it's difficult to prove that the improvement had any real effort, since any improvement in result may have just been coincident with the sporadic nature of the special cause.

When control charts are used to baseline the process, and the process is out of control, the team may quickly discover the cause. If this special cause is of the order of magnitude that justified the project, the project is on the way to rapid conclusion. In any event, this is necessary information for the Analyze stage, as common cause variation must be treated much differently than special cause variation.

At the business level, we often seek to measure the customer's perceptions of the current state of the product, product delivery, and/or service experience. Obtaining credible customer feedback is an important part of this definition. Customer surveys, focus groups, interviews, and benchmarking are all used, generally in conjunction with one another, to better understand customer needs, desires, and those items known in the Kano model as *exciters*. These topics were discussed in Part 2.

Measurement Systems Analysis

An argument can be made for asserting that quality begins with measurement. Only when quality is quantified can meaningful discussion about improvement begin. Conceptually, measurement is quite simple: measurement is the assignment of numbers to observed phenomena according to certain rules. Measurement is a *sine qua non* of any science, including management science.

Levels of Measurement

A *measurement* is simply a numerical assignment to something, usually a non-numerical element. Measurements convey certain information about the relationship between the element and other elements. Measurement involves a theoretical domain, an area of substantive concern represented as an empirical relational system, and a domain represented by a particular selected numerical relational system. There is a mapping function that carries us from the empirical system into the numerical system. The numerical system is manipulated and the results of the manipulation are studied to help the manager better understand the empirical system.

In reality, measurement is problematic: the manager can never know the "true" value of the element being measured. The numbers provide information on a certain scale, and they represent measurements of some unobservable variable of interest. Some measurements are richer than others; that is, some measurements provide more information than other measurements. The information content of a number is dependent on the scale of measurement used. This scale determines the types of statistical analyses that can be properly employed in studying the numbers. Until one has determined the scale of measurement, one cannot know if a given method of analysis is valid.

The four measurement scales are: nominal, ordinal, interval, and ratio. Harrington (1992) summarizes the properties of each scale in Table 14.1.

Numbers on a *nominal scale* aren't measurements at all, they are merely *category labels* in numerical form. Nominal measurements might indicate membership in a group (1 = male, 2 = female) or simply represent a designation (John Doe is #43 on the team). Nominal scales represent the simplest and weakest form of measurement. Nominal variables are perhaps best viewed as a form of classification rather than as a measurement scale. Ideally, categories on the nominal scale are constructed in such a way that

Scale	Definition	Example	Statistics
Nominal	Only the presence/absence of an attribute; can only count items	Go/no go; success/fail; accept/reject	Percent; proportion; chi-square tests
Ordinal	Can say that one item has more or less of an attribute than another item; can order a set of items	Taste; attractiveness	Rank-order correlation
Interval	Difference between any two successive points is equal; often treated as a ratio scale even if assumption of equal intervals is incorrect; can add, subtract, order objects	Calendar time; temperature	Correlations; t-tests; F-tests; multiple regression
Ratio	True zero point indicates absence of an attribute; can add, subtract, multiply and divide	Elapsed time; distance; weight	t-test; F-test; correlations; multiple regression

(ASQ Quality Engineering Handbook, 1992)

TABLE 14.1 Types of Measurement Scales and Permissible Statistics

all objects in the universe are members of one and only one class. Data collected on a nominal scale are called attribute data. The only mathematical operations permitted on nominal scales are = (which shows that an object possesses the attribute of concern) or ≠.

An *ordinal* variable is one that has a natural ordering of its possible values, but for which the distances between the values are undefined. An example is product preference rankings such as good, better, best. Ordinal data can be analyzed with the mathematical operators, = (equality), ≠ (inequality), > (greater than), and < (less than). There is a wide variety of statistical techniques that can be applied to ordinal data, including the Pearson correlation. Other ordinal models include odds-ratio measures, log-linear models, and logit models. In quality management, ordinal data are commonly converted into nominal data and analyzed using binomial or Poisson models. For example, if parts were classified using a poor-good-excellent ordering, the quality engineer might plot a *p* chart of the proportion of items in the poor category.

Interval scales consist of measurements where the ratios of *differences* are invariant. For example, 90°C = 194°F, 180°C = 356°F, 270°C = 518°F, 360°C = 680°F. Now, 194°F/90°C ≠ 356°F/180°C but

$$\frac{356°\text{F} - 194°\text{F}}{680°\text{F} - 518°\text{F}} = \frac{180°\text{C} - 90°\text{C}}{360°\text{C} - 270°\text{C}}$$

Conversion between two interval scales is accomplished by the transformation

$$y = ax + b, a > 0$$

For example,

$$°F = 32 + \left(\frac{9}{5} \times °C\right)$$

where $a = 9/5$ and $b = 32$. As with ratio scales, when permissible transformations are made statistical, results are unaffected by the interval scale used. Also, 0° (on either scale) is arbitrary. In this example, zero does not indicate an absence of heat.

Ratio scale measurements are so-called because measurements of an object in two different metrics are related to one another by an invariant ratio. For example, if an object's mass were measured in pounds (x) and kilograms (y), then $x/y = 2.2$ for all values of x and y. This implies that a change from one ratio measurement scale to another is performed by a transformation of the form $y = ax$, $a > 0$; for example, pounds = 2.2 × kilograms. When permissible transformations are used, statistical results based on the data are identical regardless of the ratio scale used. Zero has an inherent meaning: in this example it signifies an absence of mass.

Reliability and Validity

Fundamentally, any item measure should meet two tests:

> The item measures what it is intended to measure (i.e., it is *valid*). A remeasurement would order individual responses in the same way (i.e., it is *reliable*).

The remainder of this section describes techniques and procedures designed to ensure that measurement systems produce numbers with these properties. A good measurement system possesses certain properties. First, it should produce a number that is "close" to the actual property being measured; that is, it should be *accurate*. Second, if the measurement system is applied repeatedly to the same object, the measurements produced should be close to one another; that is, it should be *repeatable*. Third, the measurement system should be able to produce accurate and consistent results over the entire range of concern; that is, it should be *linear*. Fourth, the measurement system should produce the same results when used by any properly trained individual; that is, the results should be *reproducible*. Finally, when applied to the same items the measurement system should produce the same results in the future as it did in the past; that is, it should be *stable*. The remainder of this section is devoted to discussing ways to ascertain these properties for particular measurement systems. In general, the methods and definitions presented here are consistent with those described by the Automotive Industry Action Group (AIAG).

Definitions

Bias. The difference between the average measured value and a reference value is referred to as *bias*. The reference value is an agreed-upon standard, such as a standard traceable to a national standards body. When applied to attribute inspection, bias refers to the ability of the attribute inspection system to produce agreement on inspection standards. Bias is controlled by *calibration*, which is the process of comparing measurements with standards. The concept of bias is illustrated in Fig. 14.2.

Repeatability. AIAG defines repeatability as the variation in measurements obtained with one measurement instrument when used several times by one appraiser, while measuring the identical characteristic on the same part. Variation obtained when the measurement system is applied repeatedly under the same conditions is usually caused by conditions inherent in the measurement system. ASQ defines precision as "the closeness of agreement between randomly selected individual measurements or test results. NOTE: The standard deviation of the error of measurement is sometimes called 'imprecision.' " This is similar to what we are calling repeatability. Repeatability is illustrated in Fig. 14.3.

Reproducibility. Reproducibility is the variation in the average of the measurements made by different appraisers using the same measuring instrument when measuring the identical characteristic on the same part. Reproducibility is illustrated in Fig. 14.4.

Stability. Stability is the total variation in the measurements obtained with a measurement system on the same master or parts when

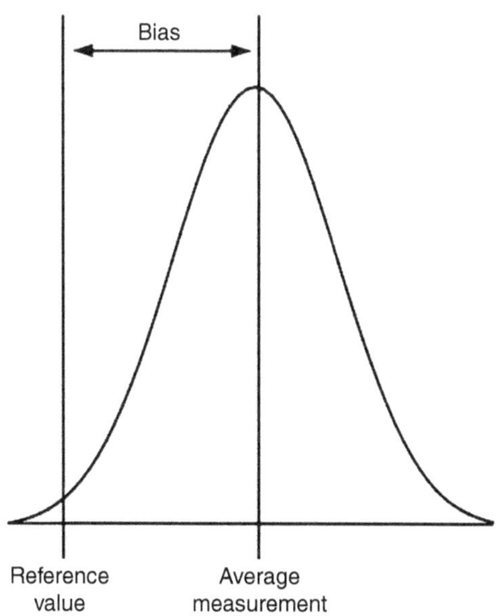

FIGURE 14.2 Bias illustrated.

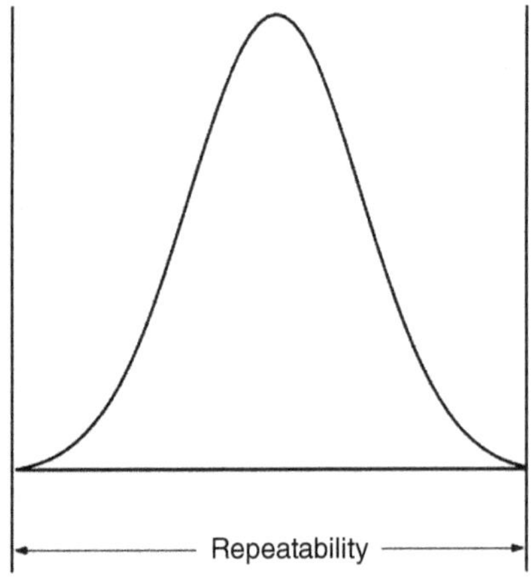

FIGURE 14.3 Repeatability illustrated.

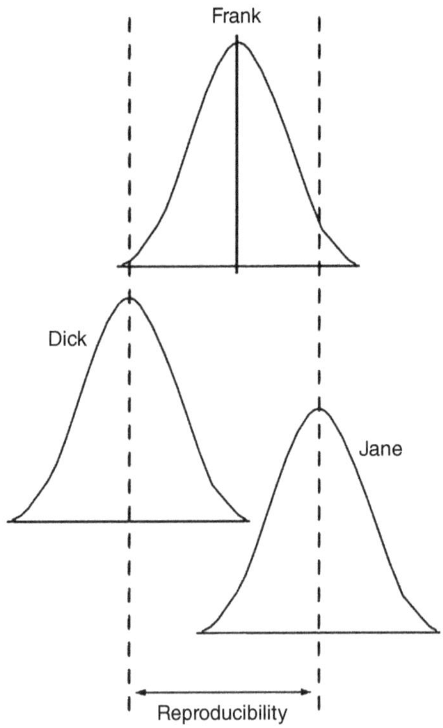

FIGURE 14.4 Reproducibility illustrated.

measuring a single characteristic over an extended time period. A system is said to be stable if the results are the same at different points in time. Stability is illustrated in Fig. 14.5.

Linearity. The difference in the bias values through the expected operating range of the gage. Linearity is illustrated in Fig. 14.6.

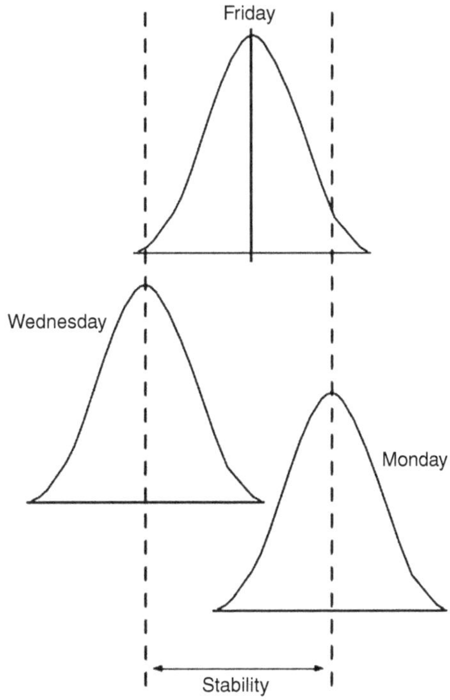

FIGURE 14.5 Stability illustrated.

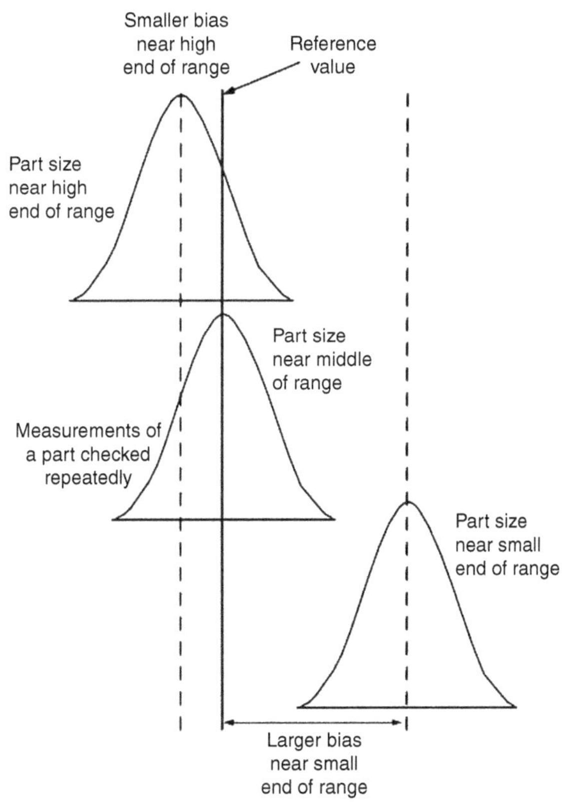

FIGURE 14.6 Linearity illustrated.

CHAPTER 15

Analyze Stage

Analyze stage objectives include (Keller, 2011a):

- Value stream analysis to determine value-producing activities
- Analyze sources of process variation
- Determine process drivers

Value Stream Analysis

Value stream analysis starts by defining the value of the product or service in the eyes of the customer. Value is alternatively defined as:

- Something the customer is willing to pay for
- An activity that changes form, fit, or function
- An activity that converts an input to an output

Value is only relevant at a specific price and point in time. One common problem in specifying value is that organizations tend to concentrate on what they are able to deliver, rather than what it the customers really want, the fallacy of "we know their needs better than they do." Of course, when they then try to improve the design or delivery process, the result can be more efficient muda, but muda none the same. The airline industry's use of hubs is a great example of this, cited frequently by Womack and Jones. The hubs serve the airlines' need to use their existing resources well, but do not provide what the customer really wants: a hassle-free journey directly from point A to point B.

Once the customer's true needs are defined, a target cost is determined, which is the cost of delivery once all the waste has been removed. If competing products or services exist in the market, the target cost can be determined by studying the customer's needs, determining the waste that exists in the current product/service, and calculating the cost of delivery once the waste has been removed. The reduced cost of delivery can lead to lower prices to the customer, or to greater margins for the business.

Many companies use *quality function deployment* (QFD), or simpler matrix diagrams, as a useful tool for defining this voice, and ensuring that the value stream is designed to meet these needs.

Once value has been determined for a given product or service, its value stream can be identified. The value stream represents the steps taken to deliver the specific product or service. (In this way it is different from a value chain, which is usually defined over broad functional areas rather than a specific product.) Value streams may be generated a number of ways. A *process map* is a useful tool for displaying the value streams, particularly when movement into functional departments is displayed via swim lanes. Once the processes are mapped out, each process step will fall into one of the following categories:

1. Steps that create value for the customer.
2. Steps that create no customer value, but are required by one or more required activities (including design, order processing, production, and delivery). These are termed type 1 muda (or more commonly *business value added activities*).
3. Steps that create no customer value. These are termed type 2 muda, and represent the proverbial low hanging fruit. They should and can be eliminated immediately.

Cycle times should also be included to assist in the analysis. Measure the "hands-on" time to complete the process step, which is best estimated using a control chart for task time. Also of interest is the "downtime" for the activity: the amount of time items sit in queue.

- If we have a dedicated process line, so that input to each process step comes only from the steps immediately preceding it in this value stream analysis, then calculate the number of items in queue, as determined through control charting.
- If the process step receives input from multiple sources, and is multitasking so that item must wait until the resource is again available, then measure the time that the item or customer waits in queue for this process step. This is best determined using a control chart for queue time.

After summing the average times, we can calculate the average process *lead time* and *velocity*. Lead time is the time needed to process all the work in progress, before new orders can be started. Velocity, sometimes known as *flow*, refers to the speed of process delivery. Speed provides flexibility and improved responsiveness to customer demands. By reducing process lead times, we can quickly respond to new orders or changes required by the customer.

Lead time is reduced, and velocity increased, when *work in progress* is reduced. (Work in progress, aka work in process, or more simply WIP, refers to order items that have been partially processed, but are not yet ready for handoff to the customer.) The rationale is simple: new orders from customers cannot be started until work (or items) in process is completed. Thus, the activity on new items is stalled. An example from a service process is a doctor's waiting room. The patients are work in progress. New patients aren't serviced by the doctor until those that arrived earlier are completed.

When a process step receives items from a single step preceding it, Little's law is used to calculate the *process lead time* by multiplying the number of items in process (in queue) by the time to complete each item. For example, if it takes 2 hours on average to complete each purchase order, and there are 10 purchase orders waiting in queue, then we need 10 times 2 equals 20 hours lead time for the process. In other words, we can't process any new orders until the 20-hour lead time has allowed the existing work in process to be completed.

When the time to complete is excessive, potential sources of delay include:

- Unnecessary process steps
- Errors requiring rework
- Non-optimal process settings
- Excessive movement of material or personnel
- Excessive wait and/or setup times

Reducing or eliminating non–value added cycle times often provides the clearest and most direct methods to reduce cycle time and lead times for better velocity. It's not uncommon for more than 50 percent of a process cycle time to consist of non–value added activities.

One of the first steps for any cycle time reduction project should be to identify and eliminate the type 2 waste; the process steps that are simply not necessary. These may include activities such as routine authorizations or approvals, or information and data collection that is not necessary.

We can force a cycle time reduction through a reduction of errors requiring rework. Practices for accomplishing this include standardization of procedures, mistake-proofing, and improvement of process capability. Each of these is covered in Chap. 16. As errors requiring rework are eliminated, the business value added inspections and approvals currently necessary may also be reduced or eliminated.

Rather than simply reducing errors, the optimization tools allow the process to operate at an improved level with respect to customer requirements. For example, the cycle time may be reduced to a point that goes beyond the elimination of complaints to the level of customer delight. These tools, discussed in Chap. 16, include designed experiments, response surface analysis, and process simulations. For example, in a service process, we might use these tools to optimize the number of trained personnel at each station within the process, based on the mix of customers arriving at a given time.

Excessive movement of material and/or personnel between process steps is non–value added, as identified in the fourth definition of waste. Efficient design of process layout reduces non–value added physical movement. For example, a lab specimen moves from the process to a laboratory, located on the other side of the plant. Once the specimen has been analyzed, the results are forwarded to the quality department, then back to the process where it is needed for the order to proceed.

A spaghetti diagram is useful to highlight the poor physical layout. The 5S tools will be used in the Improve stage to create conditions for reduced physical movement.

Waiting increases lead time by increasing both the *completion time* and the number of *items in queue*. In that regard, its impact on the lead time equation is magnified.

Process items will incur waiting when process personnel are unavailable to work on the process items. This can occur for a variety of reasons, notably:

- Multitasking
- Process steps not balanced
- Long setup times

The case of multitasking within departments, or departments that are specialized and receive process items from multiple sources, should be identified. These issues can be addressed in the Improve stage through a proper prioritization policy or dedicated personnel.

Level loading is used to balance, or match, the production rates of the process steps. When we have achieved level loading of our processes, then all work in progress (items in queue) are removed: there is no waiting as items move from one process activity to the next. This reduction in physical inventories improves cash flow and ultimately costs. The money spent on partial or completed work generates no income to the organization until the item is sold.

Inventories hide problems, such as unpredictable or low process yields, equipment failure, or uneven production levels. When inventory exists as work in progress, it prevents new orders' being processed until the WIP is completed. Although these concepts are most clearly identified with manufacturing processes, they persist in service processes, where inventory may refer to health care patients, hamburgers at the fast food, or an unfinished swimming pool under construction.

Level-loaded flow is batchless, with a shorter cycle time per unit (shorter lead time), increased flexibility, decreased response time, and an increase in the percent of value-added activities.

To balance the process steps, we first calculate the *takt* time by dividing the available resource (in units of time) by the production demand (in items). Takt is a German word meaning *metronome,* and is used to indicate the desired rhythm of the process. The takt time is posted at the cell, and the resources (machines, personnel) at each step in the process are balanced so that its cycle time equals the takt time. This level loading ensures that goods produced at each step are used immediately by the next step, ensuring a constant flow of items (or service) through the value stream. If a temporary increase in orders is received, the pace remains the same, but resources are moved to meet demand. In this way, the process steps are resourced to accommodate a pull system of management, where items are only processed when needed by the next operation.

The *lean concept of transparency,* or visual control, makes everyone aware of the current status of the process, and has been found to decrease the reaction time to waste, foster responsibility, and aid in problem solving.

While we can usually design the process and allocate standard resources for any process to meet its standard takt time, we recognize that a shift in demand will shift the takt time requirements. One way to accommodate the takt time adjustment is to shift resources.

Once personnel and equipment have been reorganized into product cells, resource allocation to these cells becomes critical. We calculate the takt time by dividing the number of hours the resource is available by the total demand. For example, if the product has an average demand of 60 units per day, and the cell works 15 hours per day (two shifts, minus breaks), then the takt time is calculated as 15 minutes (i.e., 0.25 hour per unit).

Batches are difficult to match to takt time, as they disrupt the continuous process stream. For example, a lab procedure runs 16 samples at a time through centrifuge. The first sample to reach this step waits until the 16th sample is received. If you're the patient waiting for the results of that first sample, you're not being efficiently serviced (from your perspective). The process sits idle awaiting the full 16 samples. Furthermore, the next step in the process receives all 16 samples at once, creating a large spike in demand. A better use of resources across the system is to level the load to a constant flow throughout the process.

The problem with batches is that they are not nearly as efficient, from a systems point of view and a customer's perspective, as they appear. As ironic as it may seem, a major reason our processes contain waste is because of our historical attempts to make them more efficient. One fallacy we have accepted is that we can make processes more efficient by creating specialized departments that process work in batches. These departments become efficient at what they do from a process standpoint, with economic lot quantities designed to minimize set-up time or material delivery costs, but they lack efficiency relative to specific product value streams. Waste is created in waiting for the batch to begin its departmental processing, and waste is additionally created when particular units of product, for which customers are waiting, must wait for the remainder of the batch to be processed.

The attempts to improve the departmental efficiency can create additional waste in the product value stream if the departmental efficiency produces outcomes that do not serve the customer's needs, or requires inputs that increase costs for suppliers without adding value. While standardization of product components makes the individual processes more efficient, this efficiency can come at the cost of customer value. Think about the usual new car purchase experience. You buy "the package," which includes things you are paying for but do not need, because it is more efficient for the production and delivery processes.

This batch-imposed waste is compounded if changes occur in design or customer needs, as the work in progress (WIP) or final good inventories require rework or become scrap. Note that these concepts are not limited to manufacturing; businesses in the service sector can also generate waste. Think of the hamburgers cooked in advance, waiting for an order, or checking account statements that come at the end of the month, long after you could possibly prevent an overdraw.

Three common reasons we are "forced" to consider batches are:

1. When the cost of movement of material is significant
2. When the setup time dominates the per item cycle time
3. When the process is designed for multiple items

An example of the first case is shipping a batch of items, when the customer really only wants one or a few items. The customer has to accept inventory that they do not want, or may wait until they need several items before placing an order. In *lean*, we try to reduce the space between supplier and customer to reduce costs of movement. Offshore production efficiencies may be less than perceived if true costs of consumer shipments, consumers holding unused inventory, and consumers waiting for delayed shipments are considered.

When processes are designed to produce multiple items, they may be inefficient for small batches. Most modern kitchen ovens are designed with large capacity to cook the turkey, the stuffing, and the potatoes at the same time on Thanksgiving. But what if we want to bake cookies? The boxes of prepared mixes are meant for dozens of cookies, as are the ovens, even though we would often prefer to eat only a few freshly baked cookies tonight, then a few freshly baked cookies tomorrow, and so on. In manufacturing operations, the term *monument* refers to purchased equipment that was designed for large capacity, and restricts our ability to make lean small batches.

When setup time dominates, it's natural to process as many items as feasible to spread the setup costs across the batch. This was common in the automotive industry pre-1980s, before they made concerted efforts to reduce setup times. In that same era, printing presses required elaborate setup procedures. Publishers were economically forced to order large quantities to keep unit price low. This resulted in large inventories, a disincentive to revise a book with new material. If the setup time is reduced, then smaller batch sizes would be affordable, as is now commonly practiced in the printing industry.

Setup time is defined as *the time to change from the last item of the previous order to the first good item of the next order.* When analyzing setup activities, note whether the activity is internal or external. Internal setup activities require an inactive (shut down) process, meaning that no orders can be run while the setup activity is taking place. External setup activities may be done while the process is operational. They are offline activities. Convert internal activities to external wherever possible.

Setup includes preparation, replacement, location, and adjustment activities:

- *Preparation* refers to the tasks associated with getting or storing tools or WIP needed for the process. For example, retrieving printer paper from the closet, downloading the process instructions on the computer, moving completed items to the next process step, starting up software that we need to process the order, and so on. Some suitable actions to reduce the time associated with preparation include:
  - Convert from departments to work cells to minimize the time required to move the finished product to the next process step.
  - Store tools and materials locally, such as advocated by the 5S principles.
  - Convert to *Always ready to go.* Make the software or instructions instantly accessible.
- *Replacement* refers to the tasks associated with adding or removing items or tools, for example, the movement of test fixtures, loading of new material into the hopper, and loading paper in the copy machine. Actions to reduce replacement times include:
  - Simplify setups. Reduce the number of steps required, such as through a redesign of fixtures.
  - Establish commonality of setups for product families. When we establish the same setup procedures for multiple items, we naturally have fewer instances of change required, reducing the setup time. This is the 5S tool of standardization, which will be discussed in the Improve stage, in Chap. 16. The process is simplified by reducing the number of "special items" that are processed: the higher the process complexity, the longer the cycle time. Henry Ford, in offering the first affordable automobile, realized the efficiency advantages offered by standardization.

His motto, "any color you want, so long as it's black," exemplified the notion of standardization. The modern approach to standardization is not that we need to limit options, but rather that we need to recognize the advantages of simplified processes, and seek to remove "special cases" where they provide little value at the cost of increased cycle times. Decision points and subsequent parallel paths on flowcharts provide indication of process complexities that can sometimes be avoided. By grouping parts or services into families, we can recognize that there are common methods that can be applied, thus simplifying processes and reducing overall cycle times.
  - The 5S tools of sorting and straightening also help to reduce movement and wait times.
- *Location* tasks are those associated with positioning or placement during setup. Examples include setting temperature profiles for heating, adjusting cutoff length for specific product, and placing the chunk of deli meat in the slicer. Actions to reduce the time associated with location include:
  - Poka yoke (mistake proofing the process), as discussed in Chap. 16
  - Commonality of setups as previously mentioned (the 5S tool of standardization)
- *Adjustment* refers to tasks associated with ensuring correct process settings. Examples include monitoring the temperature of a furnace, checking cutoff length, and proofing copy before printing. A suitable action to reduce adjustment time is process control. If we can improve the repeatability of the process, then the adjustments will not be necessary. Often this is achieved through robust design methods, as discussed in Chap. 16.

Although it may be your initial tendency, don't limit your value stream to the walls of your organization. Fantastic sums of money have been saved by evaluating value streams as they move from supplier to customer, often because of discovering mistaken concepts of value or attempts to achieve operational savings that diminish the customer value.

Analyze Sources of Process Variation

The sources of process variation (SPC) control charts from the Measure stage provide evidence of either a stable (i.e., in control) or unstable (i.e., out of control) process. It is critical to first differentiate between these two types of variation, as the improvement strategies are necessarily different for each. For stable processes, the common cause variation built into the process can only be reduced through a fundamental change to the system. When the process is out of control, the special cause creating the unstable condition during a specific time period must be addressed and removed to attain a stable process, which can then be improved (if needed) as noted above.

At the business level, customer data may be analyzed to establish relationships between customer satisfaction and the internal processes used to deliver the customer experience. A common tool used for this analysis is Quality Function Deployment (QFD). In many cases simplified versions of these tools are used with comparable results. This identification of key internal processes and metrics feeds into operations and process-level projects in a top-down deployment strategy. This feedback from business-level projects into the definition of operations- and process-level projects is a key to successfully harnessing the power of Six Sigma.

Quality Function Deployment

Once information about customer expectations has been obtained, techniques such as QFD can be used to link the voice of the customer directly to internal processes.

Tactical quality planning involves developing an approach to implementing the strategic quality plan. One of the most promising developments in this area has been policy deployment. Sheridan (1993) describes policy deployment as the development of a measurement-based system as a means of planning for continuous quality improvement throughout all levels of an organization. Although it was originally developed by the Japanese, American companies also use policy deployment because it clearly defines the long-range direction of company development, as opposed to short term.

QFD is a customer-driven process for planning products and services. It starts with the voice of the customer, which becomes the basis for setting requirements. QFD matrices, sometimes called "the house of quality," are graphical displays of the result of the planning process. QFD matrices vary a great deal and may show such things as competitive targets and process priorities. The matrices are created by interdepartmental teams, thus overcoming some of the barriers that exist in functionally organized systems.

QFD is also a system for design of a product or service based on customer demands, a system that moves methodically from customer requirements to specifications for the product or service. QFD involves the entire company in the design and control activity. Finally, QFD provides documentation for the decision-making process. The QFD approach involves four distinct phases (King, 1987):

- *Organization phase.* Management selects the product or service to be improved, appoints the appropriate interdepartmental team, and defines the focus of the QFD study.

- *Descriptive phase.* The team defines the product or service from several different directions such as customer demands, functions, parts, reliability, cost, and so on.
- *Breakthrough phase.* The team selects areas for improvement and finds ways to make them better through new technology, new concepts, better reliability, cost reduction, etc., and monitors the bottleneck process.
- *Implementation phase.* The team defines the new product and how it will be manufactured.

QFD is implemented through the development of a series of matrices. In its simplest form QFD involves a matrix that presents customer requirements as rows and product or service features as columns. The cell, where the row and column intersect, shows the correlation between the individual customer requirement and the product or service requirement. This matrix is sometimes called the "requirement matrix." When the requirement matrix is enhanced by showing the correlation of the columns with one another, the result is called the "house of quality." Figure 15.1 shows one commonly used house of quality layout.

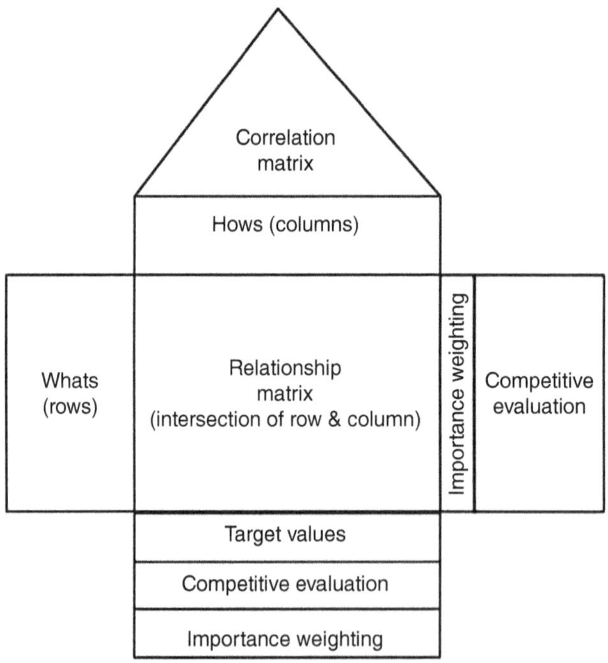

Figure 15.1 The house of quality.

The house of quality relates, in a simple graphical format, customer requirements, product characteristics, and competitive analysis. It is crucial that this matrix be developed carefully since it becomes the basis of the entire QFD process. By using the QFD approach, the customer's demands are "deployed" to the final process and product requirements.

One rendition of QFD, called the Macabe approach, proceeds by developing a series of four related matrices (King, 1987): product planning matrix, part deployment matrix, process planning matrix, and production planning matrix. Each matrix is related to the previous matrix, as shown in Fig. 15.2.

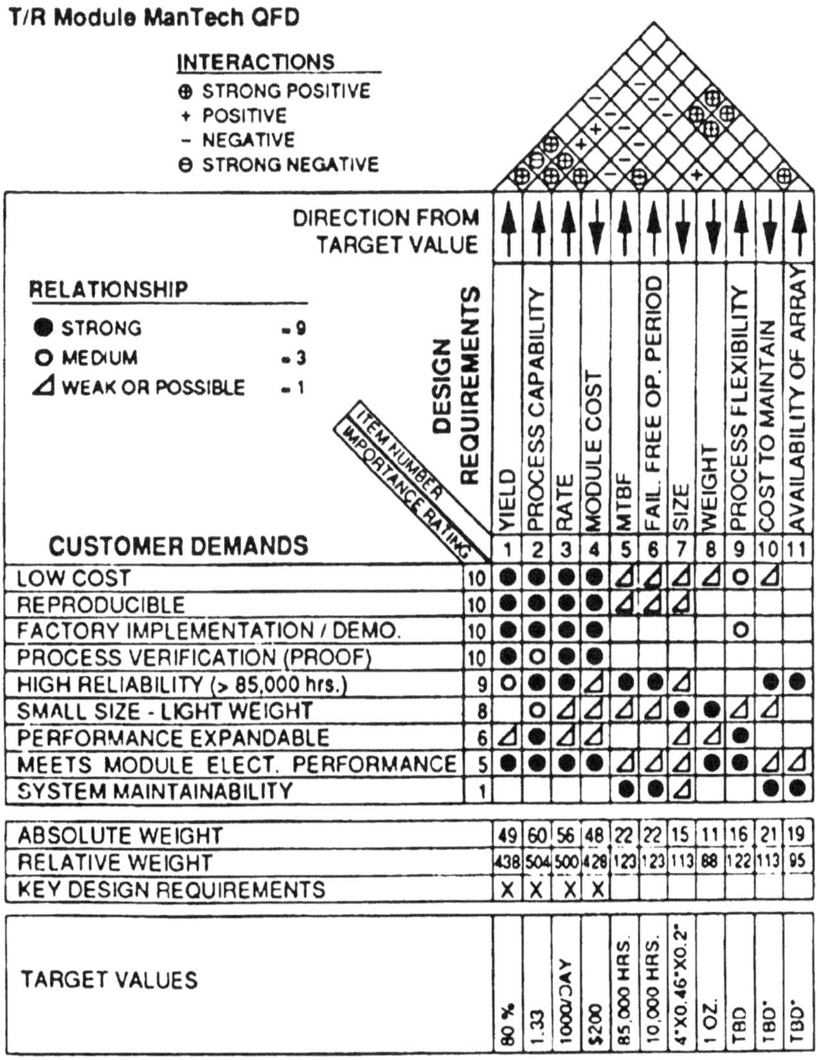

Figure 15.2 QFD matrix for an aerospace firm (Wahl and Bersbach, 1991).

Data Collection and Review of Customer Expectations, Needs, Requirements, and Specifications

Another approach to QFD is based on work done by Yoji Akao. Akao (1990, pp. 7–8) presents the following 11-step plan for developing the quality plan and quality design, using QFD.

1. First, survey both the expressed and latent quality demands of consumers in your target marketplace. Then decide what kinds of "things" to make.
2. Study the other important characteristics of your target market and make a demanded quality function deployment chart that reflects both the demands and characteristics of that market.
3. Conduct an analysis of competing products on the market, which we call a competitive analysis. Develop a quality plan and determine the selling features (sales points).
4. Determine the degree of importance of each demanded quality.
5. List the quality elements and make a quality elements deployment chart.
6. Make a quality chart by combining the demanded quality deployment chart and the quality elements deployment chart.
7. Conduct an analysis of competing products to see how other companies perform in relation to each of these quality elements.
8. Analyze customer complaints.
9. Determine the most important quality elements as indicated by customer quality demands and complaints.
10. Determine the specific design quality by studying the quality characteristics and converting them into quality elements.
11. Determine the quality assurance method and the test methods.

Cause-and-Effect Diagrams

Process improvement involves taking action on the causes of variation. With most practical applications the number of possible causes for any given problem can be huge. Dr. Kaoru Ishikawa developed a simple method of graphically displaying the causes of any given quality problem. His method is known by several names: the *Ishikawa diagram*, the *fishbone diagram*, and the *cause-and-effect diagram*.

Cause-and-effect diagrams are tools that are used to organize and graphically display all of the knowledge a group has brainstormed related to a particular problem. The brainstormed ideas are categorized into rational categories and subcategories. The cause-and-effect diagram is drawn to depict the relationships of the data in each category and each of its subcategories.

A good cause-and-effect diagram will have many "bones." If your cause-and-effect diagram doesn't have a lot of smaller branches, then the understanding of the problem is somewhat superficial.

Cause-and-effect diagrams come in several basic types. The dispersion analysis type is created by repeatedly asking "Why does this dispersion occur?" For example, we might want to know why all of our fresh peaches don't have the same color.

The production process class cause-and-effect diagram uses production processes as the main categories, or branches, of the diagram. The processes are shown joined. Other common themes in cause-and-effect diagrams include those with main branches designated using the 5M and E (machine, method, manpower, material, measurement, and environment) and the 4P (people, policy, procedure, plant).

An example is shown in Fig. 15.3.

The cause enumeration cause-and-effect diagram simply displays all possible causes of a given problem grouped according to rational categories. This type of cause-and-effect diagram lends itself readily to the brainstorming approach.

Cause-and-effect diagrams have a number of uses. Creating the diagram is an education in itself. Organizing the knowledge of the group serves as a guide for discussion and frequently inspires more ideas. The cause-and-effect diagram, once created, acts as a record of your research. Simply record your tests and results as you proceed. If the true cause is found to be something that wasn't on the original diagram, it should be added. Finally, the cause-and-effect diagram is a display of your current level of understanding. It is a good idea to post the cause-and-effect diagram in a prominent location for handy reference in the future.

A variation of the basic cause-and-effect diagram, developed by Dr. Ryuji Fukuda of Japan, is cause-and-effect diagrams with the addition of cards, or CEDAC. The main difference is that the group gathers ideas outside of the meeting room on small cards, as well as in group meetings. The cards also serve as a vehicle for gathering input from people who are not in the group; they can be distributed to anyone involved with the process. Often the cards provide more information than the brief entries on a standard cause-and-effect diagram. The cause-and-effect diagram is built by actually placing the cards on the branches.

Scatter Diagrams

A scatter diagram is a plot of one variable versus another: the *independent variable* is shown on the horizontal (bottom) axis; the *dependent variable* is shown on the vertical (side) axis.

Scatter diagrams are used to evaluate the correlation of one variable with the other. The premise is that the independent variable is causing a change in the dependent variable, although strictly speaking cause and effect cannot be proven with statistics alone. Scatter plots are used to

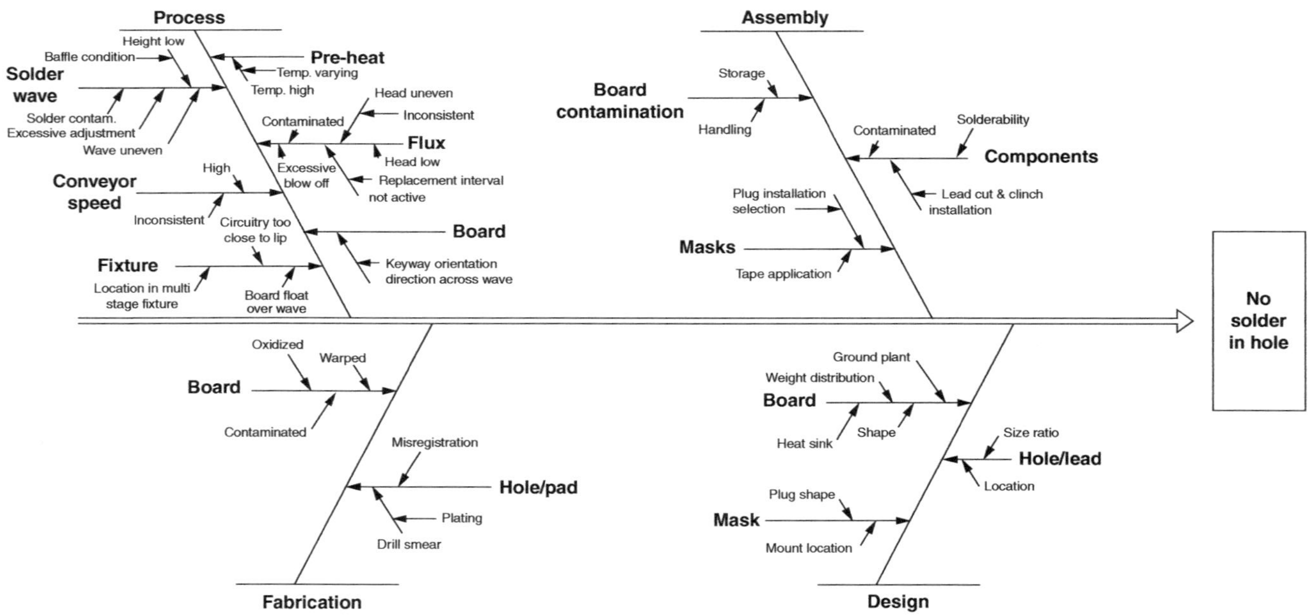

Figure 15.3 Process production class cause-and-effect diagram.

answer such questions as "Is vendor A's material machine better than vendor B's?" "Does the length of training have anything to do with the amount of scrap an operator makes?" and so on.

How to Construct a Scatter Diagram

1. Gather several paired sets of observations, preferably 20 or more. A paired set is one where the dependent variable can be directly tied to the independent variable.
2. Find the largest and smallest independent variable and the largest and smallest dependent variable.
3. Construct the vertical and horizontal axes so that the smallest and largest values can be plotted.
4. Plot the data by placing a mark at the point corresponding to each *X*–*Y* pair. If more than one classification is used, you may use different symbols to represent each group.

Example of a Scatter Diagram. The orchard manager has been keeping track of the weight of peaches on a day-by-day basis. The data is collected in pairs as shown in Table 15.1, so that for each peach its weight and the number of days on the tree were recorded.

The independent variable, *X*, is the number of days the fruit has been on the tree. The dependent variable, *Y*, is the weight of the peach. The scatter diagram is shown in Fig. 15.4.

Pointers for Using Scatter Diagrams

- Scatter diagrams display different patterns that must be interpreted; Fig. 15.5 provides a scatter diagram interpretation guide.
- Be sure that the independent variable, *X*, is varied over a sufficiently large range. When *X* is changed only a small amount, you may not see a correlation with *Y* even though the correlation really does exist.
- If you make a prediction for *Y* for an *X* value that lies outside of the range you tested, be advised that the prediction is highly questionable and should be tested thoroughly. Predicting a *Y* value beyond the *X* range actually tested is called *extrapolation*.
- Keep an eye out for the effect of variables not included in the analysis. Often, an uncontrolled variable will wipe out the effect of your *X* variable. It is also possible that an uncontrolled variable will be causing the effect and you will mistake the *X* variable you are controlling as the true cause. This problem is much less likely to occur if you choose *X* levels at random. An example of this is our peaches. It is possible that any number of variables changed

Data ID	Days on Tree	Weight (ounces)
1	75	4.5
2	76	4.5
3	77	4.4
4	78	4.6
5	79	5.0
6	80	4.8
7	80	4.9
8	81	5.1
9	82	5.2
10	82	5.2
11	83	5.5
12	84	5.4
13	85	5.5
14	85	5.5
15	86	5.6
16	87	5.7
17	88	5.8
18	89	5.8
19	90	6.0
20	90	6.1

(Pyzdek, 1990)

TABLE 15.1 Raw Data for Scatter Diagram

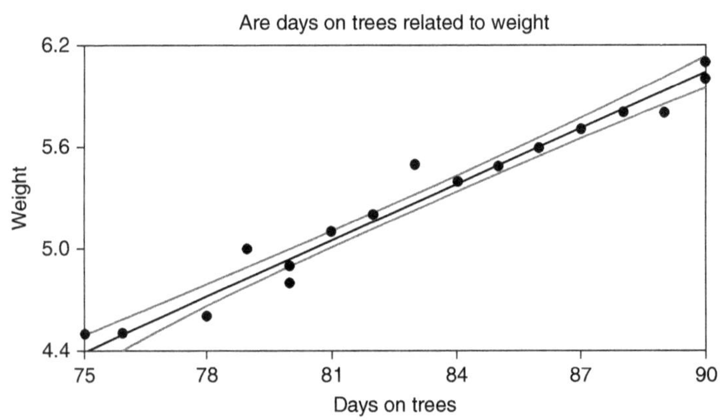

Regression function: $y = -3.847 + 0.110 * x$ Correlation coefficient: 0.980 Function explains 96.038% of variation	Regression: 1 (df) 4.907 (SS) 4.907 (MS) Residual: 18 (df) 0.202 (SS) 0.011 (MS) F: 436.296 Prob F: 0.000

FIGURE 15.4 Sample scatter diagram.

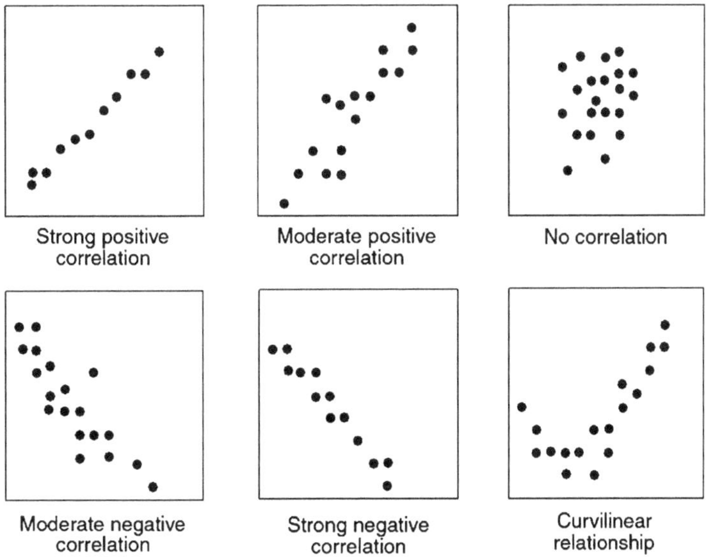

FIGURE 15.5 Scatter diagram interpretation guide (Pyzdek, 1990).

steadily over the time period investigated. It is possible that these variables, and not the independent variable, are responsible for the weight gain (e.g., was fertilizer added periodically during the time period investigated?).

- Beware of "happenstance" data! Happenstance data is data that was collected in the past for a purpose other than for constructing a scatter diagram. Since little or no control was exercised over important variables, you may find nearly anything. Happenstance data should be used only to get ideas for further investigation, never for reaching final conclusions. One common problem with happenstance data is that the variable that is truly important is not recorded. For example, records might show a correlation between the defect rate and the shift. However, perhaps the real cause of defects is the ambient temperature, which also changes with the shift.
- If there is more than one possible source for the dependent variable, try using different plotting symbols for each source. For example, if the orchard manager knew that some peaches were taken from trees near a busy highway, he could use a different symbol for those peaches. He might find an interaction; that is, perhaps the peaches from trees near the highway have a different growth rate from those from trees deep within the orchard. This technique is known as *stratification*.

Although it is possible to do advanced analysis without plotting the scatter diagram, this is generally bad practice. This misses the enormous learning opportunity provided by the graphical analysis of the data.

Determine Process Drivers

Process drivers refer to the factors that have the largest influence on the process. For any business process, there are likely to be many factors that contribute to process variation. Process improvement will require either a reduction in process variation, or a movement of the process centerline to a more favorable setting. In either case, focusing on the key process drivers will facilitate this improvement.

Correlation and Regression Analysis

Correlation analysis (the study of the strength of the linear relationships among several variables) and regression analysis (modeling the relationship between one or more independent variables and a dependent variable) are closely related to the scatter diagram. A regression problem considers the frequency distributions of one variable when another is held fixed at each of several levels. A correlation problem considers the joint variation of two variables, neither of which is restricted by the experimenter. Correlation and regression analyses are designed to assist the engineer in studying cause and effect. They may be employed in all stages of the problem-solving and planning process. Of course, statistics cannot by themselves establish cause and effect. Proving cause and effect requires sound scientific understanding of the situation at hand. The statistical methods described in this section assist in performing this analysis.

Linear Models. A simple linear model is a mathematical expression of the association between two variables, x and y. A *linear relationship* simply means that a change of a given size in x produces a proportionate change in y. Linear models have the form:

$$y = a + bx$$

where a and b are constants. The equation simply says that when x changes by one unit, y will change by b units. This relationship can be shown graphically.

In the scatter diagram shown in Fig. 15.4, $a = 3.847$ and $b = 0.110$. The term a is called the *intercept* and b is called the *slope*. When $x = 0$, y is equal to the intercept.

Many types of associations are non-linear. For example, over a given range of x values y might increase, and for other x values y might decrease. This *curvilinear relationship* is shown in Fig. 15.6.

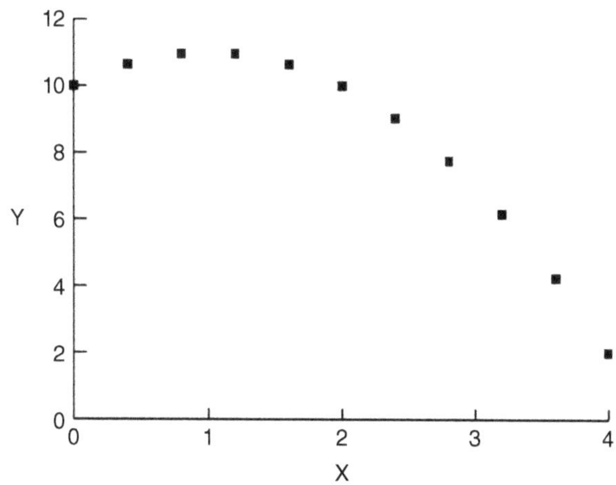

FIGURE 15.6 Scatter diagram of a curvilinear relationship.

In this case, y increases when x is less than 1, and decreases for larger values of x. A wide variety of processes produce such relationships. One common method for analyzing non-linear responses is to break the response into segments that are piecewise linear, and then analyze each piece separately. For example, in Fig. 15.6, y is roughly linear and increasing over the range $0 < x < 1$ and roughly linear and decreasing over the range $x > 1$. Of course, if you have access to powerful statistical software, non-linear forms can be analyzed directly.

When conducting regression and correlation analysis, we can distinguish two main types of variables. One type we call predictor variables or independent variables; the other, response variables or dependent variables. A predictor or independent variable can either be set to a desired variable (e.g., oven temperature), or else take values that can be observed but not controlled (e.g., outdoor ambient humidity). As a result of changes that are deliberately made, or simply take place in the predictor variables, an effect is transmitted to the response variables (e.g., the grain size of a composite material). We are usually interested in discovering how changes in the predictor variables affect the values of the response variables. Ideally, we hope that a small number of predictor variables, will "explain" nearly all of the variation in the response variables.

In practice, it is sometimes difficult to draw a clear distinction between independent and dependent variables. In many cases it depends on the objective of the investigator. For example, a quality engineer may treat ambient temperature as a predictor variable in the study of paint quality, and as the response variable in a study of clean room particulates. However, the above definitions are useful in planning quality improvement studies.

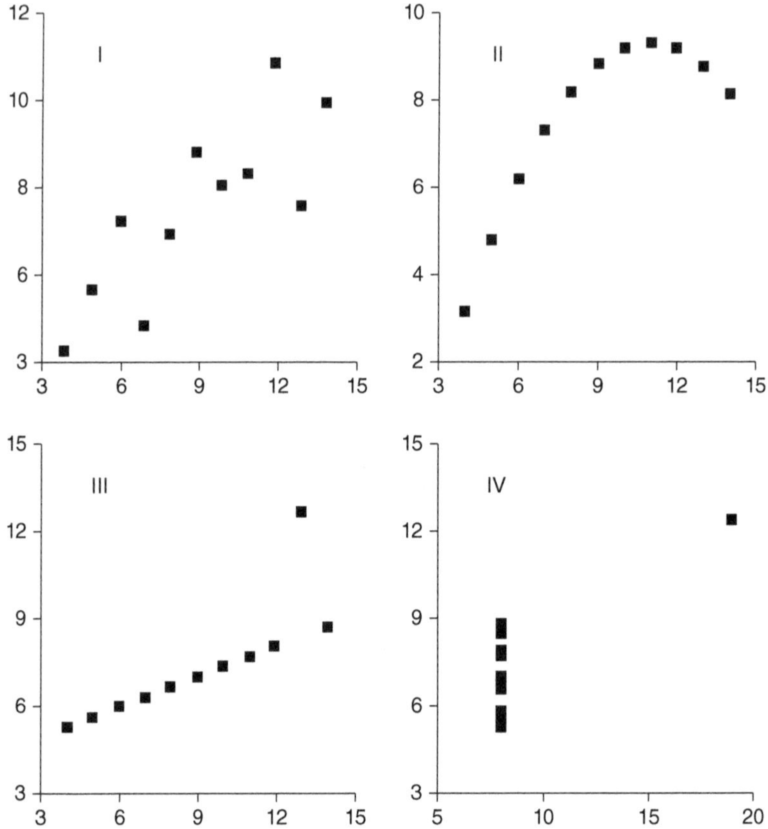

FIGURE 15.7 Illustration of the value of scatter diagrams (Tufte, 1983).

While the numerical analysis of data provides valuable information, it should always be supplemented with graphical analysis as well. Scatter diagrams are one very useful supplement to regression and correlation analysis. The four quite different scatter diagrams in Fig. 15.7 illustrates the value of supplementing numerical analysis with scatter diagrams, as the scatter diagrams have common statistical parameters.

In other words, although the scatter diagrams clearly show four distinct processes, the statistical analysis does not. In quality work, numerical analysis alone is not enough.

Least-Squares Fit

If all data fell on a perfectly straight line, it would be easy to compute the slope and intercept given any two points. However, the situation becomes more complicated when there is "scatter" around the line. That is, for a given value of x, more than one value of y appears. When this occurs, we have error in the model. Figure 15.8 illustrates the concept of error.

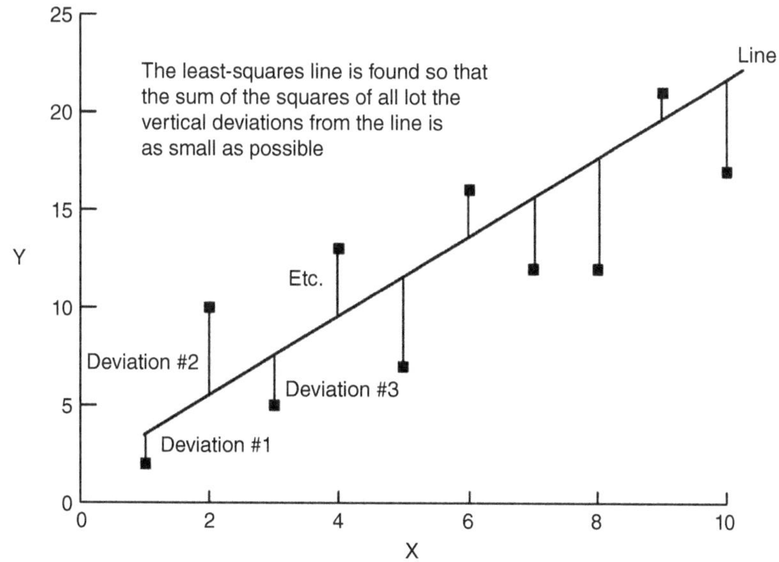

FIGURE 15.8 Error in the linear model.

The model for a simple linear regression with error is:

$$y = a + bx + \mathrm{e}$$

where e represents error. Generally, assuming the model adequately fits the data, errors are assumed to follow a normal distribution with a mean of 0 and a constant standard deviation. The standard deviation of the errors is known as the *standard error*. We discuss ways of verifying our assumptions about the error below.

When error occurs, as it does in nearly all "real-world" situations, there are many possible lines that might be used to model the data. Some method must be found that provides, in some sense, a "best-fit" equation in these everyday situations. Statisticians have developed a large number of such methods. The method most commonly used finds the straight line that minimizes the sum of the squares of the errors for all of the data points. This method is known as the "least-squares" best-fit line. In other words, the least-squares best-fit line equation is $y_i' = a + bx$, where a and b are found so that the sum of the squared deviations from the line is minimized. Most spreadsheets and scientific calculators have a built-in capability to compute a and b.

This discussion shows how a single independent variable is used to model the response of a dependent variable. This is known as *simple linear regression*. It is also possible to model the dependent variable in terms of two or more independent variables; this is known as *multiple linear regression*.

The mathematical model for multiple linear regression has additional terms for the additional independent variables, for example:

$$y = b_0 + b_1x_1 + b_2x_2 + e$$

where y is the dependent variable, x_1 and x_2 are independent variables, b_0 is the intercept, b_1 is the coefficient for x_1, b_2 is the coefficient for x_2, and e is the error. More variables can be added to the model as needed.

Example. A restaurant conducted surveys of 42 customers, obtaining customer ratings on staff service, food quality, and overall satisfaction with their visit to the restaurant. Figure 15.9 shows the regression analysis output from a spreadsheet regression function.

The data consist of two independent variables, staff and food quality, and a single dependent variable, overall satisfaction. The basic premise is that the quality of staff service and the food are *causes* and the overall satisfaction score is an *effect*.

Interpretation of Computer Output for Regression Analysis

The regression output is interpreted as follows:

Multiple R. The multiple correlation coefficient. It is the correlation between y (actual satisfaction) and y' (satisfaction estimated from the model). For the example, multiple R = 0.847, which indicates that y and y' are highly correlated, which implies that there is an association between overall satisfaction and the quality of the food and service.

SUMMARY OUTPUT

Regression statistics	
Multiple R	0.847
R square	0.717
Adjusted R square	0.703
Standard error	0.541
Observations	42

ANOVA

	df	*ss*	*ms*	*F*	*Significance F*
Regression	2	28.97	14.49	49.43	0.00
Residual	39	11.43	0.29		
Total	41	40.40			

	Coefficients	*Standard error*	*t Stat*	*P-value*	*Lower 95%*	*Upper 95%*
Intercept	-1.188	0.565	-2.102	0.042	-2.331	-0.045
Staff	0.902	0.144	6.283	0.000	0.611	1.192
Food	0.379	0.163	2.325	0.025	0.049	0.710

FIGURE 15.9 Microsoft Excel regression analysis output.

R Square. The square of multiple R, it measures the proportion of total variation about the mean explained by the regression. For the example, $R^2 = 0.717$, which indicates that the fitted equation explains 71.7 percent of the total variation about the average satisfaction level.

Adjusted R Square. A measure of R^2 "adjusted for degrees of freedom," which is necessary when there is more than one independent variable.

Standard error. The standard deviation of the residuals. The *residual* is the difference between the observed value of y and the predicted value y' based on the regression equation.

Observations. Refer to the number of cases in the regression analysis, or n.

ANOVA, or ANalysis Of Variance. A table examining the hypothesis that the variation explained by the entire regression is zero. If this is so, then the observed association could be explained by chance alone. The rows and columns are those of a standard one-factor ANOVA table. For this example, the important item is the column labeled "Significance F." The value shown, 0.00, indicates that the probability of getting these results due to chance alone is less than 0.01; that is, the association is probably not due to chance alone. Note that the ANOVA applies to the entire *model*, not to the individual variables. In other words, the ANOVA tests the hypothesis that the explanatory power of all of the independent variables combined is zero.

The next table in the output examines each of the terms in the linear model separately. The *intercept* is as described above; it corresponds to our term a in the linear equation. Our model uses two independent variables. In our terminology, staff = b_1, food = b_2. Thus, reading from the *coefficients* column, the linear model is:

$$\text{Satisfaction} = -1.188 + 0.902 * \text{staff} + 0.379 * \text{food} + \text{error}$$

The remaining columns test the hypotheses that each coefficient in the model is actually zero.

Standard error column. Gives the standard deviations of each term, that is, the standard deviation of the intercept = 0.565, etc.

t Stat column. The coefficient divided by the t statistic; that is, it shows how many standard deviations the observed coefficient is from zero.

P-value. Shows the area in the tail of a t distribution beyond the computed t value. For most experimental work a P value less that 0.05 is accepted as an indication that the coefficient is significantly different from zero.

Lower 95% and upper 95% columns. A 95 percent confidence interval on the coefficient. If the confidence interval does not include zero, we will reject the hypothesis that the coefficient is zero.

Analysis of Residuals

The experimenter should carefully examine the residuals. The residuals represent the variation "left over" after subtracting the variation explained by the model. The examination of residuals is done to answer the question "What might explain the rest of the variation?" Potential clues to the answer might arise if a pattern can be detected. For example, the experimenter might notice that the residuals tend to be associated with certain experimental conditions, or they might increase or decrease over time. Other clues might be obtained if certain residuals are *outliers*, that is, errors much larger than would be expected from chance alone. Residuals that exhibit patterns or that contain outliers are evidence that the linear model is incorrect. There are many reasons why this might be so. The response might be non-linear. The model may leave out important variables. Or, our assumptions may not be valid.

There are four common ways of plotting the residuals:

1. Overall
2. In time sequence (if the order is known)
3. Against the predicted values
4. Against the independent variables

Overall Plot of Residuals. When the assumptions are correct, we expect to see residuals that follow an approximately normal distribution with zero mean. An overall plot of the residuals, such as a histogram, can be used to evaluate this. It is often useful to plot *standardized residuals* rather than actual residuals. Standardized residuals are obtained by dividing each residual by the standard error; the result is the residual expressed in standard deviations. The standardized residuals should then be plotted on a normal probability plot to verify normality.

When performing the other three types of analysis on the list, the experimenter should look for any non-randomness in the patterns. Figure 15.10 illustrates some common patterns of residuals behavior.

Pattern #1 is the overall impression that will be conveyed when the model fits satisfactorily. Pattern #2 indicates that the size of the residual increases with respect to time or the value of the independent variable. It suggests the need for performing a transformation of the y values prior to performing the regression analysis. Pattern #3 indicates that the linear effect of the independent variable was not removed, perhaps due to an error in calculations. Pattern #4 appears when a linear model is fitted to curvilinear data. The solution is to perform a linearizing transformation of the y's, or to fit the appropriate non-linear model.

In addition to the above, the analyst should always bring his or her knowledge of the process to bear on the problem. Patterns may become

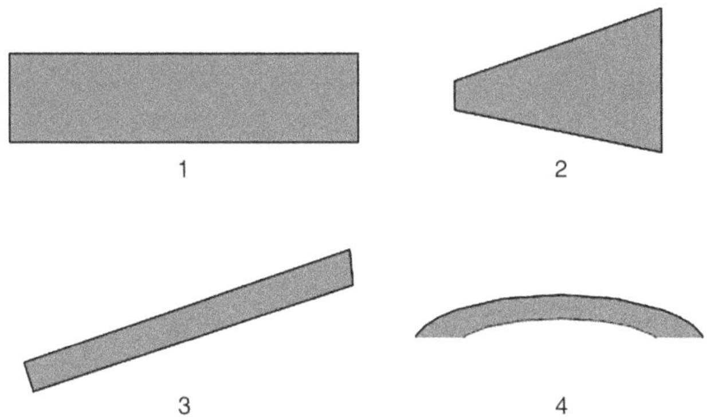

Figure 15.10 Residual patterns.

apparent when the residuals are related to known laws of science, even though they are not obvious using statistical rules only. This is especially true when analyzing outliers, that is, standardized residuals greater than 2.5 or so.

Designed Experiments

While data mining can be used for a variety of purposes, including understanding buying patterns and identifying major factors influencing costs and profitability, it cannot properly confirm cause and effect. Data mining provides a view of the seemingly complex relationships between the many factors that possibly affect outcomes, but these patterns are mere suspicions that can become the basis of additional project activity. By itself, a data mining analysis is the *happenstance data* referred to previously (in the scatter diagram discussion).

The proper tool for collecting data useful for correlation and regression analysis is the designed experiment. A project team brainstorms to produce a list of the potential process factors. From this large list, the team selects five to seven factors to include in the experiment. If it turns out an important factor was not included, the regression will include a large error term in the model and the R-square value will be low.

Once they have determined the factors, the team will conduct an experiment by varying each of the factors (i.e., the independent variables), moving several at a time, over a wide range and measure the response (or responses, i.e., the dependent variables) of the process. By manipulating the factors over a wide range they have the best chance of detecting a change in the response that may be otherwise too subtle to detect.

They then use the multiple regression techniques to estimate the effect of each factor, as well as the interactions between selected factors.

A designed experiment differs from the traditional experiment many of us learned in grade school. In the traditional experiment, one factor is

Trial	Cycle Time	Type	Method	Score
1	Low	Generic	Email	35
2	High	Generic	Email	21
3	Low	Personal	Email	28
4	Low	Generic	Phone	27

TABLE 15.2 One Factor at a Time Data Collection Scheme

varied at a time to estimate its effect. Table 15.2 lists three factors, with two possible settings for each:

- Response cycle time [low, high]
- Personalized response type [generic or personal]
- Response method [email or phone]
- An initial baseline condition is taken by measuring the response [customer satisfaction score] when each of the factors is set at its low level. A second trial is then run to estimate the effect of the first factor: response cycle time. The difference between the Trial 2 response and the baseline (Trial 1) is assumed to be the effect of the factor varied. In this case, it appears that raising the response cycle time from low to high results in a decrease in customer satisfaction of 14 units.

Likewise, the effect of a personalized response type may be estimated by comparing Trial 3 with Trial 1; the effect of response method is estimated by comparing Trial 4 with Trial 1. In this way, the effect of the personalized response is estimated as a decrease in customer satisfaction of 7 units, and the effect of phone versus email as a decrease of 8 units in customer satisfaction score.

Based on these observations, customer satisfaction may be maximized by setting the factors as follows: cycle time: low; personalized response: generic; response type: email.

The problem with this traditional one-factor-at-a-time experiment is that it ignores the effect of interactions. Figure 15.11 displays the results from a designed experiment of the same process. You can see that at the high cycle time setting (shown by the line labeled *330.000*), the satisfaction score of 21.0 was observed when the response type is set at the *generic* (*no* personal response) condition. This is Trial 2 from above, and is circled on the graph. Trial 3 is shown on the line labeled *210.000* at the *yes* personal response condition.

On the 210.000 response cycle time line, moving from *no* personalized response to *yes* personalized response (left to right along the line), there is very little change in customer satisfaction score. In other words, there was

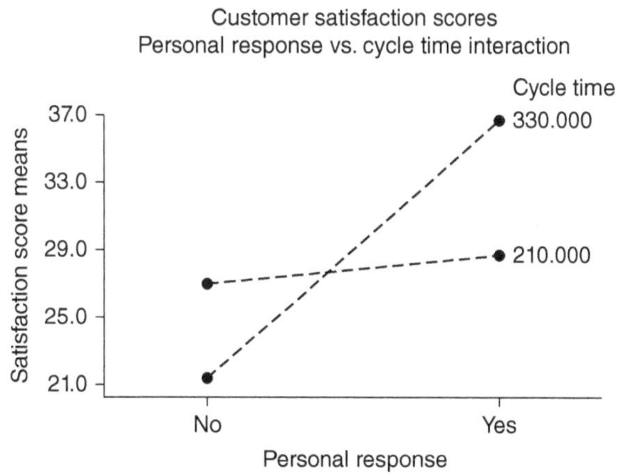

FIGURE 15.11 Results of designed experiment to estimate interaction between personal response and cycle time factors.

no observed difference between personalized response and generic response at that level of response cycle time.

On the 330.000 response cycle time line, moving from *no* personalized response to *yes* personalized response (left to right along the line), there is very *large* change in customer satisfaction score. This implies that personalized response is a significant contributor to the change in customer satisfaction score, but apparently only at the low (210) level of response cycle time.

The implication is that the estimate of the effect of personalized response changes depending on whether the effects are measured at low cycle time or high cycle time. This provides evidence of an *interaction* between personalized response and cycle time.

When interactions are ignored, haphazard results may be obtained from improvement efforts:

- A factor may appear unimportant if interacting factors are not changed at the same time, as shown in the example.
- A process improvement effort may only attain desired results for a period of time (so long as the interacting factors remain constant). If we don't vary the right factors, it appears an improvement has been made, but then it is gone. This can happen when there is another factor present that was not included in the experiment.
- The effect of one factor may be minimized by reducing the variation of another. This is the Taguchi approach to robust design, and can be seen by looking at the prior example in a slightly different way. Since the effect of personalized response was negligible at low response cycle times, if response cycle time is kept near its low setting then changes in personalized response wouldn't make much difference.

CHAPTER 16

Improve/Design Stage

The objectives of the Improve/Design stage include (Keller, 2011a):

- Propose one or more solutions to sponsor; quantify benefits of each; reach consensus on solution and implement.
- Define and mitigate failure modes for new process/design; define new operating/design conditions.

The Improve stage involves the deployment of methods to close the gap between the current process state and the desired state. Methods implemented must also be verified in this stage to ensure that the desired effects are achieved and can be maintained.

This stage is where the rubber meets the road, as it defines the improvements and cost reductions that sustain the program. As such, it is usually the make or break point, forcing the team to consider the un-considerable and become true agents of change. Management support at this point is critical.

At the business level, system-wide change results. This could include changes in policy, deployment of customer feedback mechanisms, changes to accounting systems to track quality costs and benefits of improvements, implementation of computerized systems to manage orders, and even elimination of complicated systems that promised improvement but delivered increased cycle times and/or costs.

At the process levels, designed experiments are conducted to determine process factor settings that improve process capability, resulting in a reduction in defects and increased throughput yields. Processes are redesigned, sometimes even eliminated, in the quest for improved performance in quality, cost, and scheduling.

At any level, the best and longest-lasting improvements are achieved when factors are identified that *predict* future outcomes. These critical factors can then be controlled to prevent problems before they occur.

Define New Operating/Design Conditions

The flowcharts and process maps introduced in the Define stage are useful now to develop the flow and responsibilities for the new process. Additional experimental designs can also be conducted to determine optimal operating conditions that maximize or minimize the response (as desirable).

The Lean tools to reduce or eliminate non–value added activities, including unnecessary movement of personnel or material, are also deployed at this time.

Cycle times may be improved by reducing movement of personnel or material, affecting the physical space in which the process takes place: offices may be redesigned, departments moved or reassigned, even entire factories moved closer to their customers. Movement analysis is aided by the use of spaghetti diagrams. Application of the 5S tools creates the conditions necessary to reduce the movement of material and personnel. The outcomes of space reduction include:

- *Decreased distance from "supplier" to "customer."* Both internal and external suppliers and customers are affected. This relocation reduces the wastes of unnecessary movement of material and wait time for material.
- *Less departmentalization, more multifunction work cells.* Within company walls, you may reassign individuals so they work within multifunctional work cells, rather than functional departments. This improves the flow of the process, so work is not batched up at each department. The work cell approach, in which each process step is close in location to its next step, also increases visibility of problems as the item moves from one step in the process to the next.
- *Reduced overhead costs and reduced need for new facilities.* As space is used more efficiently, overhead costs are reduced, as is the need for new facilities as new equipment is brought on.

Reducing unnecessary movement incorporates a Lean concept known as 5S, which comes from the Japanese words used to create organization and cleanliness in the workplace [Seiri (organization), Sieton (tidiness), Seiso (purity), Seiketsu (cleanliness), Shitsuke (discipline)]. The traditional 5S have been translated into the following 5S's, which are perhaps better definitions for English-speaking organizations (ReVelle, 2000):

- *Sort.* Eliminate whatever is not needed.
- *Straighten.* Organize whatever remains.
- *Shine.* Clean the work area.
- *Standardize.* Schedule regular cleaning and maintenance.
- *Sustain.* Make 5S a way of life.

The Lean methods for setup reduction and level loading are useful for improving lead and cycle times. Analysis of setup times, and categorization of activities as internal or external, and as preparation, replacement, location, or adjustment (as discussed in the Analyze stage, Chap. 15) will provide valuable input to the Improve stage activities. Prerequisites for level loading to be completely successful include:

- *Standardization of work instructions,* and cross-training, so employees can be shifted to meet increased demand or address process problems. A key strategy for eliminating barriers and creating continuous flow is through redesign of process flow into product work cells. In product work cells, all functions work together in a single close-knit cell, rather than in different departments. These cells may be defined by product or by families of like products. Product cells decrease the physical movement of the goods (whether a physical product, paperwork, or even ideas). This not only reduces the waste of waiting, it also allows the people performing each of the tasks to have visibility of the tasks, and of slowdowns, barriers, or inefficiencies that occur in the preceding or following steps. Standardized work instructions, with process costs and measurement indicators, are prominently posted. Status indicators are used so all workers can readily see if slowdowns occur. Work instructions are standardized so that fellow workers can fill in for a worker who is absent on a given day. Poka yoke (mistake proofing) is used to minimize errors.
- *Transparency.* Workers need to know about shifts in demand or process problems as soon as possible. Create visibility of the work in progress (WIP). When we see where it predominantly occurs, we have identified bottlenecks in the process. By forcing the elimination of this inventory, we can determine and resolve the root causes of the inventory build-up.

Finally, WIP is reduced when we can convert from batch processes to continuous flow of individual product or service units. "Start an item, finish an item" is the mantra. We will now look at some of the problems associated with batches, and the reasons we process in batches, so we can understand how to move away from this practice.

A four-step approach is recommended to reduce setup times (George, 2002):

1. *Classify each setup step as either internal* or *external*. Internal steps are those done while the process is inactive. External steps are done while the process is operating.
2. *Convert internal steps to external steps.* We want to reduce the time the process is non-operational, so we need to reduce the time associated with the internal steps. The quickest way to do this is to do as many of these steps as we can while the process is operational.

For example, if we can collect the money from customers as their burgers are being cooked, then the total cycle time is reduced.

3. *Reduce time for remaining internal steps.* There are some internal steps that cannot be done while the process is operational. We now want to concentrate on reducing the time required to complete those steps that require the process to be delayed. For example, since the burgers cannot be cooked until we know what the customer wants to order, we will try to reduce the time it takes to place the customer order.
4. *Eliminate adjustments.* Adjustments, as discussed in the Analyze stage, can be reduced through effective process control. Designed experiments may be used to understand the causes of process variation that precede the adjustment.

Define and Mitigate Failure Modes

Once the process flow is established, it can be evaluated for its failure modes. Understanding process failure modes allows us to define mitigation strategies to minimize the impact or occurrence of failures. These mitigation strategies may result in new process steps, optimal process settings, or process control strategies to prevent failure. In some cases, in which failure cannot be economically prevented, a strategy can be developed to minimize the occurrence of the failure and contain the damage.

The cause-and-effect diagrams discussed in the Analyze stage are again useful for brainstorming the potential causes of failures. This brainstorming activity will provide necessary input to *process decision program charts* and *failure modes and effects analysis*.

Process Decision Program Chart

The process decision program chart (PDPC) is a technique to prepare contingency plans. It is a simplified version of the reliability engineering methods of failure mode, effects, and criticality analysis (FMECA) and fault tree analysis (discussed later in this chapter). PDPC seeks to describe specific actions to be taken to prevent the problems from occurring, and to mitigate the impact of the problems if they do occur. An enhancement to classical PDPC is to assign subjective probabilities to the various problems and to use these to help assign priorities. Figure. 16.1 shows a PDPC.

Preventing Failures

Many failures occur due to human error, particularly in service processes. While the failure may result from human error, it does not necessarily imply that process or system-level solutions could not prevent its occurrence. In other words, focusing on individuals' performance (or lack of performance) will not realize an improvement.

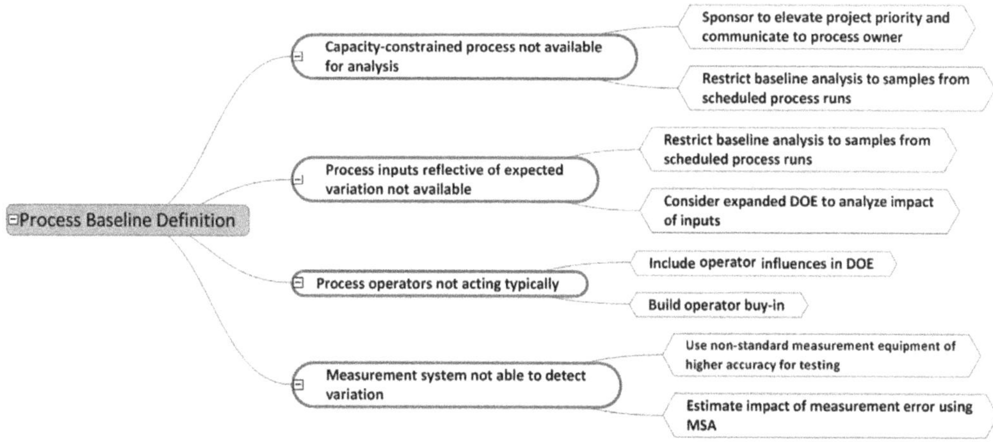

Figure 16.1 Example PDPC (Pyzdek and Keller, 2010).

There are three main categories of human errors: *inadvertent errors*, *technique errors*, and *willful errors*.

Inadvertent Errors

Many human errors occur due to lack of attention. People are notorious for their propensity to commit this type of error. Inadvertent errors have certain hallmarks:

- There is usually no advance knowledge that an error is imminent.
- The incidence of error is relatively small. That is, the task is normally performed without error.
- The occurrence of errors is random, in a statistical sense.

Examples of inadvertent errors are not hard to find. This is the type of error we all make ourselves in everyday life when we find a mistake balancing the checkbook, miss a turn on a frequently traveled route, dial a wrong number on the phone, or forget to pay a bill. At home these things can be overlooked, but in business they can have significant costs and contribute to wasted resources.

Preventing inadvertent errors may seem an impossible task. Indeed, these errors are among the most difficult of all to eliminate. As the error rate becomes small, the improvement effort becomes more difficult. Still, in most cases it is possible to make substantial improvements economically. At times it is even possible to eliminate the errors completely.

One way of dealing with inadvertent errors is foolproofing, also known as poka yoke. *Foolproofing* involves changing the design of a process or product to make the commission of a particular human error impossible. For example, a company was experiencing a sporadic problem (note: the words "sporadic problem" should raise a flag in your mind that inadvertent human error is likely!) with circuit board defects. It seems that occasionally entire orders were lost because the circuit boards were drilled wrong. A study revealed that that problem occurred because the circuit boards could be mounted backward on an automatic drill unless the manufacturing procedure was followed carefully. Most of the time there was no problem, but as people became more experienced with the drills they sometimes got careless. The problem was solved by adding an extra hole in an unused area of the circuit board panel, and then adding a pin to the drill fixture. If the board was mounted wrong, the pin wouldn't go through the hole. Result: no more orders lost. It would never happen again because *it could never happen again*.

Another method of reducing human errors is automation. People tend to commit more errors when working on dull, repetitive tasks, or when working in unpleasant environments (e.g., due to heat, odors, noise, fumes, and so on). Automation is very well suited to this type of work. A highly complicated task for a normal machine becomes a simple repetitive task for a robot. Elimination of errors is one justification for an investment in robots. On a more mundane level, simpler types of automation such as numerically controlled machining centers often produce a reduction in human errors.

Another approach to the human error problem is ergonomics, or human factors engineering. Many errors can be prevented through the application of engineering principles to design of products, processes, and workplaces. By evaluating such things as seating, lighting, sound levels, temperature change, workstation layout, etc., the environment can often be improved and errors reduced. Sometimes human factors engineering can be combined with automation to reduce errors. This involves automatic inspection and the use of alarms (lights, buzzers, etc.) that warn the employee when he's made an error. This approach is often considerably less expensive than full automation.

Technique Errors

As an example of technique errors, consider the following real-life problem with gearbox housings. The housings were gray iron castings and the problem was cracks. The supplier was made aware of the problem and their metallurgist and engineering staff had worked long and hard on the problem, but to no avail. Finally, in desperation, the customer sat down with the supplier to put together a "last-gasp" plan. If the plan failed, the customer would be forced to try an alternative source for the casting.

As might be expected, the plan was grand. The team identified many important variables in the product, process, and raw materials. Each variable was classified as either a "control variable," which would be held constant, or an "experimental variable," which would be changed in a prescribed way. The results of the experiment were to be analyzed using all the muscle of a major mainframe statistical analysis package. All of the members of the team were confident that no stone had been left unturned.

Shortly after the program began, the customer quality engineering supervisor received a call from his quality engineering representative at the supplier's foundry. "We can continue with the experiment if you really want to," he said, "but I think we've identified the problem and it isn't on our list of variables." It seems that the engineer was in the inspection room inspecting castings for our project and he noticed a loud "clanging sound" in the next room. The clanging occurred only a few times each day, but the engineer soon noticed that the cracked castings came shortly after the clanging began. Finally he investigated and found the clanging sound was a relief employee pounding the casting with a hammer to remove the sand core. Sure enough, the cracked castings had all received the "hammer treatment"!

This example illustrates a category of human error different from the inadvertent errors described earlier. Technique errors share certain common features:

- They are unintentional.
- They are usually confined to a single characteristic (e.g., cracks) or class of characteristics.
- They are often isolated to a few workers who consistently fail.

Solution of technique errors involves the same basic approaches as the solution of inadvertent errors, namely automation, foolproofing, and human factor engineering. In the meantime, unlike inadvertent errors, technique errors may be caused by a simple lack of understanding that can be corrected by developing better instructions and training.

Willful Errors (Sabotage)

This category of error is unlike either of the two previous categories. Willful errors are often very difficult to detect; however, they do bear certain trademarks:

- They are not random.
- They don't "make sense" from an engineering point of view.
- They are difficult to detect.
- Usually only a single worker is involved.
- They begin at once.
- They do not occur when an observer is present.

Another real-life example may be helpful. An electromechanical assembly suddenly began to fail on some farm equipment. An examination of the failures revealed that the wire had been broken *inside of the insulation*. However, the assemblies were checked 100 percent for continuity after the wire was installed and the open circuit should've been discovered by the test. After a long and difficult investigation, no solution had been found. However, the problem had gone away and never come back.

About a year later, the quality engineer was at a company party when a worker approached him. The worker said he knew the answer to the now infamous "broken wire mystery," as it had come to be known. The problem was caused, he said, when a newly hired probationary employee was given his two weeks' notice. The employee decided to get even by sabotaging the product. He did this by carefully breaking the wire, but not the insulation, and then pushing the broken sections together so the assembly would pass the test. However, in the field the break would eventually separate, resulting in failure. Later, the quality engineer checked the manufacturing dates and found that every failed assembly had been made during the two weeks prior to the saboteur's termination date.

In most cases, the security specialist is far better equipped and trained to deal with this type of error than quality control or engineering personnel. In serious cases, criminal charges may be brought as a result of the sabotage. If the product is being made on a government contract, federal agencies may be called in. Fortunately, willful errors are extremely rare. They should be considered a possibility only after all other explanations have been investigated and ruled out.

Failure Mode and Effects Analysis

Failure modes and effects analysis (FMEA), also known as failure modes, effects, and criticality analysis, is used to determine high-risk functions or product features based on the impact of a failure and the likelihood that a failure could occur without detection.

The methodology can be applied to products (design FMEA) or processes (process FMEA) as follows (Pyzdek and Keller, 2010):

1. Define the system to be analyzed, including a review of all functions or processes, the current performance levels for each, and a definition of failure of each process. The process and its failure modes were specified in the Define stage, and the current level of performance documented in the Measure stage; however, during the Improve stage the process was redefined, so it's possible the new process will have different failure modes. The performance levels will certainly be different, representing the fruits of the improvement effort.
2. The process map is used to define the steps and functional relationships for the new process.

3. A proper SIPOC analysis (as discussed in the Define stage, Chap. 13) ensures a thorough understanding of the process and subprocesses.
4. Step 4 is perhaps the true beginning of the FMEA process within Six Sigma DMAIC projects, since the preceding three steps have already been accomplished and serve as "inputs" at the Improve stage. In this step, we define the function of the process. The *function* provides the purpose of the step. Each step should have one or more functions, given that the step is necessary to satisfy an internal or external requirement. To identify the functions of the process step, it might be useful to consider the ramifications of removing the step. For example, in a sales process, the process step for "Enter the product ID number for each purchased item" provides the function to "Identify the item numbers that belong to the products being purchased so that they are all included in the delivery."
5. For each function, identify failure mode and its effect: *What could go wrong? What could the customer dislike?* For example, for the function "Identify the item numbers that belong to the products being purchased so that they are all included in the delivery," the failure modes might be "Product ID mistyped" and "Item numbers not correctly defined for product bundles." The second failure mode refers to products that are sold as sets. A single item number is used for the set so that the proper charge is applied for the set (discounted from the per item prices), but subsequent process steps (and subsequent processes) need the correct item numbers for each piece (such as to check inventory levels or fill the order from inventory).
6. Define the severity for each of the Failure Modes. Table 16.1 provides a good means of identifying the severity for a given failure effect. In the example given, the failure mode of mistyping the product ID, with the effect of shipping the wrong product, is given a severity of 6. From Table 16.1, severity 6 is described as "Customer will complain. Repair or return likely. Increased internal costs." Granted, defining a severity level is subjective. A severity of 5 or 7 might seem reasonable in this example. There is no one "right" answer; however, consistency between analyses is important for meaningful prioritizations.
7. Define the likelihood (or probability) of occurrence. Table 16.1 provides useful descriptions of occurrence levels from 1 to 10. Table 16.2 provides a somewhat better definition, as developed by the Automotive Industry Action Group (AIAG) based on process capability and defect rates. In the example, the failure mode of mistyping the product ID, with the effect of shipping the wrong product, is given an occurrence level of 5.

Rating	Severity (SEV)	Occurrence (OCC)	Detectability (DET)
Rating	How significant is this failure's effect to the customer?	How likely is the cause of this failure to occur?	How likely is it that the existing system will detect the cause, if it occurs? Note: p is the estimated probability of failure *not* being detected
1	Minor. Customer won't notice the effect or will consider it insignificant	Not likely	Nearly certain to detect before reaching the customer ($p \approx 0$)
2	Customer will notice the effect	Documented low failure rate	Extremely low probability of reaching the customer without detection ($0 < p \leq 0.01$)
3	Customer will become irritated at reduced performance	Undocumented low failure rate	Low probability of reaching the customer without detection ($0.01 < p \leq 0.05$)
4	Marginal. Customer dissatisfaction due to reduced performance	Failures occur from time to time	Likely to be detected before reaching the customer ($0.05 < p \leq 0.20$)
5	Customer's productivity is reduced	Documented moderate failure rate	Might be detected before reaching the customer ($0.20 < p \leq 0.50$)
6	Customer will complain. Repair or return likely. Increased internal costs (scrap, rework, etc.)	Undocumented moderate failure rate	Unlikely to be detected before reaching the customer ($0.50 < p \leq 0.70$)
7	Critical. Reduced customer loyalty. Internal operations adversely impacted	Documented high failure rate	Highly unlikely to be detected before reaching the customer ($0.70 < p \leq 0.90$)
8	Complete loss of customer goodwill. Internal operations disrupted	Undocumented high failure rate	Poor chance of detection ($0.90 < p \leq 0.95$)
9	Customer or employee safety compromised. Regulatory compliance questionable	Failures common	Extremely poor chance of detection ($0.95 < p \leq 0.99$)
10	Catastrophic. Customer or employee endangered without warning. Violation of law or regulation	Failures nearly always occur	Nearly certain that failure won't be detected ($p \approx 1$)

Table 16.1 Severity, Occurrence, and Detectability Levels

Probability of Failure	Failure Rate	Cpk	Occ.
Very high	> 1/2	< 0.33	10
	1/3	0.33	9
High (often)	1/8	0.51	8
	1/20	0.67	7
Moderate; occasional	1/80	.83	6
	1/400	1.00	5
	1/2,000	1.17	4
Low	1/15,000	1.33	3
Very low	1/150,000	1.50	2
Remote	< 1/150,000	> 1.67	1

(AIAG, 1995)

Table 16.2 AIAG Occurrence Levels

8. Define the detection method and likelihood of detection. Table 16.1 provides useful descriptions of detection levels from 1 to 10. In the example, the failure mode of mistyping the product ID, with the effect of shipping the wrong product, is given a detectability level of 4, a likely detection before reaching the customer. This is based on the detection method that has been implemented from past process improvements: the accounting clerk compares the PO with the order form as the invoice is created for shipping.
9. Calculate *risk priority number* (RPN) by multiplying the severity, occurrence, and detectability levels. In the example, the risk priority number is calculated by multiplying 6 (the severity) by 5 (the occurrence level) by 4 (the detection level), resulting in an RPN of 120.
10. Prioritize the failure modes based on the RPN.

The risk priority number will range from 1 to 1000, with larger numbers representing higher risks. Failure modes with higher RPN should be given priority for the Improve stage of DMAIC.

Some organizations use threshold values, above which preventive action must be taken. For example, the organization may require improvement for any RPN exceeding 120. Reducing the RPN requires a reduction in the severity, occurrence, and/or detectability levels associated with the failure mode. As a general rule:

- Reducing severity requires a change to the design of the product or process. For example, if the process involves a manufactured part, it may be possible to alter the design of the part so that the stated failure mode is no longer a serious problem for the customer.

- Reducing detectability level increases cost with no improvement to quality. In order to reduce the detectability level, we must improve the detection rate. We might add process steps to inspect product, approve product, or (as in the example), to double-check a previous process step. None of these activities adds value to the customer, and are hidden factory sources of waste to the organization.
- Reducing the occurrence level is often the best approach, since reducing severity can be costly (or impossible) and reducing detectability is only a costly short-term solution. Reducing the occurrence level requires a reduction in process defects, which reduces cost.

The final step in the FMEA is to re-evaluate the RPN after improvements have been implemented.

CHAPTER 17

Control/Verify Stage

The objectives of the Control/Verify stage include (Keller, 2011a):

- Standardize new procedures/product design elements.
- Continually verify project deliverables.
- Document lessons learned.

Once the process or system has been improved, we set about to control it so that the improvements are maintained. Without this critical step, old habits return and the gains are quickly lost. We must standardize on the new methodologies to sustain the improvements.

The methods discussed in Part II are used to control the process, including statistical process control as well as work instructions controlled through a document control system.

Control at the business level may be a matter of "getting the word out." Training becomes a key aspect of maintaining the improvements that were deployed, whether it was a policy change or a new computerized system for order processing.

Spreading the word at the process level involves changing process procedures, specifications, and/or statistical control charting limits. As with other levels of the organization, these changes require training for affected personnel. When these personnel understand not just how the process has changed, but also why, then further improvements may be found down the road. These aspects of training are discussed later in this chapter.

As the project team concludes its activities, it is important for project documentation to be finalized and retained. A key aspect of this is the documentation of lessons learned: What might you have done differently to achieve speedier or better results? Would these insights be useful to other teams in the organization?

Communicating these success stories to other parts of the company has proven to be an effective way to achieve greater and greater levels of performance throughout the organization. GE Capital learned from the aircraft division, and vice versa, as divergent as their businesses were.

Success breeds success. Internal Web sites, company newsletters, and Black Belt forums are effective ways to share this information.

Another important part of the team wrap-up is the recognition of their efforts.

Performance Evaluation

Evaluating team performance involves the same principles as evaluating performance in general. Often the team's performance in meeting the project's goals and objectives is a critical aspect of the evaluation. However, if meeting the goal is the sole criterion for success, management may well have a difficult time recruiting team members for anything but the slam-dunk projects of the future. Rather, teams must be praised for effort, assuming their effort was praise-worthy, even when those efforts fail to achieve the stated objectives.

Performance measures generally focus on group tasks, rather than on internal group issues. Typically, financial performance measures show a payback ratio of between 2:1 and 8:1 on team projects. Some examples of tangible performance measures are:

- Productivity
- Quality
- Cycle time
- Grievances
- Medical usage (e.g., sick days)
- Absenteeism
- Service
- Turnover
- Dismissals
- Counseling usage

Many intangibles can also be measured. Some examples of intangibles affected by teams are:

- Employee attitudes
- Customer attitudes
- Customer compliments
- Customer complaints

The performance of the team process should also be measured. Project failure rates should be carefully monitored. A p chart can be used to evaluate the causes of variation in the proportion of team projects that succeed. Failure analysis should be rigorously conducted.

Aubrey and Felkins (1988) list the following effectiveness measures:

- Leaders trained
- Number of potential volunteers
- Number of actual volunteers
- Percent volunteering
- Projects started
- Projects dropped
- Projects completed/approved
- Projects completed/rejected
- Improved productivity
- Improved work environment
- Number of teams
- Inactive teams
- Improved work quality
- Improved service
- Net annual savings

Recognition and Reward

Recognition is a form of employee motivation in which the company identifies and thanks employees who have made positive contributions to the company's success. In an ideal company, motivation flows from the employees' pride of workmanship. When employees are enabled by management to do their jobs and produce a product or service of excellent quality, they will be motivated.

The reason recognition systems are important is not that they improve work by providing incentives for achievement. Rather, they make a statement about what is important to the company. Analyzing a company's employee recognition system provides a powerful insight into the company's values in action. These are the values that are actually driving employee behavior. They are not necessarily the same as management's stated values. For example, a company that claims to value customer satisfaction but recognizes only sales achievements probably does not have customer satisfaction as one of its values in action.

Public recognition is often better for two reasons:

1. Some (but not all) people enjoy being recognized in front of their colleagues.
2. Public recognition communicates a message to all employees about the priorities and function of the organization.

The form of recognition can range from a pat on the back to a small gift to a substantial amount of cash. When substantial cash awards become an established pattern, however, it signals two potential problems:

1. It suggests that several top priorities are competing for the employee's attention, so that a large cash award is required to control the employee's choice.
2. Regular, large cash awards tend to be viewed by the recipients as part of the compensation structure, rather than as a mechanism for recognizing support of key corporate values.

Carder and Clark (1992) list the following guidelines and observations regarding recognition:

- *Recognition is not a method by which management can manipulate employees.* If workers are not performing certain kinds of tasks, establishing a recognition program to raise the priority of those tasks might be inappropriate. Recognition should not be used to get workers to do something they are not currently doing because of conflicting messages from management. A more effective approach is for management to first examine the current system of priorities. Only by working on the system can management help resolve the conflict.
- *Recognition is not compensation.* In this case, the award must represent a significant portion of the employee's regular compensation to have significant impact. Recognition and compensation differ in a variety of ways:
  - Compensation levels should be based on long-term considerations such as the employee's tenure of service, education, skills, and level of responsibility. Recognition is based on the specific accomplishments of individuals or groups.
  - Recognition is flexible. It is virtually impossible to reduce pay levels once they are set, and it is difficult and expensive to change compensation plans.
  - Recognition is more immediate. It can be given in timely fashion and therefore relate to specific accomplishments.
  - Recognition is personal. It represents a direct and personal contact between employee and manager.
- *Recognition should be personal.* Recognition should not be carried out in such a manner that implies that people of more importance (managers) are giving something to people of less importance (workers).
- *Positive reinforcement is not always a good model for recognition.* Just because the manager is using a certain behavioral criterion for

providing recognition, it doesn't mean that the recipient will perceive the same relationship between behavior and recognition.

- *Employees should not believe that recognition is based primarily on luck.* An early sign of this is cynicism. Employees will tell you that management says one thing but does another.
- *Recognition meets a basic human need.* Recognition, especially public recognition, meets the needs for belonging and self-esteem. In this way, recognition can play an important function in the workplace. According to Maslow's theory, until these needs for belonging and self-esteem are satisfied, self-actualizing needs such as pride in work, feelings of accomplishment, personal growth, and learning new skills will not come into play.
- *Recognition programs should not create winners and losers.* Recognition programs should not recognize one group of individuals time after time while never recognizing another group. This creates a static ranking system, with all of the problems discussed earlier.
- *Recognition should be given for efforts, not just for goal attainment.* According to Imai, a manager who understands that a wide variety of behaviors are essential to the company will be interested in criteria of discipline, time management, skill development, participation, morale, and communication, as well as direct revenue production. To be able to effectively use recognition to achieve business goals, managers must develop the ability to measure and recognize such process accomplishments.
- *Employee involvement is essential in planning and executing a recognition program.* It is essential to engage in extensive planning before instituting a recognition program or before changing a bad one. The perceptions and expectations of employees must be surveyed.

Principles of Effective Reward Systems

Kohn (1993) believes that nearly all existing reward systems share the following characteristics:

1. They punish the recipients.
2. They rupture relationships.
3. They ignore reasons for behavior.
4. They discourage risk-taking.

Most existing reward systems (including many compensation systems) are an attempt by management to manipulate the behaviors of employees. Kohn convincingly demonstrates, through solid academic research into the effects of rewards, that people who receive the rewards as well as those who hand them out suffer a loss of incentive—hardly the goal of the exercise!

Rather than provide cookbook solutions to the problem of rewards and incentives, Kohn offers some simple guidelines to consider when designing reward systems.

1. *Abolish incentive pay (something Deming advocated as well).* Hertzberg's hygiene theory tells us that money is not a motivator, but it can be a de-motivator. Pay people generously and equitably; then do everything in your power to put money out of the employee's mind.
2. *Reevaluate evaluation.* Review Chap. 20 for information on performance appraisals and alternatives.
3. *Create conditions for authentic motivation.* Money is no substitute for the real thing—interesting work. Here are some principles to use to make work more interesting:
   a. *Design interesting jobs.* Give teams projects that are intrinsically motivating, for example, projects that are meaningful, challenging, and achievable.
   b. *Encourage collaboration.* Help employees work together, then provide the support needed to make it possible for the teams to accomplish their goals.
   c. *Provide freedom.* Trust people to make the right choices. Encourage them when they make mistakes.

Training

When quality improvement plans are implemented, the nature of the work being done changes. People involved in or impacted by the new approach must receive two different types of training: conceptual and task-based.

Conceptual training involves explanation of the principles driving the change and a shift from an internal, product-based perspective to a customer and process-based focus. Rather than viewing their jobs in isolation, employees must be taught to see all work as a process, connected to other processes in a system. Rather than pursuing a goal of "control," where activities are done the same way indefinitely, employees learn that continuous improvement is to be the norm, with processes constantly being changed for the better. The PDCA cycle discussed in Chap. 12 is helpful. Such ideas are radically different and difficult to assimilate. Patient repetition and "walking the talk" are essential elements of such training.

Conceptual training also involves teaching employees the basics of problem-solving. Data-driven process improvement demands an understanding of the fundamentals of data collection and analysis. In Six Sigma parlance, operational employees are trained as Green Belts in the basic tools of quality, including (most importantly) SPC, which provides a means of understanding the systematic nature of process variation.

When jobs are reintegrated, the duties expected of each employee change, often radically. Task-based training is necessary to help employees acquire and maintain new skills and proficiencies. Employees are given new responsibilities for self-control of process quality. To effectively handle these new responsibilities, employees must learn to use information in ways they never did before. Often employees are asked to help design new information systems, enter data, use computer terminals to access information, read computer output, make management presentations, etc. These skills must be acquired through training and experience. Quality plans that do not include adequate employee training are commonplace, and a primary cause of the high rate of failure of quality plans.

Job Training

Job training is the vehicle through which the vast majority of training occurs. Job training involves assigning the learners to work with a more experienced employee, either a supervisor, peer, or lead hand, to learn specific tasks in the actual workplace. The learner is usually a new employee who has been recently either hired, transferred, or promoted into the position and who lacks the knowledge and skill to perform some components of her or his job. The experienced employee normally demonstrates and discusses new areas of knowledge and skill and then provides opportunities for practice and feedback. There are three common methods used in job training (Nolan, 1996):

1. *Structured on-the-job training (OJT).* Structured OJT allows the learner to acquire skills and knowledge needed to perform the job through a series of structured or planned activities at the work site. All activities are performed under the careful observation and supervision of the OJT instructor. The structured process is based on a thorough analysis of the job and the learner. The OJT instructor introduces the learner systematically to what he or she needs to know to perform competently and meet performance standards and expectations.
2. *Unstructured OJT.* Unstructured OJT often means sink or swim. Most activities in unstructured OJT have not been thought through and are done in a haphazard way. A common method of unstructured OJT is to have the learner "sit" with another employee or "follow the employee around" for a few days to see what the employee does and how she does it. This "sit-and-see" technique often leads the learner to pick up as much by trial and error as he does by any instruction given by the more experienced

employee. The learner is typically inundated with reading assignments concerning policies, procedures, and other assorted documentation which, when not put into the right context, can cause more confusion than assistance.

3. The learner is often thrust on the experienced employee without notice and is seen as a hindrance, since this training time is interrupting the experienced employee's normal work load and performance outputs. The major drawback of the unstructured approach is that objectives, expectations, and outcomes are not defined in advance and, therefore, results are unpredictable.
4. *Job instruction training.* Job instruction training was originally developed for use with World War II production workers and is based on a mechanical step procedure requiring the instructor to present the material in an orderly, disciplined manner. It is most frequently used to teach motor skills. Since it involves a systematic approach, components of it are often found in today's structured OJT.

Developing a Structured OJT Program

Structured OJT has proved to be an efficient and effective means of teaching employees about the skills required to do their jobs. Developing structured OJT programs is a process that involves the following steps:

1. *Needs analysis.* The need is established during the improvement project DMAIC cycle, including assessment of process personnel knowledge, skills and abilities (KSA), as well as attitudes.
2. *Job analysis.* Job analysis is part of the job design and employee selection processes. Job requirements are matched to the employee's knowledge, skills, and abilities during the selection process. When designing structured OJT programs, the characteristics of trainees must be examined in order to target the OJT accurately and develop effective instructional materials; for example, what works for new hires may not be best for transfers. Trainees should complete employee profile surveys to provide the instructional designer with the information needed to customize the training to each employee's needs.
3. *Course design.* This step will produce a course training plan that serves as the blueprint to be used to construct training support materials. The course training plan should include:
   - A purpose statement
   - Performance objectives
   - Criterion tests
   - Presentation

- Application and feedback methods
- Course content outline
- Lesson plan
- Training schedule

4. *Material preparation.*
5. *Validation.* After training materials have been developed for the OJT program, they must be tested to ensure they fulfill their mandate to train the individual to perform the job. This involves pilot studies with selected individuals to "shake down" the materials. Final validation can only occur by monitoring the program's effectiveness with actual trainees.
6. *Presentation.* Effective instructional presentations incorporate a systematic learning process of presentation, application, and feedback (PAF) (Nolan, 1996, p. 764). Prior to actual delivery, the OJT instructor needs to review the structured process and methods, collect all materials and tools necessary, and develop a schedule of training. Three components are necessary:

   Presentation. The OJT instructor

   - States the objective
   - Motivates the trainee
   - Overviews key steps
   - Presents tasks (tell and show)
   - Tests for understanding

   Application. The trainee applies new knowledge and skills through

   - Directed practice
   - Undirected, yet supervised, practice

   Feedback. The OJT trainer observes and communicates to the trainee

   - What was done well
   - What needs improvement
   - How to improve

7. *Evaluation.* The final step in a structured OJT program is to evaluate it. The method of evaluation is the same five-level process as described above for evaluating training in general: reaction, learning, behavior, results, and return on investment.

Instructional Games, Simulations, and Role-Plays

These techniques are based on two premises: (1) people learn better through active experience than passive listening and (2) people learn better through interacting with one another than working alone.

Instructional games. An instructional game is an activity that is deliberately designed to produce certain learning outcomes. Instructional games incorporate five characteristics (Thiagarajan, 1996):

1. *Conflict*. Games specify a goal to be achieved and throw in obstacles to its achievement. A game may involve competition among players, or it may involve player cooperation to achieve a group goal.
2. *Control*. Games are governed by rules that specify how to play the game.
3. *Closure*. Games have an ending rule, which may be a time limit, completion of a set of tasks, elimination of players from the game, etc. Most effective instructional games use multiple criteria for closure and permit different players or teams to win along different dimensions.
4. *Contrivance*. Games contain elements, such as chance, to ensure that the game retains a playful character and isn't taken too seriously.
5. *Competency base*. The game is designed to help players improve their competencies in specific areas. Learning objectives range from rote recall to complex problem-solving and may deal with motor, informational, conceptual, interpersonal, and affective domains.

Simulation games. A simulation game contains the five characteristics of instructional games, but in addition it includes a correspondence between some aspect of the game and reality. Some examples of simulation games that have been used in teaching quality concepts are:

- *Senge's "Beer Game"* in The Fifth Discipline (Senge, 1990, pp. 26–53). The beer game is designed to teach systems thinking.
- *Deming's funnel experiment*. Boardman and Boardman (1990) provide a detailed description of how to set up and conduct the funnel experiment. The funnel experiment illustrates statistical thinking and decision making.
- *"The Card Drop Shop."* The card drop shop is a small enterprise that has customers, a president, a supervisor, an inspector, a rework operator, several line operators, and an accountant. There is only one process: dropping playing cards onto a target on the floor. The customer ideally wants all cards on the target but will accept the product provided that the total deviation from the target is "not too bad." The customer also specifies that each card is to be held by its center and dropped individually. Like the funnel experiment,

the card drop shop illustrates the effects of tampering with a stable process. However, it does so using simpler tools and materials (Alloway, 1994).

- *Geometric dimensioning and tolerancing simulation* (Wearring and Karl, 1995).
- Additional applications of simulation to quality are described in Simon and Bruce (1992).

In addition to simulation games, simulation tools such as quincunxes and sampling bead boxes are commonly used in quality training. Simple simulations create a concrete link to abstract concepts, making it easier for people to understand the concept.

Role-plays. In a role-play, players spontaneously act out characters assigned to them in a scenario. Role-plays take a variety of forms, for example:

- *Media.* The scenario for a role-play may be presented through a printed handout, an audiotape, videotape, etc.
- *Characters.* The characters in a role-play may be identified in terms of job functions, personality variables, or attitudes. Some role-plays require people to play their own roles in a different situation (as in a desert survival exercise).
- *Responses.* Most role-plays involve face-to-face communication among the characters, but written or phone communication is also used.
- *Mode of usage.* Groups may be divided into smaller groups or pairs, or a group can watch as others participate. Role-players may be substituted as the role-play progresses. Coaches may be assigned to help the role-players.
- *Number of players.* Most role-plays involve two characters, but there is no fixed limit on the number of players who can be involved.
- *Replay.* The effectiveness of some role-plays can be improved through repetition. Repetition can take place after a presentation of new material by the instructor, or after changing some aspect of the game.

An excellent example of a role-play game is Deming's "Red bead experiment" (Deming, 1986, p. 346ff). The red bead experiment is designed to teach statistical thinking and the use of simple statistical tools.

PART V

Management of Human Resources

Management decisions are often classified as being judgmental, data-based, or scientific (Weaver, 1995, pp. 223–231). Management styles are often viewed from a psychological perspective. Yet before discussing specific management styles, it is fair to ponder the proper role of a manager.

Dr. W. Edwards Deming (1993) offers the following perspective:

This is the new role of a manager of people after transformation:

1. A manager understands and conveys to his people the meaning of a system. He explains the aims of the system. He teaches his people to understand how the work of the group supports these aims.
2. He helps his people to see themselves as components in a system, to work in cooperation with preceding stages and with following stages toward optimization of the efforts of all stages toward achievement of the aim.
3. A manager of people understands that people are different from each other. He tries to create for everybody interest and challenge, and joy in work. He tries to optimize the family background, education, skills, hopes, and abilities of everyone. This is not ranking people. It is, instead, recognition of differences between people, and an attempt to put everybody in position for development.
4. He is an unceasing learner. He encourages his people to study. He provides, when possible and feasible, seminars and courses for advancement of learning. He encourages continued education in college or university for people who are so inclined.
5. He is coach and counsel, not a judge.
6. He understands a stable system. He understands the interaction between people and the circumstances that they work in. He understands that the performance of anyone that can learn a skill will come to a stable state—upon which further lessons will not bring improvement of performance. A manager of people knows that in this stable state it is distracting to tell the worker about a mistake.

7. He has three sources of power: authority of office; knowledge; personality and persuasive power; tact. A successful manager of people develops Nos. 2 (knowledge) and 3 (personality and persuasive power); he does not rely on No. 1 (authority of office). He has nevertheless the obligation to use No. 1, as this source of power enables him to change the process—equipment, materials, methods—to bring improvement, such as to reduce variation in output. (Dr. Robert Klekamp). He is in authority, but lacking knowledge or personality (No. 2 or 3), must depend on his formal power (No. 1). He unconsciously fills a void in his qualifications by making it clear to everybody that he is in position of authority. His will be done.
8. He will study results with the aim to improve his performance as a manager of people.
9. He will try to discover who if anybody is outside the system, in need of special help. This can be accomplished with simple calculations, if there be individual figures on production or on failures. Special help may be only simple rearrangement of work. It might be more complicated. He in need of special help is not in the bottom 5% of the distribution of others: he is clean outside that distribution. (See Fig. V.1.)
10. He creates trust. He creates an environment that encourages freedom and innovation.
11. He does not expect perfection.
12. He listens and learns without passing judgment on him that he listens to.

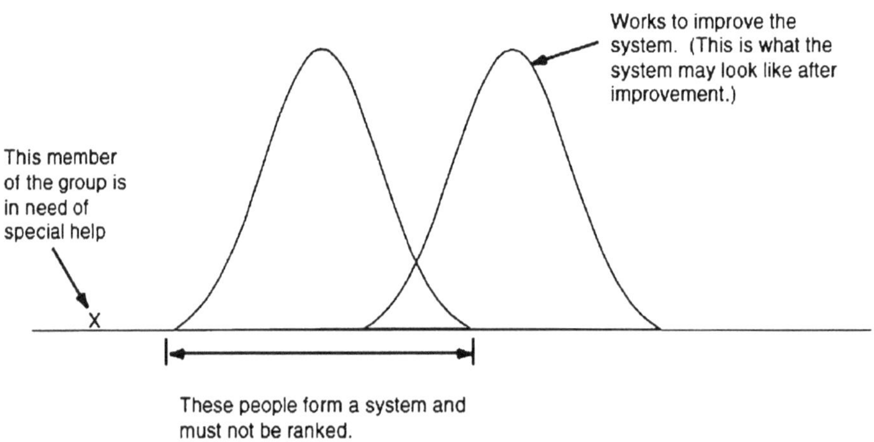

Figure V.1 Figures on production or on failures, if they exist, can be plotted. Study of the figures will show the system, and outliers if any.

13. He will hold an informal, unhurried conversation with every one of his people at least once a year, not for judgment, merely to listen. The purpose would be development of understanding of his people, their aims, hopes, and fears. The meeting will be spontaneous, not planned ahead.
14. They understand the benefits of cooperation and the losses from competition between people and between groups (Kohn, 1986).

Deming's perspective is important in that it represents a radical departure from the traditional view of the manager's role. As of this writing, it represents a set of normative guidelines that few organizations incorporate completely, and incorporates much of the modern understanding of motivational psychology.

CHAPTER 18

Motivation Theories and Principles

The science of psychology, while still in its infancy, has much to offer anyone interested in motivating people to do a better job.

Maslow's Hierarchy of Needs

Professor A.S. Maslow of Brandeis University has developed a theory of human motivation elaborated on by Douglas McGregor. The theory describes a "hierarchy of needs." Figure 18.1 illustrates this concept.

Maslow postulated that the lower needs must be satisfied before one can be motivated at higher levels. Furthermore, as an individual moves up the hierarchy the motivational strategy must be modified because *a satisfied need is no longer a motivator*; for example, how much would you pay for a breath of air right now? Of course, the answer is nothing because there is a plentiful supply of free air. However, if air were in short supply, you would be willing to pay plenty.

The hierarchy begins with physiological needs. At this level a person is seeking the simple physical necessities of life, such as food, shelter, and clothing. A person whose basic physiological needs are unmet will not be motivated with appeals to personal pride. If you wish to motivate personnel at this level, provide monetary rewards such as bonuses for good quality. Other motivational strategies include opportunities for additional work, promotions, or simple pay increases. As firms continue doing more business in underdeveloped regions of the world, this category of worker will become more commonplace.

Once the simple physiological needs have been met, motivation tends to be based on safety. At this stage issues such as job security become important. Quality motivation of workers in this stage was once difficult. However, since the loss of millions of jobs to foreign competitors who offer better quality goods, it is easy for people to see the relationship between quality, sales, and jobs.

Social needs involve the need to consider oneself as an accepted member of a group. People who are at this level of the hierarchy will respond to group situations and will work well on quality circles, employee involvement groups, or quality improvement teams.

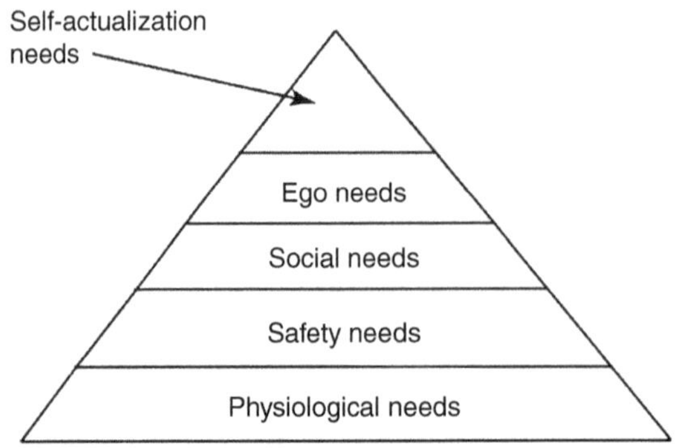

Figure 18.1 Maslow's hierarchy of needs.

The next level, ego needs, involves a need for self-respect and the respect of others. People at this level are motivated by their own craftsmanship, as well as by recognition of their achievements by others.

The highest level is that of self-actualization. People at this level are self-motivated. This type of person is characterized by creative self-expression. All you need do to "motivate" people in this group is to provide an opportunity for them to make a contribution.

Herzberg's Hygiene Theory

Frederick Herzberg is generally given credit for a theory of motivation known as the hygiene theory. The basic underlying assumption of the hygiene theory is that job satisfaction and job dissatisfaction are not opposites. Satisfaction can be increased by paying attention to "satisfiers," and dissatisfaction can be reduced by dealing with "dis-satisfiers." The theory is illustrated in Fig. 18.2.

Theories X, Y, and Z

People seem to seek a coherent set of beliefs that explain the world they see. The belief systems of managers were classified by McGregor into two categories, which he called Theory X and Theory Y.

Under Theory X, workers have no interest in work in general, including the quality of their work. Because civilization has mitigated the challenges of nature, modern man has become lazy and soft. The job of managers is to deal with this by using "carrots and sticks." The carrot is monetary incentive, such as piece rate pay. The stick is docked pay for poor quality or missed production targets. Only money can motivate the lazy, disinterested worker.

Theory Y advocates believe that workers are internally motivated. They take satisfaction in their work, and would like to perform at their

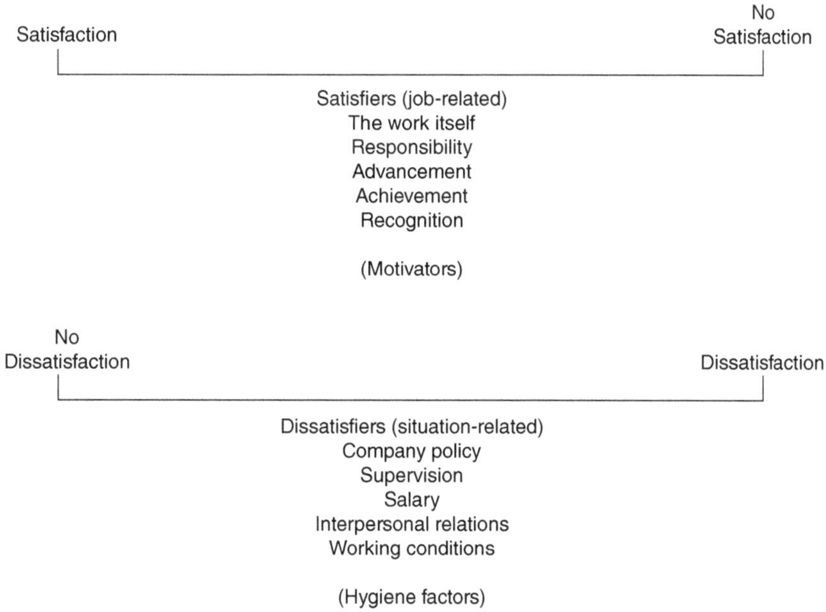

FIGURE 18.2 Herzberg's hygiene theory.

best. Symptoms of indifference are a result of the modern workplace, which restricts what workers can do and separates them from the final results of their efforts. It is management's job to change the workplace so that the workers can, once again, recapture their pride of workmanship. Elements of Theory Y are evident in Deming's discussion of the role of a manager of people, presented in the introduction to Part V.

Theories X and Y have been around for decades. Much later, in the 1980s, Theory Z came into vogue. Z organizations have consistent cultures in which relationships are holistic, egalitarian, and based on trust. Since the goals of the organization are obvious to everyone, and integrated into each person's belief system, self-direction is predominant. In the Z organization, Theories X and Y become irrelevant. Workers don't need the direction of Theory X management, nor does management need to work on the removal of barriers since there are none.

CHAPTER 19

Management Styles

Judgmental Management Style

If the traditional organizational hierarchy is viewed as a "chain-of-command," then ultimate authority resides in the top-most position on the organization chart. The individual occupying this position delegates authority to subordinates who may, in turn, delegate authority to employees further down in the hierarchy. In this system, managers are expected to use their authority to get the work done via command-and-control. Action is based on the manager's judgment. This system effectively requires that managers possess complete knowledge of the work being done by their subordinates—how else could the manager "command-and-control" the work? Of course, this omniscience does not exist. Thus, managers who attempt to follow this metaphor too closely find themselves making decisions based on guesswork to a greater or lesser degree. This results in mistakes, for which the managers are held accountable. Managers who make too many mistakes may be fired, demoted, or disciplined. A natural response to this threat is fear, which may result in the managers blaming their subordinates for their "failures." Because of the authoritarian mind-set, problems are ascribed to individuals, not systems. This produces the classic approach to performance appraisal, including ranking of employees, merit pay, etc. Another outcome is acting only when it is absolutely necessary. Since actions are based on judgments, judgments can lead to mistakes, and mistakes are punished; managers who can minimize action will minimize the chance that mistakes will occur that can be blamed on them. Of course, this tendency is partially offset by the threat of being blamed for not meeting goals set by higher authorities.

Data-Based Management Style

One reaction to the obvious shortcomings of the judgmental management style has been to try to improve the judgments by relying on "facts." Managers solicit feedback from employees and review data in reports before making a decision. Ostensibly, this "data-based approach" changes the basis for action from the manager's judgment to data. Results are marginally better than with the purely judgmental approach. However, data are always incomplete and the element of judgment can never be completely removed. To the extent that managers abdicate their responsibility for making a judgment, the quality of the decisions will suffer. Another problem is the time involved in collecting data. The time (and expense) required increases exponentially to the extent that managers wish to remove all judgment from the decision and insist on "complete" data.

Combination Data-Based/Judgment Management Style

Most experts in management advocate making management decisions based on a combination of the manager's judgment and reasonable amounts of data analysis. Managers, working with all parties impacted, formulate a coherent model of the system. The model (or theory) is used to predict the outcome that would result from operating the system in a certain manner. The system is operated and data is collected on the results obtained. The results are compared with the results predicted by the model, and the theory and systems are updated accordingly. This is the classic Shewhart Plan-Do-Check-Act (PDCA) cycle, or Deming Plan-Do-Study-Act (PDSA) cycle. It closely resembles the scientific method, hypothesize-experiment-test analyze. It is used extensively in organizations that have adopted the Six Sigma DMAIC approach to problem solving, where management uses focused Six Sigma projects to execute data-driven decision-making.

With this management style *systems* are evaluated rather than *people*. The change in focus is fundamental and profound. Here judgment is a source of generating hypotheses about systems or problems, and data is used to evaluate the quality of the hypotheses. People are asked to work to stabilize, then improve, the systems and the organization as a whole.

Participatory Management Style

The premise of the participatory management style is the belief that workers can make a contribution to the design of their own work, based on McGregor's Theory Y.

Managers who practice the participatory style of management tend to engage in certain types of behavior. To engage the workers they establish and communicate the purpose and direction of the organization. This is used to help develop a shared vision of what the organization should be, which is used to develop a set of shared plans for achieving the vision. The managers' role is that of a leader. By their actions and words they show the way to their employees. They are also coaches, evaluating the results of their people's efforts and helping them use the results to improve their processes. They work with the leaders above them in the organization to improve the organization's systems and the organization as a whole.

Autocratic Management Style

The premise of the autocratic management style is McGregor's Theory X: the belief that in most cases workers *cannot* make a contribution to their own work, and that even if they could, they wouldn't. Theory X practitioners would favor the autocratic management style. Autocratic managers attempt to control work to the maximum extent possible. A major threat to control is complexity; complex jobs are more difficult to learn and workers who master such jobs are scarce and possess a certain amount of control over how the job is done. Thus, autocratic managers attempt to simplify work to gain maximum control. Planning of work, including quality planning, is centralized. A strict top-down, chain-of-command approach to management is practiced. Procedures are maintained in exquisite detail and enforced by frequent audits. Product and process requirements are recorded in equally fine detail and in-process and final inspection are used to control quality.

Management by Wandering Around

Peters and Austin (1985, p. 8) call Management by Wandering Around (MBWA) "the technology of the obvious." MBWA addresses a major problem with modern managers: lack of direct contact with reality. Many, perhaps most, managers don't have enough direct contact with their employees, their suppliers, or, especially, their customers. They maintain superficial contact with the world through meetings, presentations, reports, phone calls, email, and a hundred other ways that don't engage all of their senses. This is not enough. Without more intense contact managers simply can't fully internalize the other person's experience. They need to give reality a chance to make them *really* experience the world. The difference between reality and many managers' perception of reality is as great as the difference between an icy blast of arctic air piercing thin indoor clothing and watching a weather report of a blizzard from a sunny beach in the Bahamas.

MBWA is another, more personal way, to collect data. Statistical purists disdain and often dismiss data obtained from opportunistic encounters or unstructured observations. But the information obtained from listening to employees or customers pour their heart out is no less "scientifically valid" than a computer printout of customer survey results. And MBWA data is of a different type. Science has yet to develop reliable instruments for capturing the information contained in angry or excited voice pitch, facial expressions, the heavy sigh—but humans have no trouble understanding the meaning these convey in the context of a face-to-face encounter. It may be that nature has hardwired us to receive and understand these signals through eons of evolution.

The techniques employed by managers who practice MBWA are as varied as the people themselves. The important point is to establish direct contact with the customer, employee, or supplier, up close and personal. This may involve visiting customers at their place of business, or bringing them to yours, manning the order desk or complaint line every month, spontaneously sitting down with employees in the cafeteria, either one-on-one or in groups, or inviting the supplier's truck driver to your office for coffee. Use your imagination. One tip: be sure to schedule regular MBWA time. If it's not on your calendar, you probably won't do it.

Fourth Generation Management

In his book *Fourth Generation Management*, Brian Joiner develops four categories of management styles, which he calls "generations" (Joiner, 1994, pp. 8–9):

- *1st Generation—management by doing.* We simply do the task ourselves. Assuming we possess the necessary knowledge, skills, abilities, and technology, this is an effective way of ensuring that tasks are done to our personal requirements. Its main problem is, of course, limited capacity. As individuals we lack the needed prerequisites to do all but a limited range of tasks, as well as the time to do more than a few things. Of course, there will always be some tasks performed using this approach.
- *2nd Generation—management by directing (micromanagement).* People found that they could expand their capacity by telling others exactly what to do and how to do it: a master craftsman giving detailed directions to apprentices. This approach allows experts to leverage their time by getting others to do some of the work, and it maintains strict compliance to the experts' standards. Although the capacity of this approach is better than 1st generation management, micromanagement still has limited capacity.
- *3rd Generation—management by results.* People get sick and tired of your telling them every detail of how to do their jobs and say, "Just tell me what you want by when, and leave it up to me to figure out how to do it." So you say, "OK. Reduce inventories by 20 percent this year. I'll reward you or punish you based on how well you do it. Good luck." This is the current approach to management practiced by most modern organizations, with all of the problems of suboptimization discussed earlier. Suboptimizing

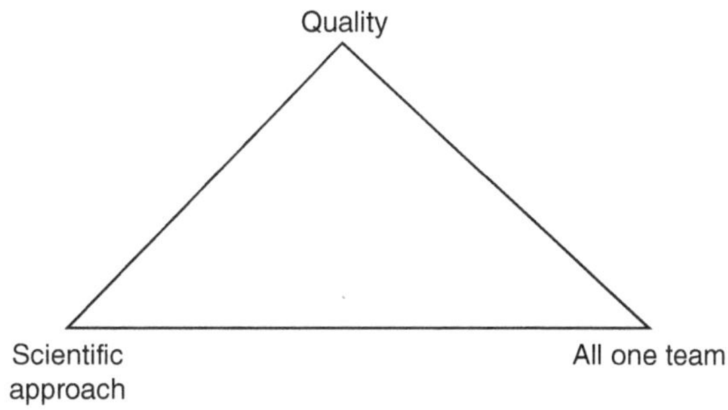

FIGURE 19.1 The Joiner triangle (Joiner, 1994, by permission).

a single department, division, or other unit is the most commonly observed problem with this approach.

- *4th Generation—systems approach.* The core elements of 4th generation management are shown in Fig. 19.1. The quality corner of the triangle represents an obsession with customer-perceived quality. The organization seeks to delight its customers, not to merely satisfy them. The scientific-approach corner indicates learning to manage the organization as a system, developing process thinking, basing decisions on data, and understanding data. "All one team" means believing in people; treating everyone with dignity, trust, and respect; and working toward win-win for customers, employees, shareholders, suppliers, and the communities in which we live.

The Fifth Discipline

Senge (1990) defines five key disciplines for organizational success:

1. *Systems thinking.* The ability to recognize interrelationships between the many actions occurring within systems, rather than to focus on linear snap-shots of simple cause and effect relationships.
2. *Personal mastery.* The ability of individuals to continually seek higher levels of proficiency and excellence, much like Maslow's highest level of self actualization.
3. *Mental models.* Individuals often have preconceived notions, perhaps unknown to themselves, that influence their perceptions and outlook. Critical thinking can only occur when organizations foster discussions to uncover and influence mental models.
4. *Building shared vision.* Even well-articulated organizational visions will flounder if they are not shared by the organization's members. Creating and executing a shared vision requires deeper

commitment to the vision and the development of personnel and systems to sustain the vision.

5. *Team learning*. Senge describes teams as the "fundamental learning unit" in an organization. Learning results from effective dialogue, a word which Senge notes has its origins in the free exchange of ideas, as well as the recognition and remedy of counter-productive communication (e.g., defensiveness).

A critical aspect of Senge's approach is the systems integration of the five elements; individually, they will not suffice.

The reader should recognize the systems level approach advocated by Senge and Joiner is woven throughout this book in a variety of contexts. Deming strongly advocated systems thinking in problem solving and management and specifically warned of the dangers in localized process optimization at the expense of system-wide improvements. Womack and Jones (1996) recommend implementation of the lean principles across the complete supply chain from raw material to final customer use. Six Sigma programs are designed at the organizational level, implementing cross-functional projects, attacking issues critical to cost, quality and schedule, impacting key stakeholder groups including customers, suppliers, employees, and shareholders.

It should be clear that all aspects of quality management, including assurance, planning and improvement, require a systems approach. Anything short of that, such as departmental-level "grass-roots" process improvement activities, risks suboptimization of the system at the expense of the local process optimization. It is the responsibility of management to effectively harness the organizational inertia for improvement, and guide it toward a systematic solution of issues. Anything less is a failure of management.

CHAPTER 20

Resource Requirements to Manage the Quality Function

The modern approach to quality requires an investment of time and resources throughout the entire organization—for many people, the price is about 10 percent of their time (Juran and Gryna, 1993, p. 129). Eventually, this investment yields time savings that become available for quality activities and other activities. In the short run, however, these resources are diverted from other, urgent organizational priorities. Upper management has the key role in providing resources for quality activities. This is accomplished through the quality councils mentioned earlier, as well as through routinely funding the activities of the quality department. One alternative is to add resources, but this is seldom feasible in a highly competitive environment. More commonly, priorities are adjusted to divert existing resources to quality planning, control, and improvement. This means that other work must be eliminated or postponed until the payback from quality efforts permits its completion.

Before resources can be requested, their usage must be determined. This should be done using a rational process that can be explained to others. The exercise will also provide a basis for budgeting. Figure 20.1 illustrates the approach used by a large integrated health care organization to determine their quality resource needs. The approach is used by the entire organization, not just the quality department. In fact, the organization doesn't have a "quality department" in the traditional sense of a centralized quality control activity.

People assigned to quality teams should be aware of the amount of time that will be required while they are on a team. If time is a problem, they should be encouraged to propose changes in their other priorities before the team activity starts. Resources for project teams will be made available only if pilot teams demonstrate benefits by achieving tangible results. As teams compile records of success, it will become easier to secure resources for the quality function. Ideally, quality improvement funding will become a routine part of the budgeting process, much like R&D.

Quality cost reports provide a valuable source of information for securing resources for quality improvement. Quality costs indicate the total impact of nonconformance on the company's bottom line. If the

The following questions have been outlined in an effort to provide a recipe to enable each area of HealthPartners of Southern Arizona to identify, illustrate, and detail the service requirements and costs of delivering services and/or products to our customers. Please answer each of the following questions in as much detail as necessary.

I. Customer/Customer requirements

A. What are the key products or services you provide to your customers?
B. Who are your customers? (Who receives the output of your products or services?)
C. What are your customers' expectations of the products or services you provide to them?
D. What attributes of these services or products are most important to your customers (e.g., timely, comfortable, cost effective)?
E. What measurement techniques do you use to confirm your customers' requirements?
F. What are the major factors that make your products or services necessary?

II. Structure and processes

A. Create a top-down or detailed flowchart of your key processes. (Key processes are defined as those processes that deliver services or products directly to your customers.)
B. Identify skill levels required for each step of your process. (E.g., an RN starts an IV or a Health Records Analyst completes the coding process for the chart.)

1. Identify the number of skill levels required at each step. (Steps may involve more than one skill level and/or multiples of one skill level.)
2. How much time is required by each skill level to complete each step of the process?
3. What is the average time to complete one cycle of the process? Does an industry standard time to complete this process exist? How does your process compare to any industry standards? If there is no comparable standard, how long should this process take as compared to the actual time required to complete the process? (Hint: If you were to measure this process assuming there were no delays, bottlenecks, shortage of resources, etc., how long would it take?)
4. Identify any other resources required to support the entire process (e.g., reusable equipment, disposable supplies, purchased services).

III. Cost assignment

A. What are the total costs each time the process is completed and the product or service is delivered? (That is, what are the total costs per unit of service?)
B. What are the cycle time costs for each step of the process?

IV. Process improvements

A. What are the causes, delays or roadblocks in your process that create non–value added cost?
B. What level of variation and complexity exist in your process? How do you measure the variation and complexity of your process? (Hint: What does a control chart tell you about the process variability?)
C. If you could do one or two things to reduce time delays, waste, or complexity, what would you do?
D. If you were successful in reducing time delays, waste, errors, and complexity how would this improve the product or service in the eyes of your customer? (What improved level of satisfaction would result?)

Figure 20.1 Process for identifying quality resource needs. HealthPartners of Southern Arizona; Decision Package. Courtesy of HealthPartners of Southern Arizona. Used with permission.

allocation for preventing these costs is small relative to total quality costs, management can see for themselves the potential impact of spending in the area of quality cost reduction. Likewise, if external failure costs are high relative to appraisal costs, additional appraisal expenditures may be justified.

A shortcoming of using quality costs to determine resource requirements for the quality function is that the highest costs are difficult to determine with a high degree of accuracy. What is the cost of someone *not* buying your product? As difficult as it is to measure the value of keeping an existing customer, it is more difficult to know when a prospective customer didn't consider your product because of a poor reputation for quality. It is also hard to estimate the cost of future failures with new products or processes. For example, a company decided to begin manufacturing their own gear boxes. The quality department's request for an expensive gear-checker was turned down because they couldn't precisely quantify the benefit of the equipment (the cost was easy to determine, $50,000). Two years later a field problem with the gears lead to a $1.5 million field repair program. Afterwards, the purchase of the gear-checker was quickly approved.

Another problem with quality costs is that they measure negatives, rather than the lack of positives. Quality is not only the absence of negatives (e.g., defects); it is the presence of desirable features. Quality costs measure only the cost of the former. While techniques exist for estimating the cost of lost business due to a lack of desirable features, they are not as well defined or standardized as quality costs. Thus, it is difficult to make a general statement regarding their use in securing resources for the quality function.

Performance Evaluation

Few subjects raise so much ire as performance appraisals. There are strong feelings on both sides of the issue. We will discuss the traditional employee appraisal process, some criticisms of the approach, and some alternatives.

Traditional Performance Appraisals

Performance appraisal systems typically include:

- *Standards of performance.* The standards are usually both qualitative and quantitative. Both the supervisor and the employee must know what the standards are and, usually, both agree to the standards in advance. Often both employees and supervisors work together to develop the performance standards.
- *Evaluation period.* Usually the evaluation period corresponds to the budget cycle, that is, 1 year. This is because performance appraisals

are often used to determine who gets what share of the budget "pie" allocated to salary increases. Also, a year is deemed to be a sufficiently long period to accomplish the performance goals set forth when the standards were developed.

- *The assessment.* Traditionally, the assessment is one-way: the supervisor evaluating the employee. Employees are sometimes invited to provide observations of their own.
- *Meeting.* A private, face-to-face meeting is held between the supervisor and the employee to discuss the assessment.

Companies that use the traditional approach feel that it provides such benefits as:

- Giving feedback to employees
- Giving direction to employees
- Identifying training needs
- Fostering communication between manager and employee
- Providing evidence for promotion decisions
- Providing a basis for compensation decisions
- Serving as a defense in legal cases associated with promotions or terminations

The "deliverable" of the performance appraisal is often a ranking of employees, that is, employee #1, #2, etc. Another, related deliverable is the "employee rating." Labels such as "outstanding," "satisfactory," or "low" are applied to each employee. Some firms place people into groups defined in other ways. The rankings, ratings, and groupings are then used to determine promotions, pay increases, or even disciplinary actions and dismissals.

Criticisms of Traditional Employee Appraisals

Let me preface this section by stating that the literature criticizing employee appraisals is so vast that we can provide but a brief summary of it here. Benneyan (1994) provides a bibliography of papers critical to performance appraisal that contains almost 300 references dating back to 1932. Deming lists performance appraisals as the "third deadly disease" of Western management:

> *Personal review system, or evaluation of performance, merit rating, annual review, or annual appraisal, by whatever name, for people in management, the effects … are devastating. Annual merit rating is destructive to long-term planning, nourishes short-term performance, annihilates teamwork, and demoralizes employees. Management by objective, on a go, no-go basis, without a method for accomplishment of the objective, is the same thing by another name. Management by fear would be still better.*

Joiner (1994) calls rating, ranking, and grouping of employees "the three great demoralizers" and recommends:

> *Abolish them tomorrow! These three do much harm and no good; the remedy is simple and swift.*

The primary criticism of performance appraisals is that the practice is inconsistent with, and even contradictory of, the role of the manager in the modern workplace. Deming's quality philosophy maintains that the negative consequences of annual merit review systems are absolutely devastating to any organization, and they remain an impassable barrier to meaningful process improvement. Benneyan summarizes the arguments against annual merit review as follows:

> Contrary to the desires of an organization focused on quality, these ineffective management processes encourage short-term "safe" performance at the expense of long-term planning. Additionally they
>
> - Annihilate teamwork and trust
> - Demoralize employees and destroy staff satisfaction
> - Instill fear
> - Discourage risk-taking and research
> - Foster mediocrity
> - Increase process variability
> - Encourage rivalry, competition, and politics
>
> The net result is to discourage meaningful and maximum process improvement. The negative effects far outweigh any *perceived* value for, without their removal, desired levels of quality may not ever be achievable.
>
> Numerous additional arguments exist for rigorously driving such practices from all quality organizations. For example, it is essentially impossible to design a system wherein people are evaluated only on events under their control. The very notion that all, or for that matter any, managers possess, or could ever possess, the skill necessary to judge the value of an employee is both somewhat preposterous and insulting. If anything, what is considered an evaluation of the employee is more a reflection of the system of management of that employee!
>
> Merit ratings reward people who do well within the current system. Contrary to modern quality philosophies, the focus is primarily on *measurable* goals. Ultimately, such motivation tactics do not reward attempts to improve the system, and *the organization is the loser*. Moreover, the sincere attempt to evaluate the performance of an individual based on another's personal observation and input from others introduces unavoidable biases that compromise even the best intentions.

Other criticisms that have been leveled against traditional performance appraisal include:

- Traditional performance appraisal focuses on the individual, despite solid evidence that the variance in performance is predominantly due to the system.
- It ignores, or at best, devalues, cooperative efforts.
- The goals it sets tend to be static.
- The nature of the appraisal (boss evaluating employee) emphasizes the hierarchical status, at the expense of a process-and-customer orientation.
- It reinforces command and control behavior and de-emphasizes initiative.
- The multiple ratings/rankings/groupings are not statistically valid.
- Few people are ever classified as "average."
- It causes "high" performers to slack off, and "average" or "poor" performers to brood.
- Appraisals are a detection-oriented technique.
- It leads to tampering with the system.

Alternatives to Traditional Appraisals

One alternative has already been mentioned: stop it now. This is the action recommended by Deming, Joiner, Scholtes, and many others. It's a way to stop the damage while you explore the options described below, or invent your own alternative.

Alternative #1: Fix What's Broken

Prince (1994) believes that the above criticisms of traditional performance appraisals can be summarized as arguing that "performance appraisal systems cannot accomplish what they are designed to achieve and inevitably do more harm than good." He believes that this argument is simply saying that bad systems are worse than no system at all. As a middle ground, Prince offers the following broad guidelines to firms who wish to design reward and appraisal systems compatible with a quality improvement strategy:

1. Rating scales used should have few, not many, rating categories. Most performers should be in the middle category. (Your author suggests three categories: below the system, within the system, and above the system. Unless extremely strong evidence is available, performers should be rated as within the system.)

2. Integrate subordinate, peer, customer, and self-evaluations with supervisory ratings into the process. This approach is sometimes called a 360-degree appraisal process.
3. Use continuous improvement, quality, and customer satisfaction as key evaluative criteria as well as traditional outcome or behavioral criteria.
4. Require work team or group evaluations that are at least equal in emphasis to individual-focused evaluations.
5. Use review procedures, particularly the appraisal meeting, that include the supervisor, focal employee or team, peer and/or work team representatives, and possibly customers.
6. Require more frequent performance reviews that have a dominant emphasis on future performance planning and problem solving.
7. Promotion decisions should be made by an independent administrative process that draws on performance-in-current-job data from the individual appraisal system where appropriate but also independently addresses employee's performance potential for the new job.
8. Adjustments to individual base salary should be skill based rather than performance based.
9. Include performance-based rewards with a mix of individual and team (or plant) bonuses, with the latter generally being the larger of the two.
10. Require supervisors to have primary responsibility for addressing work system constraints on performance uncovered in the performance review session.

Alternative #2: Customer-Supplier Appraisals

Eckes (1994) also believes that performance appraisals are a necessary part of business. He points out that customers constantly conduct performance appraisals. When a customer doesn't like a hotel's service, the result of his or her appraisal of that establishment is to not stay there again. More businesses are now obtaining performance evaluations through customer satisfaction surveys. Thus, as more companies are recognizing their internal and external customers, it is logical that the appraisal function should take into account how well employees satisfy the customers' requirements. Eckes describes a six-step process for customer-supplier appraisals:

- Identify the customers. This includes external customers, interdepartmental internal customers, and intradepartmental internal customers (department employees, department manager, and technical advisors).

- Identify customer requirements. Meet with the customers to determine what they expect.
- Determine metrics for current performance. Metrics should be determined for each requirement.
- Identify areas for improvement.
- Form teams to develop improvement plans.
- Develop tools (surveys, interviews, etc.) for measuring customer-perceived performance improvement.

Alternative #3: Process Appraisals

In a process appraisal the focus is not on outcomes, where the employee's level of control is debatable. Rather, the appraiser focuses on key process behaviors that are under the employee's direct control. For example, suppose the output of a process is successful supplier projects. Rather than concentrating on this factor in the performance appraisal, the focus would be on key success factors associated with individuals who consistently produce successful supplier projects (e.g., communication skills). These factors can not only form the developmental plans for each employee but can also improve personnel practices: candidates who already exhibit these traits can be hired.

Alternative #4: Managing Personal Growth

In response to the criticisms of traditional performance appraisals, Dow adopted a process called Managing Personal Growth (MPG) to help employees and managers clarify what each manager expects from his or her employees and what the employees' personal strengths and development needs are. It is used by managers and their direct reports at all levels and in all functions. The MPG process has four steps:

- *Pre-workshop assignment.* Each participant and his or her manager independently define the participant's job responsibilities, important skills, and personal capabilities. Employees are asked to consider and rate their personal and professional values. They are encouraged to think of their career growth and accept responsibility for their success or failure.
- *Workshop.* Participants attend a 1½- or 2-day workshop at which they clarify their values and personal motivators and develop ideas for increasing job satisfaction. They also compare their priorities and skills with their managers' assessments of the same. Then they outline specific actions to increase job satisfaction and improve performance.
- *Development discussion.* The participant and manager review the employee's development plan and agree on job priorities, areas for

development, and the best ways to use the employee's strengths. The open discussion (typically 60 to 90 minutes) creates a partnership that helps promote individual and organizational success. It differs from a performance appraisal by focusing on future development and encouraging two-way communication.

- *Continuing feedback.* The manager and employee continue to meet two or three times annually to talk about progress and plan for continuing development.

Alternative #5: The Boss-less Performance Review

Fitzsimmons (1996) describes a performance review process developed by a clerical organization for clerics in Baltimore, Maryland. The process used is applicable to any professional and management position.

The organization was concerned with gathering systematic and reliable information about its ministers' performance that could be used to provide feedback and make future assignments. The director of the Pastoral Personnel Services Staff in the Baltimore archdiocese, along with three ministers, identified the key roles of the parish minister along with associated behaviors that are central to fulfilling each role. Based on these roles, a preliminary assessment instrument was field tested for validity. The responses and discussions from the field test led to collapsing the four roles into three. At the same time, two versions of the assessment instrument evolved: one for pastors and one for associates. Further field tests of the revised document were conducted until the staff and pastors were satisfied that the instrument was valid and reliable.

Two members of the Pastoral Personnel Services Staff were made part of the committee to bring their experiences and insights from the development of the assessment instrument. The committee, with the assistance of an outside consultant, then began to focus its attention on a process for performance assessment. After brainstorming some ideas, hopes, and preferences, it developed a sequence of activities that would make up the assessment process:

- Identify who is to be assessed. (Each minister would be evaluated once every 5 years.)
- Prepare survey forms and letters.
- Develop a plan for administering the survey to the assessee's constituents.
- Conduct the survey. The survey instrument is used to collect assessments from parish constituencies, such as a sampling of the total congregation, church support staff, and the leadership of various committees and special groups within the parish. This is akin to conducting a survey of customers, suppliers, supervisors, and colleagues.

- Review responses.
- Summarize the data into a printout.
- Discuss the summary with the assessee during a feedback session.
- With the assessee, write a memo of understanding (a joint agreement on the significant points made by respondents) and a summary for the personnel office.
- With the assessee, create an individual growth and development plan.
- Conduct a follow-up 1 year later.

The person who fills the role traditionally filled by the "boss" in this boss-less process is the *confidant*. The confidant is the person who conducts the survey, reviews the responses, and helps create the memo, summary, and development plan. This individual, a respected peer of the assessee, attends a 2-day certification program. The training focuses on the process, the administrative activities associated with each step, the skills of objectivity and joint decision making, and the format and guidelines for creating a development plan. Affirmation and mutual respect are stressed throughout the training, as well as the need to protect the anonymity of the survey respondents.

Use of the Boss-less Review in Any Organization Fitzsimmons offers the following guidelines for using this peer review process for performance evaluations with any professional:

- Create a process that will enable each professional to learn how his or her behaviors are viewed by customers.
- Focus on professional behaviors exhibited by top performers that are validated through testing.
- Gather information, as rated by the customers or constituents of the professional's work, on that person's behavior through a general sampling methodology that assures respondents' anonymity.
- Use a peer, trained in coaching skills and the administrative procedures, to oversee the process, conduct the assessment review, help create a growth and development plan, and conduct the follow-up session.
- Ensure confidentiality throughout the process by storing the information in a special file and limiting access to only certain individuals. This helps build trust and confidence in the process.
- Involve professionals in the decisions that affect them. Allow them to validate the behaviors on which they will be assessed, to review the sampling process for respondents, to participate

in identifying peer confidants and selecting their own peer reviewer, and to be part of building a support system of resources that will help them learn and grow toward professional mastery.

Professional Development

Professional development is the set of activities associated with obtaining and maintaining professional credentials and of expanding one's knowledge in one's chosen field. Here are a few examples of how some companies encourage their employees to continue their professional development:

Granite Rock, 1992 Baldrige Award winner, conducts annual professional development reviews with every employee.

Virginia Beach Ambulatory Surgery Center (outpatient surgery) gives cash awards and newsletter recognition to people who pass certification exams.

Grumman and *IBM* both have "Quality Colleges" that offer employees the equivalent of a college degree program in quality-related subjects.

Credentials

A *credential* is that which entitles one to a claim of authority or expertise in a certain area. More concretely, a credential provides evidence that one has a right to such a claim.

One class of credentials is *compliance credentials*. Compliance credentials include licenses required by regulatory agencies for jobs that involve public health and/or safety. In addition to legally required credentials, employers or customers may require credentials for such jobs as a condition of the contract. Credentials are sometimes required to permit one to perform certain tasks that require a level of skill that can't be easily determined by after-the-fact inspection of the work, for example, welding or reading of X-ray images. Such credentials are often highly task-specific, for example, certified to perform a particular surgical procedure or to weld nuclear reactors. Since skills can deteriorate, periodic recertification is usually required to maintain the credential. Examples of jobs requiring compliance credentials are surgeons, certain engineering professions (including, in some states, quality engineering), nurses, midwifes, radiology technicians, food and drug workers, nuclear inspectors, nuclear welders, etc.

Professional Certification

It is possible to obtain certification in many of the broad categories of jobs in the quality field. Certification is formal recognition by one's peers (ASQ) that an individual has demonstrated a proficiency within and a comprehension of a specified body of knowledge at a point in time. Peer recognition is not registration or licensing.

Professional Development Courses

Ongoing professional development can be obtained through in-house or public seminars. In-house training is typically conducted by staff personnel, or by purchasing the service from outside sources. In-house courses can be scheduled at the company's convenience and can be tailored to the company's specific needs. Because all of the attendees of in-house training programs are from the same organization, discussions can focus on issues that are relevant to the company. Also, the participants bring a common background to the sessions that reduces the amount of time spent explaining the examples to "outsiders." In-house courses can be significantly lower in terms of per-person registration fees, and the savings in travel expenses can be considerable. To be cost effective, a minimum attendance is usually necessary.

In-house training also has some disadvantages. If it is conducted on-site, participants may be interrupted. The fact that all participants are from the same company limits the opportunity to share ideas and experiences with others from different backgrounds. Unless there are a significant number of attendees, it may be less cost-effective than public seminars.

Goals and Objectives

Performance goal setting is an important activity that is closely related to quality improvement. Goals set for departments, teams, or individuals should be linked to the organization's mission, purpose, and strategic plans. Goals should not be set in a vacuum. The test of the value of a particular goal is that it moves the entire organization toward its mission. A goal that, if met by one area of the organization, causes difficulty in another area, is not valid, for example, a purchasing department that sets a goal of reducing the cost of purchased material without regard to its impact on quality, production schedules, etc. One way to safeguard against inadvertently setting such goals is to involve the customer in the goal-setting process. Goal setting is also integral to performance evaluation. To be useful, goals should conform to certain guidelines (Johnson, 1993b):

- *Goals should be specific and measurable.* Vague goals mean different things to different people. "Improve customer satisfaction" is a vague goal. "Improve customer satisfaction as indicated by a significant improvement in item #30 on the monthly customer survey" is better.
- *Goals should be challenging, yet realistic.* All parties should agree that the goal is attainable, and that they would derive personal satisfaction from having attained it.
- *Goals should be written.* Goals should be consistent—goals should not contradict other goals. It should be possible to attain the entire set of goals simultaneously.

- *Goals should be accepted.* The best goal is one set by oneself, even though the supervisor provides guidance and assistance in the goal-setting process.

Achieving the Goals

Goals should be accompanied by a detailed plan of how each goal will be achieved. Goals without plans are little more than wishes. The plan will detail the steps that will be taken to reach the goal: who will be responsible for each step, resources that will be required, a timetable. The supervisor should assist the employee(s) with the planning process, and he should agree to the plan. If the plan is carried out and the goals not reached, the problem is in the plan, not in the employee. Remember, Plan-Do-Check-Act is a process of continuous improvement. If the plan doesn't achieve the result, improve the plan and try it again. Since progress will be monitored on an ongoing basis, and the plan will include a timetable, lack of progress should be evident well before the time the goal is to be accomplished. The lack of progress is a signal to revise the plan.

The responsibility of achieving the goals belongs to both the supervisor and the employee, as well as everyone on the staff. It's a team effort. It's a companywide effort. The supervisor should work with the employee and the staff to identify ways that the supervisor can assist them in meeting their goals. Progress toward the goals should be monitored constantly.

Coaching

In modern, quality-focused organizations supervisors may spend up to 60 percent of their time coaching. On a one-on-one basis, coaching refers to the process of helping a single employee improve some aspect of his or her performance. On a group level, coaching is a process of developing effective teams and work groups. The successful coach needs the following skills:

- Communication skills
- Listening skills
- Analysis skills
- Negotiation skills
- Conflict resolution skills

The coach must also possess sufficient subject matter knowledge to assist the employee in achieving results. Expert knowledge of the task enables the coach to describe and demonstrate the desired behavior and to observe the employee performing the task and give feedback. Just as an NBA coach must be an expert in basketball, business coaches must be experts on the subjects they are coaching.

Coaches must also understand how adults learn and grow. Brookfield (1986) identified six factors that influence adult learning:

1. More learning takes place if learning is seen as voluntary, self-initiated activity.
2. More learning takes place in a climate of mutual respect.
3. The best learning takes place in an environment characterized by a spirit of collaboration.
4. Learning involves a balance between action and self-reflection.
5. People learn best when their learning is self-directed.

Coaches should be aware of the essentials of motivation theory, such as Maslow's needs hierarchy, McGregor's theory X and theory Y models of management, and Herzberg's hygiene theory, as discussed in Chap. 18.

There are many variations on the coaching process, but all involve similar steps (Finnerty, 1996):

- Define performance goals.
- Identify necessary resources for success.
- Observe and analyze current performance.
- Set expectations for performance improvement.
- Plan a coaching schedule.
- Meet with the individual or team to get commitment to goals, demonstrate the desired behavior, and establish boundaries.
- Give feedback on practice and performance.
- Follow up to maintain goals.

Situations That Require Coaching to Improve Performance

Coaching discussions may be initiated as a result of an administrative situation or because the manager has become aware of an event or incident of concern in relation to a task or project. In their publication, *Coaching Skills*, The Center for Management and Organization Effectiveness lists the following situations in which coaching may be required:

Administrative situations

- Setting objectives
- Performance reviews
- Salary discussions
- Career planning-development discussions
- Job posting and bidding discussions

Project or task situations

- A specific project or assignment problem: for instance, delays, quality problems, quantity problems, lack of follow-through on commitments
- Absenteeism and/or tardiness
- Deficiency in effort or motivation
- Behavior that causes problems, for example, aggressiveness
- Training: opportunity or assignment
- When someone new joins your group or team
- Conflicts between employees or within groups
- Communication problems or breakdowns
- When your own supervisor makes you aware that one of your employees has a problem

Forms of Coaching

Coaching can take on a variety of forms. The traditional form is, as described above, a supervisor coaching a subordinate or a sponsor coaching a team. However, other forms do occur on occasion and should probably occur more often than they do. A number of the more common alternative forms of coaching are described here (Finnerty, 1996):

- *Mentoring.* Mentoring involves a relationship between a senior manager and a less-experienced employee. The mentor is a trusted friend as well as a source of feedback. Mentors provide employees with information and feedback they may not otherwise receive. It was stated earlier that coaches must be subject-matter experts in the area they are coaching. Mentors are experts on the *organization itself.* Because of their superior knowledge of the written and unwritten rules of the organization, the mentor can help the employee traverse the often perplexing maze that must be negotiated to achieve success. Knowledge of the organization's informal leadership, norms, values, and culture can usually only be acquired from either a mentor, or by experience. *"Experience is a stern school, 'tis a fool that will learn in no other."*
- *Peer coaching.* Traditional, vertically structured organizations have been the model for mentoring. Typically, the mentors or coaches are above the employees in the organizational hierarchy and have formal command authority over them. Peer coaching relationships are those where all parties are approximately the same level within the organization or, at least, where no party in the relationship has

command authority over any other party. For all the talk of formal training, policy, and procedure manuals, etc., the truth is that in the real world, most "training" takes the form of a co-worker telling the employee "the way it's *really* done around here . . ." Peer coaching recognizes that this approach has tremendous value. Among the advantages is that there is less hesitancy in asking a peer for help for fear of revealing ignorance or bothering one's superior with a minor problem. Peer coaching can be facilitated by putting people together for the purpose of learning from one another. Peer coaches are provided with training in the skills of coaching.

- *Executive coaching.* By its nature a command hierarchy discourages upward communication of "bad news." The result of this is that the higher one's position in the hierarchy, the less feedback one obtains on their performance. This is especially true for negative feedback. Employees quickly learn that it is not in their best interest to criticize the boss. While these problems are partially alleviated by such innovations as anonymous 360-degree performance assessments, it may also be useful for the executive to have a coach. The executive coach is usually a consultant hired from outside of the firm. The coaching role is typically to act as coach to the executive team, rather than to a particular executive.

APPENDIX A

Control Chart Constants

	Chart for Average			Chart for Standard Deviations					
	Factors for Control Limits			Factors for Central Line		Factors for Control Limits			
Observations in Sample, n	A	A_2	A_3	c_4	$1/c_4$	B_3	B_4	B_5	B_6
2	2.121	1.880	2.659	0.7979	1.2533	0	3.267	0	2.606
3	1.732	1.023	1.954	0.8862	1.1284	0	2.568	0	2.276
4	1.500	0.729	1.628	0.9213	1.0854	0	2.266	0	2.088
5	1.342	0.577	1.427	0.9400	1.0638	0	2.089	0	1.964
6	1.225	0.483	1.287	0.9515	1.0510	0.030	1.970	0.029	1.874
7	1.134	0.419	1.182	0.9594	1.0423	0.118	1.882	0.113	1.806
8	1.061	0.373	1.099	0.9650	1.0363	0.185	1.815	0.179	1.751
9	1.000	0.337	1.032	0.9693	1.0317	0.239	1.761	0.232	1.707
10	0.949	0.308	0.975	0.9727	1.0281	0.284	1.716	0.276	1.669
11	0.905	0.285	0.927	0.9754	1.0252	0.321	1.679	0.313	1.637
12	0.866	0.266	0.886	0.9776	1.0229	0.354	1.646	0.346	1.610
13	0.832	0.249	0.850	0.9794	1.0210	0.382	1.618	0.374	1.585
14	0.802	0.235	0.817	0.9810	1.0194	0.406	1.594	0.399	1.563
15	0.775	0.223	0.789	0.9823	1.0180	0.428	1.572	0.421	1.544
16	0.750	0.212	0.763	0.9835	1.0168	0.448	1.552	0.440	1.526
17	0.728	0.203	0.739	0.9845	1.0157	0.466	1.534	0.458	1.511
18	0.707	0.194	0.718	0.9854	1.0148	0.482	1.518	0.475	1.496
19	0.688	0.187	0.698	0.9862	1.0140	0.497	1.503	0.490	1.483
20	0.671	0.180	0.680	0.9869	1.0133	0.510	1.490	0.504	1.470
21	0.655	0.173	0.663	0.9876	1.0126	0.523	1.477	0.516	1.459
22	0.640	0.167	0.647	0.9882	1.0119	0.534	1.466	0.528	1.448
23	0.626	0.162	0.633	0.9887	1.0114	0.545	1.455	0.539	1.438
24	0.612	0.157	0.619	0.9892	1.0109	0.555	1.445	0.549	1.429
25	0.600	0.153	0.606	0.9896	1.0105	0.565	1.435	0.559	1.420

Chart for Ranges							
Observations in Sample, n	Factors for Central Line			Factors for Control Limits			
	d_2	$1/d_2$	d_3	D_1	D_2	D_3	D_4
2	1.128	0.8865	0.853	0	3.686	0	3.267
3	1.693	0.5907	0.888	0	4.358	0	2.574
4	2.059	0.4857	0.880	0	4.698	0	2.282
5	2.326	0.4299	0.864	0	4.918	0	2.114
6	2.534	0.3946	0.848	0	5.078	0	2.004
7	2.704	0.3698	0.833	0.204	5.204	0.076	1.924
8	2.847	0.3512	0.820	0.388	5.306	0.136	1.864
9	2.970	0.3367	0.808	0.547	5.393	0.184	1.816
10	3.078	0.3249	0.797	0.687	5.469	0.223	1.777
11	3.173	0.3152	0.787	0.811	5.535	0.256	1.744
12	3.258	0.3069	0.778	0.922	5.594	0.283	1.717
13	3.336	0.2998	0.770	1.025	5.647	0.307	1.693
14	3.407	0.2935	0.763	1.118	5.696	0.328	1.672
15	3.472	0.2880	0.756	1.203	5.741	0.347	1.653
16	3.532	0.2831	0.750	1.282	5.782	0.363	1.637
17	3.588	0.2787	0.744	1.356	5.820	0.378	1.622
18	3.640	0.2747	0.739	1.424	5.856	0.391	1.608
19	3.689	0.2711	0.734	1.487	5.891	0.403	1.597
20	3.735	0.2677	0.729	1.549	5.921	0.415	1.585
21	3.778	0.2647	0.724	1.605	5.951	0.425	1.575
22	3.819	0.2618	0.720	1.659	5.979	0.434	1.566
23	3.858	0.2592	0.716	1.710	6.006	0.443	1.557
24	3.895	0.2567	0.712	1.759	6.031	0.451	1.548
25	3.931	0.2544	0.708	1.806	6.056	0.459	1.541

APPENDIX B

Control Chart Equations

	np Chart	*p* Chart
LCL	$LCL = n\bar{p} - 3\sqrt{n\bar{p}(1-\bar{p})}$ **Or 0 if LCL is negative**	$LCL = \bar{p} - 3\sqrt{\frac{\bar{p}(1-\bar{p})}{n}}$ **Or 0 if LCL is negative**
Center Line	$n\bar{p} = \frac{\text{sum of subgroup defective counts}}{\text{number of subgroups}}$	$\bar{p} = \frac{\text{sum of subgroup defective counts}}{\text{sum of subgroup sizes}}$
UCL	$UCL = n\bar{p} + 3\sqrt{n\bar{p}(1-\bar{p})}$ **or n if UCL is greater than n**	$UCL = \bar{p} + 3\sqrt{\frac{\bar{p}(1-\bar{p})}{n}}$ **or 1 if UCL is greater than 1**

	c Chart	*u* Chart
LCL	$LCL = \bar{c} - 3\sqrt{\bar{c}}$ **Or 0 if LCL is negative**	$LCL = \bar{u} - 3\sqrt{\frac{\bar{u}}{n}}$ **Or 0 if LCL is negative**
Center Line	$\bar{c} = \frac{\text{sum of subgroup occurrences}}{\text{number of subgroups}}$	$\bar{u} = \frac{\text{sum of subgroup occurrences}}{\text{number of units in all subgroups}}$
UCL	$UCL = \bar{c} + 3\sqrt{\bar{c}}$	$UCL = \bar{u} + 3\sqrt{\frac{\bar{u}}{n}}$

	X Chart
LCL	$LCL = \bar{X} - 2.66 \times \bar{R}$
Center Line	$\bar{X} = \frac{\text{sum of measurements}}{\text{number of measurements}}$
UCL	$UCL = \bar{X} + 2.66 \times \bar{R}$

	R Chart	*X*-bar Chart
LCL	$LCL = D_3\bar{R}$ Or 0 if LCL is negative	$LCL = \bar{\bar{X}} - A_2\bar{R}$
Center Line	$\bar{R} = \frac{\text{sum of subgroup ranges}}{\text{number of subgroups}}$	$\bar{\bar{X}} = \frac{\text{sum of subgroup averages}}{\text{number of subgroups}}$
UCL	$UCL = D_4\bar{R}$	$UCL = \bar{\bar{X}} + A_2\bar{R}$

	S Chart	*X*-bar Chart
LCL	$LCL = B_3\bar{S}$ Or 0 if LCL is negative	$LCL = \bar{\bar{X}} - A_3\bar{S}$
Center Line	$\bar{S} = \frac{\text{sum of subgroup sigmas}}{\text{number of subgroups}}$	$\bar{\bar{X}} = \frac{\text{sum of subgroup averages}}{\text{number of subgroups}}$
UCL	$UCL = B_4\bar{S}$	$UCL = \bar{\bar{X}} + A_3\bar{S}$

APPENDIX C

Area under the Standard Normal Curve

z	0.00	0.01	0.02	0.03	0.04	0.05	0.06	0.07	0.08	0.09
–3.4	0.0003	0.0003	0.0003	0.0003	0.0003	0.0003	0.0003	0.0003	0.0003	0.0002
–3.3	0.0005	0.0005	0.0005	0.0004	0.0004	0.0004	0.0004	0.0004	0.0004	0.0003
–3.2	0.0007	0.0007	0.0006	0.0006	0.0006	0.0006	0.0006	0.0005	0.0005	0.0005
–3.1	0.0010	0.0009	0.0009	0.0009	0.0008	0.0008	0.0008	0.0008	0.0007	0.0007
–3.0	0.0013	0.0013	0.0013	0.0012	0.0012	0.0011	0.0011	0.0011	0.0010	0.0010
–2.9	0.0019	0.0018	0.0018	0.0017	0.0016	0.0016	0.0015	0.0015	0.0014	0.0014
–2.8	0.0026	0.0025	0.0024	0.0023	0.0023	0.0022	0.0021	0.0021	0.0020	0.0019
–2.7	0.0035	0.0034	0.0033	0.0032	0.0031	0.0030	0.0029	0.0028	0.0027	0.0026
–2.6	0.0047	0.0045	0.0044	0.0043	0.0041	0.0040	0.0039	0.0038	0.0037	0.0036
–2.5	0.0062	0.0060	0.0059	0.0057	0.0055	0.0054	0.0052	0.0051	0.0049	0.0048
–2.4	0.0082	0.0080	0.0078	0.0075	0.0073	0.0071	0.0069	0.0068	0.0066	0.0064
–2.3	0.0107	0.0104	0.0102	0.0090	0.0096	0.0094	0.0091	0.0089	0.0087	0.0084
–2.2	0.0139	0.0136	0.0132	0.0129	0.0125	0.0122	0.0119	0.0116	0.0113	0.0110
–2.1	0.0179	0.0174	0.0170	0.0166	0.0162	0.0158	0.0154	0.0150	0.0146	0.0143
–2.0	0.0228	0.0222	0.0217	0.0212	0.0207	0.0202	0.0197	0.0192	0.0188	0.0183
–1.9	0.0287	0.0281	0.0274	0.0268	0.0262	0.0256	0.0250	0.0244	0.0239	0.0233
–1.8	0.0359	0.0351	0.0344	0.0336	0.0329	0.0322	0.0314	0.0307	0.0301	0.0294
–1.7	0.0446	0.0436	0.0427	0.0418	0.0409	0.0401	0.0392	0.0384	0.0375	0.0367
–1.6	0.0548	0.0537	0.0526	0.0516	0.0505	0.0495	0.0485	0.0475	0.0465	0.0455
–1.5	0.0668	0.0655	0.0643	0.0630	0.0618	0.0606	0.0594	0.0582	0.0571	0.0559
–1.4	0.0808	0.0793	0.0778	0.0764	0.0749	0.0735	0.0721	0.0708	0.0694	0.0681
–1.3	0.0968	0.0951	0.0934	0.0918	0.0901	0.0885	0.0869	0.0853	0.0838	0.0823
–1.2	0.1151	0.1131	0.1112	0.1093	0.1075	0.1056	0.1038	0.1020	0.1003	0.0985
–1.1	0.1357	0.1335	0.1314	0.1292	0.1271	0.1251	0.1230	0.1210	0.1190	0.1170
–1.0	0.1587	0.1562	0.1539	0.1515	0.1492	0.1469	0.1446	0.1423	0.1401	0.1379

(Continued)

z	0.00	0.01	0.02	0.03	0.04	0.05	0.06	0.07	0.08	0.09
–0.9	0.1841	0.1814	0.1788	0.1762	0.1736	0.1711	0.1685	0.1660	0.1635	0.1611
–0.8	0.2119	0.2090	0.2061	0.2033	0.2005	0.1977	0.1949	0.1922	0.1894	0.1867
–0.7	0.2420	0.2389	0.2358	0.2327	0.2296	0.2266	0.2236	0.2206	0.2177	0.2148
–0.6	0.2743	0.2709	0.2676	0.2643	0.2611	0.2578	0.2546	0.2514	0.2483	0.2451
–0.5	0.3085	0.3050	0.3015	0.2981	0.2946	0.2912	0.2877	0.2843	0.2810	0.2776
–0.4	0.3446	0.3409	0.3372	0.3336	0.3300	0.3264	0.3228	0.3192	0.3156	0.3121
–0.3	0.3821	0.3783	0.3745	0.3707	0.3669	0.3632	0.3594	0.3557	0.3520	0.3483
–0.2	0.4207	0.4168	0.4129	0.4090	0.4052	0.4013	0.3974	0.3936	0.3897	0.3859
–0.1	0.4602	0.4562	0.4522	0.4483	0.4443	0.4404	0.4364	0.4325	0.4286	0.4247
–0.0	0.5000	0.4960	0.4920	0.4880	0.4840	0.4801	0.4761	0.4721	0.4681	0.4641
0.0	0.5000	0.5040	0.5080	0.5120	0.5160	0.5199	0.5239	0.5279	0.5319	0.5359
0.1	0.5398	0.5438	0.5478	0.5517	0.5557	0.5596	0.5636	0.5675	0.5714	0.5753
0.2	0.5793	0.5832	0.5871	0.5910	0.5948	0.5987	0.6026	0.6064	0.6103	0.6141
0.3	0.6179	0.6217	0.6255	0.6293	0.6331	0.6368	0.6406	0.6443	0.6480	0.6517
0.4	0.6554	0.6591	0.6628	0.6664	0.6700	0.6736	0.6772	0.6808	0.6844	0.6879
0.5	0.6915	0.6950	0.6985	0.7019	0.7054	0.7088	0.7123	0.7157	0.7190	0.7224
0.6	0.7257	0.7291	0.7324	0.7357	0.7389	0.7422	0.7454	0.7486	0.7517	0.7549
0.7	0.7580	0.7611	0.7642	0.7673	0.7704	0.7734	0.7764	0.7794	0.7823	0.7852
0.8	0.7881	0.7910	0.7939	0.7967	0.7995	0.8023	0.8051	0.8078	0.8106	0.8133
0.9	0.8159	0.8186	0.8212	0.8238	0.8264	0.8289	0.8315	0.8340	0.8365	0.8389
1.0	0.8413	0.8438	0.8461	0.8485	0.8508	0.8531	0.8554	0.8577	0.8599	0.8621
1.1	0.8643	0.8665	0.8686	0.8708	0.8729	0.8749	0.8770	0.8790	0.8810	0.8830

1.2	0.8849	0.8869	0.8888	0.8907	0.8925	0.8944	0.8962	0.8980	0.8997	0.9015
1.3	0.9032	0.9049	0.9066	0.9082	0.9099	0.9115	0.9131	0.9147	0.9162	0.9177
1.4	0.9192	0.9207	0.9222	0.9236	0.9251	0.9265	0.9279	0.9292	0.9306	0.9319
1.5	0.9332	0.9345	0.9357	0.9370	0.9382	0.9394	0.9406	0.9418	0.9429	0.9441
1.6	0.9452	0.9463	0.9474	0.9484	0.9495	0.9505	0.9515	0.9525	0.9535	0.9545
1.7	0.9554	0.9564	0.9573	0.9582	0.9591	0.9599	0.9608	0.9616	0.9625	0.9633
1.8	0.9641	0.9649	0.9656	0.9664	0.9671	0.9678	0.9686	0.9693	0.9699	0.9706
1.9	0.9713	0.9719	0.9726	0.9732	0.9738	0.9744	0.9750	0.9756	0.9761	0.9767
2.0	0.9772	0.9778	0.9783	0.9788	0.9793	0.9798	0.9803	0.9808	0.9812	0.9817
2.1	0.9821	0.9826	0.9830	0.9834	0.9838	0.9842	0.9846	0.9850	0.9854	0.9857
2.2	0.9861	0.9864	0.9868	0.9871	0.9875	0.9878	0.9881	0.9884	0.9887	0.9890
2.3	0.9893	0.9896	0.9898	0.9901	0.9904	0.9906	0.9909	0.9911	0.9913	0.9916
2.4	0.9918	0.9920	0.9922	0.9925	0.9927	0.9929	0.9931	0.9932	0.9934	0.9936
2.5	0.9938	0.9940	0.9941	0.9943	0.9945	0.9946	0.9948	0.9949	0.9951	0.9952
2.6	0.9953	0.9955	0.9956	0.9957	0.9959	0.9960	0.9961	0.9962	0.9963	0.9964
2.7	0.9965	0.9966	0.9967	0.9968	0.9969	0.9970	0.9971	0.9972	0.9973	0.9974
2.8	0.9974	0.9975	0.9976	0.9977	0.9977	0.9978	0.9979	0.9979	0.9980	0.9981
2.9	0.9981	0.9982	0.9982	0.9983	0.9984	0.9984	0.9985	0.9985	0.9986	0.9986
3.0	0.9987	0.9987	0.9987	0.9988	0.9988	0.9989	0.9989	0.9989	0.9990	0.9990
3.1	0.9990	0.9991	0.9991	0.9991	0.9992	0.9992	0.9992	0.9992	0.9993	0.9993
3.2	0.9993	0.9993	0.9994	0.9994	0.9994	0.9994	0.9994	0.9995	0.9995	0.9995
3.3	0.9995	0.9995	0.9995	0.9996	0.9996	0.9996	0.9996	0.9996	0.9996	0.9997
3.4	0.9997	0.9997	0.9997	0.9997	0.9997	0.9997	0.9997	0.9997	0.9997	0.9998

APPENDIX D

Simulated Certification Exam Questions

This handbook is designed for quality professionals wishing to improve their understanding of the quality management body of knowledge that they can apply in their day-to-day work.

Readers should note that questions that appear on any given certification exam can come from any source. No single book can hope to specifically deal with every possible question. Furthermore, it is our belief that preparing for the exam by focusing on questions is a flawed strategy. Frankly, the student who is only interested in passing an exam deserves to fail. The goal should be mastery of the subject matter, not a passing score on an exam. Quality management is a serious business and it should be practiced by people who have in-depth knowledge of the subject. The customer's safety and the organization's viability may depend on the expertise of the quality manager. There is no room for a quality manager whose sole objective is to know enough about quality management to answer some questions correctly on a given day.

By mastering the quality management body of knowledge, you will be ready for whatever problems come your way. The problems may be in the form of questions on a certification exam, or decisions you will make on the job. Mastery implies understanding, not memorization of a few facts. When you understand the subject, you will have grasped the principles of quality management. The fundamental principles will be your guiding light in times of confusion. Such wisdom can never be attained by focusing on exam questions that ask, for example, whether Ishikawa referred to his system of management as "total quality control" or "company-wide quality control." Knowing such trivia may make you seem erudite at a cocktail party, but it will not help you make better decisions regarding quality.

What is involved in "mastery?" Mastery involves a firm grasp of fundamental principles not as floating abstractions, but as a means of understanding reality. The principle "reducing variation results in improved quality" is understood if it immediately creates in your mind the image of a customer who knows that she can depend on the quality of your products. This book presents fundamental principles, along with numerous examples of the principles in practice. Of course, you will also need to provide your own examples from your own experience. The point is that the focus is on fundamental principles of quality management, not memorizing answers to potential exam questions.

This handbook was originally written with chapter headings and subheadings matching those of the ASQ Body of Knowledge. Frankly, this resulted in a flawed product. The body of knowledge was not written, nor intended, as a rational flow of concepts and ideas that build in a readable fashion to develop the reader's skills. This second edition handbook was reformatted to group like topics that will build on the concepts discussed.

While you may find it useful to "weight" your study time toward the more prevalent test topics, consider the earlier advice: It is nearly impossible to "master" the Body of Knowledge by studying test questions. When you review the material covered in a given section, note that some sections cover a broad Body of Knowledge, sometimes at an advanced (Evaluation, Analysis) cognitive level. As a result, many questions are needed to adequately test these Body of Knowledge topics, even though a given exam may ask only a few questions requiring less breadth or depth of understanding.

Regardless, we feel the materials presented, if properly reviewed, will offer you great opportunity in passing a certification exam and (more importantly) help you to develop exceptional skills in quality management. Best wishes on your success!

Answers may be downloaded at www.mhprofessional.com/HQM2

1. Under the Kano model, which of the following is most likely to occur as a result of competitive pressure?
   a. A basic quality feature will become an expected quality feature.
   b. An exciting quality feature will become an expected quality feature.
   c. An expected quality feature will become a basic quality feature.
   d. Choices b and c.

2. The graph shown is interpreted as follows:

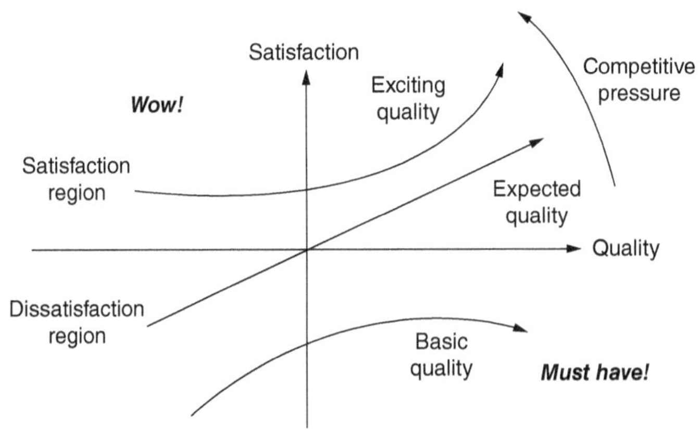

a. Customer satisfaction is determined solely by the quantity of the product or service delivered.
b. Customer wants can be determined once and for all and used to design high-quality products and services.
c. Customer wants, needs, and expectations are dynamic and must be monitored continually. Providing products or services that match the customers' expectations is not enough to ensure customer satisfaction.
d. Customers will be satisfied if you supply them with products and services that meet their needs at a fair price.

3. The primary reason for evaluating and maintaining surveillance over a supplier's quality program is to:
   a. perform product inspection at source.
   b. eliminate incoming inspection cost.
   c. motivate suppliers to improve quality.
   d. make sure the supplier's quality program is functioning effectively.

4. Incoming-material inspection is based most directly on:
   a. design requirements.
   b. purchase order requirements.
   c. manufacturing requirements.
   d. customer use of the end product.

5. The most important step in vendor certification is to:
   a. obtain copies of the vendor's handbook.
   b. familiarize the vendor with quality requirements.
   c. analyze the vendor's first shipment.
   d. visit the vendor's plant.

6. The advantage of a written procedure is:
   a. it provides flexibility in dealing with problems.
   b. unusual conditions are handled better.
   c. it is a perpetual coordination device.
   d. coordination with other departments is not required.

7. A vendor quality control plan has been adopted. Which of the following provisions would you advise top management to be the least effective?
   a. Product audits
   b. Source inspection
   c. Certificate of analysis
   d. Certificate of compliance
   e. Pre-award surveys

8. The most desirable method of evaluating a supplier is:
   a. history evaluation.
   b. survey evaluation.
   c. questionnaire.
   d. discussing with quality manager on phone.
   e. all of the above

9. When purchasing materials from vendors, it is sometimes advantageous to choose vendors whose prices are higher because:
   a. materials that cost more can be expected to be better, and "you get what you pay for."
   b. such vendors may become obligated to bestow special favors.
   c. such a statement is basically incorrect. Always buy at lowest bid price.
   d. the true cost of purchased materials, which should include items such as sorting, inspection, contacting vendors, and production delays, may be lower.

10. A quality audit program should begin with:
    a. a study of the quality documentation system.
    b. an evaluation of the work being performed.
    c. a report listing findings, the action taken, and recommendations.
    d. a charter of policy, objectives, and procedures.
    e. a follow-up check on the manager's response to recommendations.

11. Auditing of a quality program is most effective on a:
    a. quarterly basis, auditing all characteristics on the checklist.
    b. periodic unscheduled basis, auditing some of the procedures.
    c. monthly basis, auditing selected procedures.
    d. continuing basis, auditing randomly selected procedures.
    e. continually specified time period basis, frequency adjustable, auditing randomly selected procedures.

12. An inspection performance audit is made of eight inspectors in an area of complex assembly, all doing similar work. Seven inspectors have an average monthly acceptance rate of 86 to 92 percent; one inspector has an average rate of 72 percent with approximately four times the daily variation as the others. As inspection supervisor you should, based on this audit,
    a. promote the 72 percent inspector, as he is very conscientious.
    b. discipline the 72 percent inspector, as he is creating needless rework and wasted time.
    c. initiate a special investigation of inspection and manufacturing performance.
    d. discipline the other seven inspectors as they are not "cracking down."

13. The quality audit could be used to judge all of the following except:
    a. a prospective vendor's capability for meeting quality standards.
    b. the adequacy of a current vendor's system for controlling quality.
    c. the application of a specification to a unique situation.
    d. the adequacy of a company's own system for controlling quality.

14. Audit inspectors should report to someone who is independent from:
    a. middle management.
    b. marketing.
    c. inspection supervision.
    d. production staff.

15. The term "quality audit" can refer to the appraisal of the quality system of:
    a. an entire plant or company.
    b. one product.
    c. one major quality activity.
    d. any of the above

16. You would normally not include data from which of the following investigations in quality auditing?
    a. Examination of all items produced
    b. Examination of customer needs and the adequacy of design specifications in reflecting these needs
    c. Examination of vendor product specifications and monitoring procedures
    d. Examination of customer quality complaints and adequacy of corrective action

17. In order to be effective, the quality audit function should ideally be:
    a. an independent organizational segment in the quality control function.
    b. an independent organizational segment in the production control function.
    c. an independent organizational segment in manufacturing operations function.
    d. all of the above

18. The following are reasons why an independent audit of actual practice versus procedures should be performed periodically:
    1. Pressures may force the supervisor to deviate from approved procedures.
    2. The supervisor may not have time for organized follow-up or adherence to procedures.
    3. Supervisors are not responsible for implementing procedures.

a. 1 and 2 only
b. 1 and 3 only
c. 2 and 3 only
d. 1, 2, and 3

19. A vendor quality survey:
    a. is used to predict whether a potential vendor can meet quality requirements.
    b. is an audit of a vendor's product for a designated period of time.
    c. is always conducted by quality control personnel only.
    d. reduces cost by eliminating the need for receiving inspection of the surveyed vendor's product.

20. A quality control program is considered to be:
    a. a collection of quality control procedures and guidelines.
    b. a step-by-step list of all quality control checkpoints.
    c. a summary of company quality control policies.
    d. a system of activities to provide quality of products and service.

21. A technique whereby various product features are graded as to relative importance is called:
    a. classification of defects.
    b. quality engineering.
    c. classification of characteristics.
    d. feature grading.

22. Much managerial decision making is based on comparing actual performance with:
    a. personnel ratio.
    b. cost of operations.
    c. number of complaints.
    d. standards of performance.

23. Which of the following is not a legitimate audit function?
    a. Identify function responsible for primary control and corrective action.
    b. Provide no surprises.
    c. Provide data on worker performance to supervision for punitive action.
    d. Contribute to a reduction in quality cost.
    e. None of the above.

24. In many programs, what is generally the weakest link in the quality auditing program?
    a. Lack of adequate audit check lists
    b. Scheduling of audits (frequency)

c. Audit reporting
d. Follow-up of corrective action implementation

25. What item(s) should be included by management when establishing a quality audit function within their organization?
   a. Proper positioning of the audit function within the quality organization
   b. A planned audit approach, efficient and timely audit reporting, and a method for obtaining effective corrective action
   c. Selection of capable audit personnel
   d. Management objectivity toward the quality program audit concept
   e. All of the above

26. Which of the following best describes the "specific activity" type of audit?
   a. Customer-oriented sampling of finished goods
   b. Evaluation for contractual compliance of quality system
   c. Assessment or survey of potential vendor
   d. An inspection performance audit
   e. None of the above

27. Which of the following techniques would not be used in a quality audit?
   a. Select samples only from completed lots.
   b. Examine samples from the viewpoint of a critical customer.
   c. Audit only those items that have caused customer complaints.
   d. Use audit information in future design planning.
   e. Use economic and quality requirements to determine frequency of audit.

28. During the pre-award survey at a potential key supplier, you discover the existence of a quality control manual. This means:
   a. that a quality system has been developed.
   b. that a quality system has been implemented.
   c. that the firm is quality conscious.
   d. that the firm has a quality manager.
   e. all of the above

29. Which of the following quality system provisions is of the least concern when preparing an audit checklist for the upcoming branch operation quality system audit?
   a. Drawing and print control
   b. Makeup of the MRB (material review board)
   c. Engineering design change control
   d. Control of special processes
   e. Calibration of test equipment

30. An audit will be viewed as a constructive service to the function that is audited when it:
    a. is conducted by non-technical auditors.
    b. proposes corrective action for each item uncovered.
    c. furnishes enough detailed facts to determine the necessary action.
    d. is general enough to permit managerial intervention.

31. Which of the following is not a responsibility of the auditor?
    a. Prepare a plan and checklist.
    b. Report results to those responsible.
    c. Investigate deficiencies for cause and define the corrective action that must be taken.
    d. Follow up to see if the corrective action was taken.

32. To ensure success of a quality audit program, the most important activity for a quality supervisor is:
    a. setting up audit frequency.
    b. maintenance of a checking procedure to see that all required audits are performed.
    c. getting corrective action as a result of audit findings.
    d. checking that the audit procedure is adequate and complete.

33. It is generally considered desirable that quality audit reports be:
    a. stated in terms different from those of the function being audited.
    b. simple but complete.
    c. sent to the general manager in all cases.
    d. quantitative in all cases.

34. Classification of defects is most essential as a prior step to a valid establishment of:
    a. design characteristics to be inspected.
    b. vendor specifications of critical parts.
    c. process control points.
    d. economical sampling inspection.
    e. a product audit checklist.

35. Classification of characteristics:
    a. is the same as classification of defects.
    b. can only be performed after product is produced.
    c. must have tolerances associated with it.
    d. is independent of defects.

36. Characteristics are often classified (critical, major, etc.) so that:
    a. equal emphasis can be placed on each characteristic.
    b. punitive action against the responsible individuals can be equitably distributed.

c. an assessment of quality can be made.
d. a quality audit is compatible with management desires.

37. A classification of characteristics makes it possible to:
   a. separate the "vital few" from the "trivial many" kinds of defects.
   b. direct the greatest inspection effort to the most important quality characteristics.
   c. establish inspection tolerances.
   d. allow the inspector to choose what to inspect and what not to inspect.

38. One defective is:
   a. an item that is unacceptable to the inspector.
   b. the same as one defect.
   c. a characteristic that may be unacceptable for more than one reason.
   d. an item that fails to meet quality standards and specifications.

39. In recent months, several quality problems have resulted from apparent change in design specifications by engineering, including material substitutions. This has only come to light through quality engineering's failure analysis system. You recommend which of the following quality system provisions as the best corrective action?
   a. Establishing a formal procedure for initial design review
   b. Establishing a formal procedure for process control
   c. Establishing a formal procedure for specification change control (sometimes called an ECO or SCO system)
   d. Establishing a formal system for drawing and print control
   e. Establishing a formal material review (MRB) system

40. When giving instructions to those who will perform a task, the communication process is completed:
   a. when the worker goes to his work station to do the task.
   b. when the person giving the instruction has finished talking.
   c. when the worker acknowledges these instructions by describing how he will perform the task.
   d. when the worker says that he understands the instructions.

41. Studies have shown that the most effective communications method for transferring information is:
   a. oral only.
   b. written only.
   c. combined written and oral.
   d. bulletin board.

42. The most important reason for a checklist in a process control audit is to:
    a. ensure that the auditor is qualified.
    b. minimize the time required for audit.
    c. obtain relatively uniform audits.
    d. notify the audited function prior to audit.

43. Effective supervisors:
    a. see their role primarily as one of making people happy.
    b. sometimes do a job themselves because they can do it better than others.
    c. have objectives of growth and increased profit by working through other people.
    d. assume the functions of planning, decision-making, and monitoring performance, but leave personnel development to the personnel department.

44. Essential to the success of any quality control organization is the receipt of:
    a. adequate and stable resources.
    b. clear and concise project statements.
    c. delegation of authority to accomplish the objective.
    d. all of the above

45. The "quality function" of a company is best described as:
    a. the degree to which the company product conforms to a design or specification.
    b. that collection of activities through which "fitness for use" is achieved.
    c. the degree to which a class or category of product possesses satisfaction for people generally.
    d. all of the above

46. The quality assurance function is comparable to which of the following other business functions in concept?
    a. General accounting
    b. Cost accounting
    c. Audit accounting
    d. All of the above

47. The prime use of a control chart is to:
    a. detect assignable causes of variation in the process.
    b. detect nonconforming product.
    c. measure the performance of all quality characteristics of a process.
    d. detect the presence of random variation in the process.

48. In preparing a product quality policy for your company, you should do all of the following except:
    a. specify the means by which quality performance is measured.
    b. develop criteria for identifying risk situations and specify whose approval is required when there are known risks.
    c. include procedural matters and functional responsibilities.
    d. state quality goals.

49. The first and most important step in establishing a good corporate quality plan is:
    a. determining customer requirements.
    b. determining manufacturing process capabilities.
    c. evaluating vendor quality system.
    d. ensuring quality participation in design review.

50. The most important measure of outgoing quality needed by managers is product performance as viewed by:
    a. the customer.
    b. the final inspector.
    c. production.
    d. marketing.

51. In planning for quality, an important consideration at the start is:
    a. the relation of the total cost of quality to the net sales.
    b. the establishment of a company quality policy or objective.
    c. deciding precisely how much money is to be spent.
    d. the selling of the quality program to top management.

52. A useful tool to determine when to investigate excessive variation in a process is:
    a. MIL-STD-105E.
    b. a control chart.
    c. Dodge-Romig AOQL sampling table.
    d. process capability study.

53. Shewhart X-bar control charts are designed with which one of the following objectives?
    a. Reduce sample size.
    b. Fix the risk of accepting poor product.
    c. Decide when to hunt for causes of variation.
    d. Establish an acceptable quality level.

54. A quality program has the best foundation for success when it is initiated by:
    a. a certified quality engineer.
    b. contractual requirements.

c. the chief executive of company.
d. production management.
e. an experienced quality manager.

55. There are two basic aspects of product quality:
    a. in-process and finished product quality.
    b. appraisal costs and failure costs.
    c. quality of design and quality of conformance.
    d. impact of machines and impact of men.

56. Establishing the quality policy for the company is typically the responsibility of:
    a. the marketing department.
    b. top management.
    c. quality control.
    d. the customer.

57. Complaint indices should:
    a. recognize the degree of dissatisfaction as viewed by the customer.
    b. provide a direct input to corrective action.
    c. not necessarily be based on field complaints or dollar values of claims paid or on service calls.
    d. ignore life cycle costs.

58. For a typical month, 900D Manufacturing Company identified and reported the following quality costs:
    - Inspection wages $12,000
    - Quality planning $4000
    - Source inspection $2000
    - In-plant scrap and rework $88,000
    - Final product test $110,000
    - Retest and troubleshooting $39,000
    - Field warranty cost $205,000
    - Evaluation and processing of deviation requests $6000

    What is the total failure cost for this month?

    a. $244,000
    b. $151,000
    c. $261,000
    d. $205,000
    e. $332,000

59. If prevention costs are increased to pay for engineering work in quality control, and this results in a reduction in the number of product defects, this yields a reduction in:
    a. appraisal costs.
    b. operating costs.

c. quality costs.
d. failure costs.
e. manufacturing costs.

60. The cost of writing instructions and operating procedures for inspection and testing should be charged to:
   a. prevention costs.
   b. appraisal costs.
   c. internal failure costs.
   d. external failure costs.

61. Which of the following activities is not normally charged as a preventive cost?
   a. Quality training
   b. Design and development of quality measurement equipment
   c. Quality planning
   d. Laboratory acceptance testing

62. In selecting a base for measuring quality costs, which of the following should be considered?
   a. Is it sensitive to increases and decreases in production schedules?
   b. Is it affected by mechanization and the resulting lower direct labor costs?
   c. Is it affected by seasonal product sales?
   d. Is it oversensitive to material price fluctuations?
   e. All of the above.

63. Which of the following quality cost indices is likely to have the greatest appeal to top management as an indicator of relative cost?
   a. Quality cost per unit of product
   b. Quality cost per hour of direct production labor
   c. Quality cost per unit of processing cost
   d. Quality cost per unit of sales
   e. Quality cost per dollar of direct production labor

64. Review of purchase orders for quality requirements falls into which one of the following quality cost segments?
   a. Prevention
   b. Appraisal
   c. Internal failures
   d. External failures

65. Failure costs include costs due to:
   a. quality control engineering.
   b. inspection setup for tests.
   c. certification of special-process suppliers.
   d. supplier analysis of nonconforming hardware.

66. The basic objective of a quality cost program is to:
    a. identify the source of quality failures.
    b. interface with the accounting department.
    c. improve the profit of your company.
    d. identify quality control department costs.

67. Cost of calibrating test and inspection equipment would be included in:
    a. prevention cost.
    b. appraisal cost.
    c. failure cost.
    d. material-procurement cost.

68. In some instances, the ordinary cost-balance formula is not valid and cannot be applied because of the presence of vital intangibles. Such an intangible involves:
    a. safety of human beings.
    b. compliance with legislation.
    c. apparatus for collection of revenue.
    d. credit to marketing as new sales for warranty replacements.
    e. none of the above

69. What is the standard deviation of the following sample 3.2, 3.1, 3.3, 3.3, 3.1?
    a. 3.2
    b. 0.0894
    c. 0.1
    d. 0.0498
    e. 0.2

70. Which of the following is most important when calibrating a piece of equipment?
    a. Calibration sticker
    b. Maintenance history card
    c. Standard used
    d. Calibration interval

71. Which one of the following best describes machine capability?
    a. The total variation of all cavities of a mold, cavities of a die cast machine, or spindles of an automatic assembly machine
    b. The inherent variation of the machine
    c. The total variation over a shift
    d. The variation in a short run of consecutively produced parts

72. Machine capability studies on four machines yielded the following information:

Machine	Average (X)	Capability (6s)
#1	1.495	.004″
#2	1.502	.006″
#3	1.500	.012″
#4	1.498	.012″

The tolerance on the particular dimension is 1.500 ± .005″. If the average value can be readily shifted by adjustment to the machine, then the best machine to use is:

a. Machine #1.
b. Machine #2.
c. Machine #3.
d. Machine #4.

73. How should measurement standards be controlled?
   1. Develop a listing of measurement standards with nomenclature and number for control.
   2. Determine calibration intervals and calibration sources for measurement standards.
   3. Maintain proper environmental conditions and traceability of accuracy to National Bureau of Standards.

a. 1 and 2 only
b. 1 and 3 only
c. 2 and 3 only
d. 1, 2, and 3

74. When making measurements with test instruments, precision and accuracy mean:
a. the same.
b. the opposite.
c. consistency and correctness, respectively.
d. exactness and traceability, respectively.
e. none of the above

75. Calibration intervals should be adjusted when:
a. no defective product is reported as acceptable due to measurement errors.
b. few instruments are scrapped during calibration.
c. the results of previous calibrations reflect few "out of tolerance" conditions during calibration.
d. a particular characteristic on the gauge is consistently found out of tolerance.

76. Random selection of a sample:
    a. theoretically means that each item in the lot had an equal chance to be selected in the sample.
    b. ensures that the sample average will equal the population average.
    c. means that a table of random numbers was used to dictate the selection.
    d. is a meaningless theoretical requirement.

77. An X-bar and R chart was prepared for an operation using 20 samples with five pieces in each sample. X was found to be 33.6 and R was 6.2. During production a sample of five was taken and the pieces measured 36, 43, 37, 34, and 38. At the time this sample was taken:
    a. both average and range were within control limits.
    b. neither average nor range was within control limits.
    c. only the average was outside control limits.
    d. only the range was outside control limits.
    e. The information given is not sufficient to construct an X-bar and R chart using tables usually available.

78. A chart for number of defects is called:
    a. *np* chart.
    b. *p* chart.
    c. X chart.
    d. *c* chart.

79. A process is checked at random by inspection of samples of four shafts after a polishing operation, and X and R charts are maintained. A person making a spot check measures two shafts accurately, and plots their range on the R chart. The point falls just outside the control limit. He advises the department foreman to stop the process. This decision indicates that:
    a. the process level is out of control.
    b. the process level is out of control but not the dispersion.
    c. the person is misusing the chart.
    d. the process dispersion is out of control.

80. A process is in control with $p = 0.10$ and $n = 100$. The three-sigma limits of the *np*-control chart are:
    a. 1 and 19.
    b. 9.1 and 10.9.
    c. 0.01 and 0.19.
    d. 0.07 and 0.13.

81. A *p* char:
    a. can be used for only one type of defect per chart.
    b. plots the number of defects in a sample.

c. plots either the fraction or percent detective in order of time.
d. plots variations in dimensions.

82. The control chart that is most sensitive to variations in a measurement is:
a. *p* chart.
b. *np* chart.
c. *c* chart.
d. X-bar and R chart.

83. A *p* chart is a type of control chart for:
a. plotting bar-stock lengths from receiving inspection samples.
b. plotting fraction defective results from shipping inspection samples.
c. plotting defects per unit from in-process inspection samples.
d. answers a, b, and c above.
e. answers a and c only.

84. When subgroups are outside of the control limits and we wish to set up a control chart for future production:
a. more data are needed.
b. discard those points falling outside the control limits for which you can identify an assignable cause, and revise the limits.
c. check with production to determine the true process capability.
d. discard those points falling outside the control limits and revise the limits.

85. You have just returned from a 2-week vacation. You and your QC manager are going over the control charts that have been maintained during your absence. He calls your attention to the fact that one of the X charts shows the last 50 points to be very near the centerline. In fact, they all seem to be within about one sigma of the centerline. What explanation would you offer him?
a. Somebody "goofed" in the original calculation of the control limits.
b. The process standard deviation has decreased during the time the last 50 samples were taken and nobody thought to re-compute the control limits.
c. This is a terrible situation. I'll get on it right away and see what the trouble is. I hope we haven't produced too much scrap.
d. This is fine. The closer the points are to the centerline, the better our control.

86. In which one of the following is the use of an X and R chart liable to be helpful as a tool to control a process?
a. The machine capability is wider than the specification.
b. The machine capability is equal to the specification.
c. The machine capability is somewhat smaller than the specification.
d. The machine capability is very small compared to the specification.

87. Quality cost trend analysis is facilitated by comparing quality costs with:
    a. manufacturing costs over the same time period.
    b. appropriate measurement bases.
    c. cash flow reports.
    d. the QC department budget.

88. Of the following, which are typically appraisal costs?
    a. Vendor surveys and vendor faults
    b. Quality planning and quality reports
    c. Drawing control centers and material dispositions
    d. Quality audits and final inspection
    e. None of the above

89. Which of the following cost elements is normally a prevention cost?
    a. Receiving inspection
    b. Outside endorsements or approvals
    c. Design of quality measurement equipment
    d. All of the above

90. When analyzing quality cost data gathered during the initial stages of a new management emphasis on quality control and corrective action as part of a product improvement program, one normally expects to see:
    a. increased prevention costs and decreased appraisal costs.
    b. increased appraisal costs with little change in prevention costs.
    c. decreased internal failure costs.
    d. decreased total quality costs.
    e. all of these

91. Quality costs are best classified as:
    a. cost of inspection and test, cost of quality engineering, cost of quality administration, and cost of quality equipment.
    b. direct, indirect, and overhead.
    c. cost of prevention, cost of appraisal, and cost of failure.
    d. unnecessary.
    e. none of the above

92. Which of the following bases of performance measurement (denominators), when related to operating quality costs (numerator), would provide reliable indicator(s) to quality management for overall evaluation of the effectiveness of the company's quality program? Quality costs per:
    a. total manufacturing costs
    b. unit produced
    c. total direct labor dollars
    d. only one of the above
    e. any two of the above

93. Quality cost data:
    a. must be maintained when the end product is for the government.
    b. must be mailed to the contracting officer on request.
    c. is often an effective means of identifying quality problem areas.
    d. all of the above

94. Operating quality costs can be related to different volume bases. An example of volume base that could be used would be:
    a. direct labor cost.
    b. standard manufacturing cost.
    c. processing cost.
    d. sales.
    e. all of the above

95. When operating a quality cost system, excessive costs can be identified when:
    a. appraisal costs exceed failure costs.
    b. total quality costs exceed 10 percent of sales.
    c. appraisal and failure costs are equal.
    d. total quality costs exceed 4 percent of manufacturing costs.
    e. There is no fixed rule; management experience must be used.

96. Quality cost systems provide for defect prevention. Which of the following elements is primary to defect prevention?
    a. Corrective action
    b. Data collection
    c. Cost analysis
    d. Training

97. Quality cost analysis has shown that appraisal costs are apparently too high in relation to sales. Which of the following actions probably would not be considered in pursuing this problem?
    a. Work sampling in inspection and test areas
    b. Adding inspectors to reduce scrap costs
    c. Pareto analysis of quality costs
    d. Considering elimination of some test operations
    e. Comparing appraisal costs with bases other than sales—for example direct labor, value added, etc.

98. Analyze the cost data below:
    - $10,000—equipment design
    - $150,000—scrap
    - $180,000—re-inspection and retest
    - $45,000—loss or disposition of surplus stock
    - $4000—vendor quality surveys
    - $40,000—repair

Considering only the quality costs shown above, we might conclude that:

a. prevention costs should be decreased.
b. internal failure costs can be decreased.
c. prevention costs are too low a proportion of the quality costs shown.
d. appraisal costs should be increased.
e. nothing can be concluded.

99. This month's quality cost data collection shows the following:
   - Returned material processing $1800
   - Adjustment of customer complaints $4500
   - Rework and repair $10,700
   - Quality management salaries $25,000
   - Warranty replacement $54,500
   - Calibration and maintenance of test equipment $2500
   - Inspection and testing $28,000

   For your "action" report to top management, you select which one of the following as the percentage of "External Failure" to "Total Quality Costs" to show the true impact of field problems?
   a. 20 percent
   b. 55 percent
   c. 48 percent
   d. 24 percent
   e. 8 percent

100. You have been assigned as a quality manager to a small company. The quality control manager desires some cost data and the accounting department reported that the following information is available. Cost accounts are production inspection, $14,185; test inspection, $4264; procurement inspection, $2198; shop labor, $141,698; shop rework, $1402; first article, $675; engineering analysis (rework), $845; repair service (warrantee), $298; quality engineering, $2175; design engineering salaries, $241,451; quality equipment, $18,745; training, $275; receiving laboratories, $385; underwriters laboratories, $1200; installation service cost, $9000: scrap, $1182; and calibration service, $794. What are the preventive costs?
   a. $3727
   b. $23,701
   c. $23,026
   d. $3295
   e. $2450

101. The percentages of total quality cost are distributed as follows:
   - Prevention: 12 percent
   - Appraisal: 28 percent

- Internal failure: 40 percent
- External failure: 20 percent

We conclude:

a. we should invest more money in prevention.
b. expenditures for failures are excessive.
c. the amount spent for appraisal seems about right.
d. nothing.

102. One of the following is not a factor to consider in establishing quality information equipment cost:
a. debugging cost
b. amortization period
c. design cost
d. replacement parts and spares
e. book cost

103. One method to control inspection costs even without a budget is by comparing as a ratio to productive machine time to produce the product.
a. Product cost
b. Company profit
c. Inspection hours
d. Scrap material

104. A complete Quality Cost Reporting System would include which of the following as part of the quality cost?
a. Test time costs associated with installing the product at the customer's facility prior to turning the product over to the customer
b. The salary of a product designer preparing a deviation authorization for material produced outside of design specifications
c. Cost of scrap
d. All of the above
e. None of the above

105. When prevention costs are increased to pay for the right kind of engineering work in quality control, a reduction in the number of product defects occurs. This defect reduction means a substantial reduction in:
a. appraisal costs.
b. operating costs.
c. prevention costs.
d. failure costs.
e. manufacturing costs.

106. The quality cost of writing instructions and operating procedures for inspection and testing should be charged to:
   a. appraisal costs.
   b. internal failure costs.
   c. prevention costs.
   d. external failure costs.

107. When analyzing quality costs, a helpful method for singling out the highest cost contributors is:
   a. a series of interviews with the line foreman.
   b. the application of the Pareto theory.
   c. an audit of budget variances.
   d. the application of break-even and profit volume analysis.

108. Included as a "prevention quality cost" would be:
   a. salaries of personnel engaged in the design of measurement and control equipment that is to be purchased.
   b. capital equipment purchased.
   c. training costs of instructing plant personnel to achieve production standards.
   d. sorting of nonconforming material that will delay or stop production.

109. The modern concept of budgeting quality costs is to:
   a. budget each of the four segments: prevention, appraisal, internal failure, and external failure.
   b. concentrate on external failures; they are important to the business since they represent customer acceptance.
   c. establish a budget for reducing the total of the quality costs.
   d. reduce expenditures on each segment.

110. The percentages of total quality cost are distributed as follows:
   - Prevention: 2 percent
   - Appraisal: 33 percent
   - Internal failure: 35 percent
   - External failure: 30 percent

   We can conclude:

   a. expenditures for failures are excessive.
   b. nothing.
   c. we should invest more money in prevention.
   d. the amount spent for appraisal seems about right.

111. Assume that the cost data available to you for a certain period are limited to the following:
   - $20,000—Final test
   - $350,000—Field warranty costs

- $170,000—Re-inspection and retest
- $45,000—Loss on disposition of surplus stock
- $4000—Vendor quality surveys
- $30,000—Rework

The total of the quality costs is:

a. $619,000
b. $574,000
c. $615,000
d. $570,000

112. Assume that the cost data available to you for a certain period are limited to the following:
   - $20,000—Final test
   - $350,000—Field warranty costs
   - $170,000—Re-inspection and retest
   - $45,000—Loss on disposition of surplus stock
   - $4000—Vendor quality surveys
   - $30,000—Rework

   The total failure cost is:

   a. $550,000
   b. $30,000
   c. $350,000
   d. $380,000

113. A goal of a quality cost report should be to:
   a. get the best product quality possible.
   b. be able to satisfy MIL-Q-9858A.
   c. integrate two financial reporting techniques.
   d. indicate areas of excessive costs.

114. The concept of quality cost budgeting:
   a. involves budgeting the individual elements.
   b. replaces the traditional profit and loss statement.
   c. does not consider total quality costs.
   d. considers the four categories of quality costs and their general trends.

115. Sources of quality cost data do not normally include:
   a. scrap reports.
   b. labor reports.
   c. salary budget reports.
   d. capital expenditure reports.

116. When one first analyzes quality cost data, he might expect to find that, relative to total quality costs:
   a. costs of prevention are high.
   b. costs of appraisal are high.
   c. costs of failure are high.
   d. all of the above

117. Quality costs should not be reported against which one of following measurement bases:
   a. direct labor.
   b. sales.
   c. net profit.
   d. unit volume of production.

118. The basic objective of a quality cost program is to:
   a. identify the source of quality failures.
   b. determine quality control department responsibilities.
   c. utilize accounting department reports.
   d. improve the profit posture of your company.

119. Accuracy is:
   a. getting consistent results repeatedly.
   b. reading to four decimals.
   c. using the best measuring device available.
   d. getting an unbiased true value.

120. Measurement error:
   a. is the fault of the inspector.
   b. can be determined.
   c. is usually of no consequence.
   d. can be eliminated by frequent calibrations of the measuring device.

121. Precision is:
   a. getting consistent results repeatedly.
   b. reading to four or more decimals.
   c. distinguishing small deviations from the standard value.
   d. extreme care in the analysis of data.

122. If a distribution is skewed to the left, the median will always be:
   a. less than the mean.
   b. between the mean and the mode.
   c. greater than the mode.
   d. equal to the mean.
   e. equal to the mode.

123. Consumer risk is defined as:
   a. accepting an unsatisfactory lot as satisfactory.
   b. passing a satisfactory lot as satisfactory.
   c. an alpha risk.
   d. a 5 percent risk of accepting an unsatisfactory lot.

124. When an initial study is made of a repetitive industrial process for the purpose of setting up a Shewhart control chart, information on the following process characteristic is sought.
   a. Process capability
   b. Process performance
   c. Process reliability
   d. Process conformance

125. Which one of the following would most closely describe machine process capability?
   a. The process variation
   b. The total variation over a shift
   c. The total variation of all cavities of a mold, cavities of a die cast machine, or spindles of an automatic assembly machine
   d. The variation in a very short run of consecutively produced parts

126. Recognizing the nature of process variability, the process capability target is usually:
   a. looser than product specifications.
   b. the same as product specifications.
   c. tighter than product specifications.
   d. not related to product specifications.

127. A variable measurement of a dimension should include:
   a. an estimate of the accuracy of the measurement process.
   b. a controlled measurement procedure.
   c. a numerical value for the parameter being measured.
   d. an estimate of the precision of the measurement process.
   e. all of the above

128. When specifying the "10:1 calibration principle" we are referring to what?
   a. The ratio of operators to inspectors
   b. The ratio of quality engineers to metrology personnel
   c. The ratio of main scale to vernier scale calibration
   d. The ratio of calibration standard accuracy to calibrated instrument accuracy
   e. None of the above

129. Measuring and test equipment are calibrated to:
   a. comply with federal regulations.
   b. ensure their precision.
   c. determine and/or ensure their accuracy.
   d. check the validity of reference standards.
   e. accomplish all of the above.

130. A basic requirement of most gage calibration system specifications is:
   a. all inspection equipment must be calibrated with master gage blocks.
   b. gages must be color coded for identification.
   c. equipment shall be labeled or coded to indicate date calibrated by whom, and date due for next calibration.
   d. gages must be identified with a tool number.
   e. all of the above

131. What four functions are necessary to have an acceptable calibration system covering measuring and test equipment in a written procedure?
   a. Calibration sources, calibration intervals, environmental conditions, and sensitivity required for use
   b. Calibration sources, calibration intervals, humidity control, and utilization of published standards
   c. Calibration sources, calibration intervals, environmental conditions under which equipment is calibrated, controls for unsuitable equipment
   d. List of standards, identification report, certificate number, and recall records

132. Select the non-hygienic motivator, as defined by Maslow.
   a. Salary increases
   b. Longer vacations
   c. Improved medical plan
   d. Sales bonuses
   e. Performance recognition

133. Which one of these human management approaches has led to the practice of job enrichment?
   a. Skinner
   b. Maslow
   c. Herzberg's "Hygiene Theory"
   d. McGregor

134. Which of the following is not a management-initiated error?
   a. The imposition of conflicting priorities
   b. The lack of operator capacity

c. Management indifference or apathy
d. Conflicting quality specifications
e. Work space, equipment, and environment

135. Which of the following does not generate product-quality characteristics?
    a. Designer
    b. Inspector
    c. Machinist
    d. Equipment engineer

136. Extensive research into the results of quality motivation has shown that:
    a. the supervisor's attitude toward his or her people is of little long term consequence.
    b. motivation is too nebulous to be correlated with results.
    c. motivation is increased when employees set their own goals.
    d. motivation is increased when management sets challenging goals slightly beyond the attainment of the better employees.

137. McGregor's theory X manager is typified as one who operates from the following basic assumption about people working for him or her. (Select the one best answer.)
    a. Performance can be improved through tolerance and trust.
    b. People have a basic need to produce.
    c. Status is more important than money.
    d. Self-actualization is the highest order of human need.
    e. People are lazy and are motivated by reward and punishment.

138. Quality motivation in industry should be directed at:
    a. manufacturing management.
    b. procurement and engineering.
    c. the quality assurance staff.
    d. the working force.
    e. all the above.

139. Who has the initial responsibility for manufactured product quality?
    a. The inspector
    b. The vice president
    c. The operator
    d. The quality manager

140. A fully developed position description for a quality engineer must contain clarification of:
    a. responsibility.
    b. accountability.
    c. authority.

d. answers a and c above.
e. answers a, b, and c above.

141. One of the most important techniques in making a training program effective is to:
a. give people meaningful measures of performance.
b. transmit all of the information that is even remotely related to the function.
c. set individual goals instead of group goals.
d. concentrate only on developing knowledge and skills needed to do a good job.

142. In order to instill the quality control employee with the desire to perform to his or her utmost and optimum ability, which of the following recognition for sustaining motivation has been found effective for most people?
a. Recognition by issuance of monetary award
b. Public verbal recognition
c. Private verbal recognition
d. Public recognition, plus non-monetary award
e. No recognition; salary he or she obtains is sufficient motivation

143. Which of the following methods used to improve employee efficiency and promote an atmosphere conducive to quality and profit is the most effective in the long run?
a. Offering incentives such as bonus, praise, profit sharing, etc.
b. Strict discipline to reduce mistakes, idleness, and sloppiness
c. Combination of incentive and discipline to provide both reward for excellence and punishment for inferior performance
d. Building constructive attitudes through development of realistic quality goals relating to both company and employee success
e. All of the above provided emphasis is placed on attitude motivation, with incentive and discipline used with utmost caution

144. An essential technique in making training programs effective is to:
a. set group goals.
b. have training classes that teach skills and knowledge required.
c. feed back to the employee meaningful measures of his performance.
d. post results of performance before and after the training program.
e. set individual goals instead of group goals.

145. In the pre-production phase of quality planning, an appropriate activity would be to:
a. determine responsibility for process control.
b. determine the technical depth of available manpower.
c. establish compatible approaches for accumulation of process data.
d. conduct process capability studies to measure process expectations.

146. Systems are improved:
   a. by improving each of the processes within the system to its best level of performance.
   b. by improving the process that is most important to the customer.
   c. by considering how the processes work within the system.
   d. sometimes at the expense of processes that operate within the system. In other words, the performance of some processes may improve, and others degrade, to achieve maximum system performance.
   e. choices c and d
   f. choices a and b

147. Deming says that the responsibility for optimizing a system rests with:
   a. the team leaders assigned to that project.
   b. the process workers who know the system the best.
   c. management.
   d. none of the above

148. How are the number of constraints in a system determined?
   a. Since they know the problems in the system, the personnel working in the system are generally able to identify the constraints in a brainstorming exercise.
   b. Each task on the critical path is a constraint, so sum the number of tasks on the critical path.
   c. Sum the number of critical tasks on the critical path.
   d. There is only one system constraint at a time (in each independent chain).

149. Constraint management is:
   a. a descriptive theory.
   b. a prescriptive theory.
   c. a hygiene theory.
   d. none of the above

150. Constraint management theory:
   a. explains how a constraint impacts a system.
   b. provides a definition for a constraint.
   c. provides management direction for dealing with a constraint.
   d. all of the above
   e. choices a and b

151. Constraints may be described as:
   a. anything that limits a system in reaching its goal.
   b. the weak link in a chain.
   c. the poorest performing process in a system.
   d. all of the above
   e. choices a and b

152. Examples of constraints include:
    a. insufficient demand for your product.
    b. an internal policy that slows response time to customer demand.
    c. a process operation that acts as a bottleneck, slowing the delivery of product to the customer.
    d. insufficient training resources, preventing adequate job skills.
    e. all of the above

153. Which of the following is NOT a basic assumption related to Constraint Management?
    a. Systems have goals and corresponding necessary conditions required to achieve the goals. We must identify these to effectively improve the system.
    b. We must maximize the performance of each link in the chain to improve the system.
    c. The performance of the system is dictated by the "weakest link" in the system.
    d. None of the above.

154. Constraint Management's focusing steps, in order, are:
    a. Identify, Elevate, Subordinate, Exploit, Repeat
    b. Plan, Do, Check, Act
    c. Identify, Exploit, Subordinate, Elevate, Repeat
    d. Identify, Elevate, Subordinate, Exploit

155. Once we have identified the constraint, and taken actions to make the most of its resources, we should:
    a. define parameters for other system elements to complement the constraint's needs.
    b. improve the capacity of downstream processes.
    c. regularly verify that the constraint has not moved.
    d. all of the above
    e. choices a and c

156. Buffers are used to:
    a. increase capacity.
    b. decrease cycle time of a constraint.
    c. ensure that the constraint is not waiting for materials.
    d. increase the system cycle time.

157. Buffers should:
    a. be used at each stage of the process.
    b. be used at the constraint, to protect against upstream cycle time variation.
    c. never be used in Constraint Management.
    d. protect the constraint from being overworked.

158. In constraint management, buffers:
   a. are specified in units of time, such as requiring delivery of material 24 hours before it is needed.
   b. are specified in production units, such as requiring that 10 percent more product be delivered to the process step than is necessary for the customer order.
   c. are not recommended, since they produce inefficiencies.
   d. none of the above

159. In Goldratt's Drum-Buffer-Rope, the rope:
   a. causes backlogs at various stages, in order to improve the efficiency of the constraint.
   b. prevents resources from being allocated faster than they can be used by the constraint.
   c. acts to pull material through the system, in contrast to push production systems.
   d. prevents a critical resource from being without material.

160. Critical chain refers to constraint management applied to:
   a. critical processes.
   b. repetitive production.
   c. projects.
   d. choices a and b
   e. none of the above
   f. all of the above

161. The critical chain approach accounts for which of the following behaviors?
   a. Estimating task time longer than necessary to avoid late completion
   b. Starting the task just prior to the scheduled completion date
   c. Using all of allotted task time, regardless of how long it really takes
   d. Assigning a given person multiple tasks with deadlines
   e. All of the above

162. Critical chain management:
   a. ensures on-time project completion by maintaining on-time completion of each activity on the critical chain.
   b. replaces time estimate padding for activities with time buffers at key points on the critical chain.
   c. should be used with caution if a resource has more than one task assigned to it.
   d. should not be applied to managing multiple projects at once.

163. A key difference between critical chain management and PERT/CPM critical path is:
   a. the critical path will always be shorter.
   b. tasks feeding the critical path have buffers imposed.

c. the critical chain considers dependent activities in series, as well conflicting resource needs.
d. there is no fundamental difference between the two.

164. Throughput may be expressed:
  a. as the rate at which a system generates money.
  b. as the marginal contribution of sales to profit.
  c. for an entire company.
  d. as sales revenue minus variable cost.
  e. all of the above

165. In constraint management, operating expense:
  a. is the amount of money spent converting inventory into throughput.
  b. includes labor costs.
  c. includes fixed costs and overhead.
  d. all of the above
  e. choices a and c

166. According to constraint management theory, as inventory costs increase:
  a. Net profit decreases.
  b. Return on investment decreases.
  c. Net profit remains the same.
  d. choices a and b
  e. choices b and c

167. Six Sigma methodologies:
  a. can only be applied to companies that produce goods with large volume.
  b. concentrate on cost savings rather than customer needs.
  c. have not been successfully applied to service companies.
  d. all of the above
  e. none of the above

168. A Six Sigma level of quality:
  a. implies 99.73 percent of the output will meet customer requirement.
  b. equates to a capability index of 1.33.
  c. represents 3.4 defects per million opportunities.
  d. provides half the defects of a 3 Sigma level of quality.
  e. all of the above

169. As an organization's sigma level increases:
  a. the cost of quality increases.
  b. the cost of quality decreases.
  c. the cost of quality is not affected.
  d. none of the above

170. As a company moves from 3 Sigma level of quality to 4 and 5 Sigma levels of quality, they tend to:
   a. spend more money on prevention costs.
   b. spend less money on appraisal costs.
   c. spend less money on failure costs.
   d. improve customer satisfaction, which can lead to increased sales.
   e. all of the above

171. Companies that successfully implement Six Sigma are likely to:
   a. initially spend a lot of money on training, but receive benefits that might be hard to quantify over the course of time.
   b. realize decreased quality costs and improved quality.
   c. see a reduction in critical defects and cycle times.
   d. all of the above
   e. choices b and c only

172. Project sponsors:
   a. ensure that the Six Sigma projects are defined with clear deliverables.
   b. help clear roadblocks encountered by the project team.
   c. are generally members of management.
   d. all of the above

173. In a typical deployment, Green Belts:
   a. are full-time change agents.
   b. maintain their regular role in the company, and work Six Sigma projects as needed.
   c. receive extensive training in advanced statistical methods.
   d. all of the above

174. Examples of Six Sigma projects might include:
   a. reducing cost of product shipments.
   b. reducing customs delays for international shipments.
   c. reducing the design cycle for a new product.
   d. increasing the market share of a particular product through improved marketing.
   e. all of the above

175. In the Six Sigma project methodology acronym DMAIC, the "I" stands for:
   a. Integrate.
   b. Investigate.
   c. Improve.
   d. Ignore.

176. The Define stage of DMAIC:
   a. is linked with the project charter and provides input to the Measure stage.
   b. stands alone in the methodology, but is always necessary.
   c. is not necessary when the project is mandated by top management.
   d. none of the above

177. The Control stage of DMAIC:
   a. is only used when you need to define control chart parameters.
   b. allows the improvements to be maintained and institutionalized.
   c. is only needed if you have ISO 9000 certification.
   d. none of the above

178. A top-down deployment of Six Sigma projects:
   a. is discouraged because projects get delayed by other management commitments.
   b. undermines the project sponsors.
   c. emphasizes projects that line workers feel are important.
   d. ensures that projects are aligned with the top-level business strategy.

179. Which of the following are NOT parts of Deming's system of profound knowledge?
   a. Constantly evaluate all employees
   b. Appreciation for a system
   c. Knowledge about variation
   d. Theory of knowledge

180. Which of the following are NOT one of Deming's 14 points?
   a. Institute a vigorous program of education and self-improvement.
   b. Put everybody in the company to work to accomplish the transformation. The transformation is everybody's job.
   c. Eliminate slogans, exhortations, and targets asking for zero defects or new levels of productivity.
   d. None of the above.

181. Examples of hidden factory losses include all of the following except:
   a. capacity losses due to reworks and scrap.
   b. stockpiling of raw material to accommodate poor yield.
   c. rush deliveries.
   d. All of the above are examples of hidden factory losses.

182. An invoicing process generates only ten to fifteen orders per month. In establishing the statistical control of the invoice process:
   a. use a subgroup size of one.
   b. use a subgroup size of five, which is generally the best size.
   c. use a subgroup size of ten.
   d. use a subgroup size of fifteen.

183. Effective leaders:
   a. share many of the same traits and responsibilities as effective managers.
   b. have a vision for the organization.
   c. are best suited for designing the processes and systems for daily operations.
   d. all of the above

184. Effective managers:
   a. share many of the same traits and responsibilities as effective leaders.
   b. have a vision for the organization.
   c. are best suited for designing the processes and systems for daily operations.
   d. all of the above

185. Management training should include:
   a. coaching skills.
   b. theory and practice of organizational systems.
   c. conflict resolution skills.
   d. all of the above

186. Which of the following is NOT one of Deming's 14 points?
   a. Drive out fear.
   b. Create constancy of purpose.
   c. Hold management accountable for meeting the business's numerical goals.
   d. Improve constantly and forever each process for planning, production, and service.

187. The management team has decided that there are three criteria for choosing projects, with their relative importance weighting shown in parenthesis:
   - Financial benefit/cost ratio (0.3)
   - Perceived customer benefit (0.5)
   - Safety benefit (0.2)

   Four projects have been reviewed relative to these criteria, with the scores for each criterion shown in the following table.

Project	Fin. Benefit/Cost Ratio	Customer Benefit	Safety Benefit
A	120	140	30
B	80	200	25
C	100	100	45
D	140	70	65

Which project should be selected?

a. Project A
b. Project B
c. Project C
d. Project D

188. Joe's project seemed to be going along well until the project team started to implement the solution. At that point, a department that hadn't been involved, but will be affected, starting raising objections and pointing out problems with the proposed solution. This indicates:
    a. the team should immediately get the sponsor involved to settle the disagreements.
    b. the department is afraid of change and needs to be told to accept the team's findings.
    c. the department should have been identified as stakeholders early on and included in the project team or the team's problem solving sessions.
    d. choices a and b

189. Bob, a team leader, is having trouble getting buy-in from various members of the team. In one particular problem-solving meeting, these team members didn't seem to listen to any of Bob's ideas, and were insistent that their ideas were more credible. Some reasonable advice to Bob would be:
    a. Replace the team members with those more willing to work as team players.
    b. Work on his communication skills, display interest for others' ideas, and use data to determine which ideas have the most merit.
    c. Ask their managers or the project sponsor to persuade them to get on board.
    d. choices a and c

190. A particular project has many stakeholder groups. In an attempt to keep the team size at a reasonable level, some of the non-key stakeholder groups were not included in the team. As a result:
    a. the team leader can boost buy-in from these groups by bringing credible group members into the problem-solving as ad hoc team members.
    b. the team leader should also restrict the distributed progress reports to only the key groups represented in the team to prevent confusion and interference on the part of the non-key groups.
    c. the sponsor should ensure that the concerns of the non-key groups are met by the team recommendations.
    d. all of the above

191. Bill's team is having a hard time agreeing on a plan for data gathering. There are three general suggestions that have been offered: one by a process expert and two by other team members. To decide which plan to deploy, the team should:
   a. accept the idea offered by the process expert. Doing otherwise is an insult to his or her expertise.
   b. vote on the different proposals, with the plan receiving the highest vote being deployed.
   c. develop new plans that take the best parts of the proposed plans with compromises on the conflicting aspects of the various plans. Vote on the resulting plans.
   d. try to reach a compromise on the various plans, with a resulting plan that everyone can live with, even if it's not perfect to any of the parties.

192. Jill is the team leader for a project aimed at reducing the cycle time for invoices. The team has reached an impasse on generating potential root causes of process failure; only a few ideas have been offered by only a few of the team members. As team leader, Jill should:
   a. request that the current team be dissolved, and a new team formed with process experts.
   b. report the impasse to the sponsor, and suggest the team meet again in a month or two when they have a fresh perspective.
   c. use brainstorming tools.
   d. end the meeting, and work on developing the list herself.

193. Joan is a Black Belt and project team leader. Her team includes, amongst other members, a manager and a clerk from different departments. Early in today's team meeting, after the clerk had offered an idea during the brainstorming session, the manager made a joke about the feasibility of the idea. Just now, the manager has stifled the clerk's comments by asserting the clerk lacked the experience to suggest potential causes of the problem under investigation. Joan should:
   a. wait until the end of the meeting and discuss the issue separately with the clerk and the manager.
   b. immediately remind the manager of the team ground rule of "respectful communication" and the general rules for brainstorming.
   c. give it some time and allow personalities to gel. Perhaps the manager is having a bad day and will be more agreeable in future meetings.
   d. do nothing. The manager should be given respect for his or her position in the company, and the point is well taken on the clerk's experience. Furthermore, the earlier joke really gave the team something to chuckle about, easing tension.

194. In an initial team meeting, the team should:
    a. establish ground rules and review member responsibilities.
    b. agree on project purpose, scope, plan, and timeline.
    c. establish workable meeting times and locations.
    d. all of the above

195. With regard to team dynamics,
    a. initial meetings are generally friendly, with the team leader exercising control.
    b. conflict is common, and can indicate that team members are becoming involved.
    c. the team leader should move the members toward thinking independently.
    d. all of the above

196. A conflict has developed among team members regarding a proposed solution. Joan, the team leader, should:
    a. insist that the team members behave and stop disagreeing.
    b. allow each member to explain his or her point of view, then take a vote to see which proposal wins.
    c. use problem solving tools to determine the true causes of dissension, then use that information to guide their solution.
    d. all of the above

197. At the team's third meeting, its leader, John, is starting to feel a bit uncomfortable. He had established ground rules for the team, and some of its members are starting to question those rules. John should:
    a. exercise his authority as team leader and insist that the team follow the rules.
    b. lead the team to establish its own rules.
    c. defer to the team sponsor.
    d. none of the above

198. In team meetings, Jane seems to listen to whoever is speaking, but then has questions for the speaker. John, the team leader, senses that a few team members are frustrated with Jane, thinking she takes up too much time. John should:
    a. politely ask Jane, after the meeting, to try to keep her questions to a minimum.
    b. politely ask Jane during the meeting to try to keep her questions to a minimum.
    c. thank Jane publicly for asking relevant questions of team members, so that issues and opinions are clearly understood.
    d. ignore the frustrations. Personalities don't always mesh.

199. Jim is assembling a team to improve quality of a process with two stakeholder groups that have a history of fixing blame on one another. Jim would like to avoid getting "stuck in the middle" of these two factions. Jim can reduce the likelihood of this by:
   a. asking only one group (the one providing the most value to the team) to be part of the team on a full-time basis, with the other group represented only as needed for input.
   b. asking the sponsor, who oversees both groups, to attend meetings so that she can settle the disagreements.
   c. discussing team ground rules at the first team meeting, and firmly enforcing these rules throughout. These ground rules would include respectful communication between team members and decisions on basis of data (rather than opinion).
   d. asking the two groups to each recommend someone who can co-lead the team.

200. Project charters help to prevent the occurrence of which of the following reducers to stakeholder buy-in?
   a. Unclear goals
   b. No accountability
   c. Insufficient resources
   d. All of the above

References

AIAG (1995). *Potential Failure Modes and Effects Analysis Reference Guide.* Automotive Industry Action Group.

——— (1995). *MSA Reference Manual.* Automotive Industry Action Group.

Akao, Y. editor (1990). *Quality Function Deployment: Integrating Customer Requirements into Product Design.* Cambridge, MA: Productivity Press.

Alloway, J.A., Jr. (1994). "The card drop shop," *Quality Progress*, July.

ASQ (1981). *Product Recall Planning Guide*. Milwaukee: ASQ Press.

ASQ (1992). *Quality Engineering Handbook*. Milwaukee: ASQ Press.

Athey, T.H. (1982). *The Systematic Systems Approach*. Englewood Cliffs: Prentice-Hall, Inc.

Aubrey, C.A. and Felkins, P.K. (1988). *Teamwork: Involving People in Quality and Productivity Improvement*. Milwaukee, WI: ASQ Quality Press.

Benneyan, J.C. (1994). "The merits of merit review...?: Bibliography of recommended study and summary of arguments against annual merit review," Amherst, MA: University of Massachusetts, Industrial Engineering/Operations Research. benneyan@ecs.umass.edu.

Blanchard, K. and Johnson, S. (1982). *The One-Minute Manager*. New York: William Morrow & Co.

Boardman, T.J. and Boardman, E.C. (1990). "Don't touch that funnel!" *Quality Progress*, December.

Brainard, E.H. (1966). "Just how good are vendor surveys?" *Quality Assurance*, August, pp. 22–26.

Brassard, M. (1989). *The Memory Jogger Plus+*. Methuen, MA: GOAL/QPC.

Brookfield, D. (1986). *Understanding and Facilitating Adult Learning*. San Francisco, CA: Jossey-Bass.

Burt, D.N. and Doyle, M.F. (1993). *The American Keiretsu: A Strategic Weapon for Global Competitiveness*. Homewood, IL: Times Mirror Books, Business One Irwin.

Buzzell, R.D. and Gale, B.L. (1987). *The PIMS Principles: Linking Strategy to Performance*. New York: The Free Press.

Camp, R.C. (1989). *Benchmarking: The Search for Industry Best Practices that Lead to Superior Performance*. Milwaukee, WI: ASQ Quality Press and White Plains, NY: Quality Resources.

Campanella, J. editor (1990). *Principles of Quality Costs*, 2nd edition. Milwaukee, WI: ASQ Quality Press.
Carder, B. and Clark, J.D. (1992). "The theory and practice of employee recognition," *Quality Progress*, December.
Carlzon, J. (1987). *Moments of Truth*. New York: Harper and Row.
Coase, R. (1937). "The nature of the firm," *Economica*, 4, pp. 386–405.
Corbett, T. (1998). *Throughput Accounting*. Great Barrington, MA: North River Press.
Crain, Nicole V. and Crain, W. Mark. (2010). "The impact of regulatory costs on small firms," Washington, DC: U.S. Small Business Administration, Office of Advocacy. (http://www.sba.gov/advo/).
Deming, W.E. (1975). "On probability as a basis for action," *The American Statistician*, 29(4), pp. 146–152.
Deming, W.E. (1986). *Out of the Crisis*. Cambridge, MA: MIT Center for Advanced Engineering Study.
Deming, W.E. (1993). *The New Economics for Industry, Government, Education*. Cambridge, MA: MIT Center for Advanced Engineering Study.
DeToro, I. (1995). "The 10 pitfalls of benchmarking," *Quality Progress*, January, pp. 61–63.
Dettmer, H.W. (1997). *Goldratt's Theory of Constraints: A Systems Approach to Continuous Improvement*. Milwaukee, WI: ASQ Quality Press.
Dettmer, H.W. (1998). *Breaking the Constraints to World-Class Performance*. Milwaukee, WI: ASQ Quality Press.
Dillman, D.A. (1983). "Mail and other self-administered questionnaires," in *Handbook of Survey Research*. Rossi, P., Wright, J. and Anderson, A. editors. New York: Academic Press, Inc., pp. 359–377.
Drucker, P.F. (1974). *Management: Tasks, Responsibilities, Practices*. New York: Harper and Row.
Eckes, G. (1994). "Practical alternatives to performance appraisals," *Quality Progress*, November, pp. 57–60.
Edosomwan, J.A. (1993). *Customer and Market-Driven Quality Management*. Milwaukee, WI: ASQ Quality Press.
Feigenbaum, A.J. (1951, 1983). *Total Quality Control*, 3rd edition. New York: McGraw-Hill.
Finnerty, M.F. (1996). "Coaching for growth and development," *The ASTD Training & Development Handbook: A Guide to Human Resources Development*, in Craig, R.L. editor-in-chief. New York: McGraw-Hill, pp. 415–436.
Fitzsimmons, C.F. (1996). "The bossless performance review," *Quality Progress*, June, pp. 77–81.
Fivars, G. (1980). "The Critical Incident Technique: A Bibliography," Research and Publication Service, American Institutes for Research ED 195 681.
Flanagan, J.C. (1954). "The critical incident technique," *Psychological Bulletin*, 51(4), July, pp. 327–358.

Forum Corporation (1996). Annual Report. Available at: www.forum.com/publications/esub_archive/sales.html.

Forsha, H.I. (1992). *The Pursuit of Quality Through Personal Change*. Milwaukee, WI: ASQ Quality Press.

Galloway, D. (1994). *Mapping Work Processes*. Milwaukee, WI: ASQ Quality Press.

GAO (1986). "Developing and Using Questionnaires—Transfer Paper 7," Washington, D.C.: United States General Accounting Office.

George, M.L. (2002). *Lean Six Sigma*. New York: McGraw-Hill.

GOAL/QPC (1990). "Cross-Functional Management: Research Report #90-12-01," Methuen, MA.

Goldratt, E.M. (1986). *The Goal*. Croton-on-Hudson, New York: The North River Press.

Goldratt, E.M. (1990). *The Haystack Syndrome: Sifting Information Out of the Data Ocean*. Croton-on-Hudson, New York: The North River Press, p. 53.

Goldratt, E.M. (1997). *Critical Chain*. Great Barrington, MA: The North River Press.

Goldratt, E.M. and Fox, R.E. (1986). *The Race*. Croton-on-Hudson, New York: The North River Press.

Hagan, John T. , editor (1990). *Principles of Quality Costs*. Milwaukee: ASQ Quality Press.

Hammer, M. and Champy, J. (1993). *Reengineering the Corporation: A Manifesto for Business Revolution*. New York: HarperCollins Publishers.

Harrington, H.J. (1992). "Probability and statistics," *Quality Engineering Handbook*, in Pyzdek, T. and Berger, R.W. editors, Milwaukee, WI: ASQ Quality Press, pp. 513–577.

Harry, M. and Schroeder, R. (2000). *Six Sigma*. New York: Doubleday.

Hayes, Bob E. (1992). *Measuring Customer Satisfaction: Development and Use of Questionnaires*. Milwaukee, WI: ASQ Quality Press.

Hillier, F.S. and Lieberman, G.J. (1980). *Introduction to Operations Research*, 3rd edition. San Francisco, CA: Holden-Day, Inc.

Hutton, D.W. (1994). *The Change Agent's Handbook: A Survival Guide for Quality Improvement Champions*. Milwaukee, WI: ASQ Quality Press.

Imai, M. (1986). *Kaizen*. New York: Random House.

Ishikawa, K. (1985). *What Is Total Quality Control the Japanese Way?* Englewood Cliffs, NJ: Prentice-Hall.

Johnson, R.S. (1993a). *TQM: Leadership for the Quality Transformation*. Milwaukee, WI: ASQ Quality Press.

Johnson, R.S. (1993b). *TQM: Management Processes for Quality Operations*. Milwaukee, WI: ASQ Quality Press.

Joiner, B.L. (1994). *Fourth Generation Management: The New Business Consciousness*. New York: McGraw-Hill.

Juran, J.M. (1994). *The Upcoming Century of Quality*. Wilton, CT: Juran Institute, Inc.

Juran, J.M. (1995). *A History of Managing for Quality*. Milwaukee: ASQC Quality Press.

Juran, J.M. and DeFeo, J.A. (2010). *Juran's Quality Handbook*, 6th edition. New York: McGraw-Hill.

Juran, J.M. and Gryna, F.M. (1988). *Juran's Quality Control Handbook*, 4th edition. New York: McGraw-Hill.

Juran, J.M. and Gryna, F.M. (1993). *Quality Planning and Analysis*, 3rd edition. New York: McGraw-Hill.

Kano, N. (1993). "A perspective on quality activities in American firms," *California Management Review*, 35(3), Spring 1993, pp. 12–31.

Kaplan, R.S. and Norton, D.P. (1992). "The balanced scorecard—measures that drive performance," *Harvard Business Review*, January-February, pp. 71–79.

Keller, P. (2011a). *Six Sigma Demystified*, 2nd edition, New York: McGraw Hill.

Keller, P. (2011b). *Statistical Process Control Demystified*. New York: McGraw Hill.

Keeney, K.A. (1995). *The Audit Kit*. Milwaukee, WI: ASQ Quality Press.

Kendrick, J.J. (1992). "How to spot total quality leaders and predict success of quality programs," *Quality*, April, p. 13.

King, B. (1987). *Better Designs in Half the Time: Implementing QFD in America.* Methuen, MA: Goal/QPC.

Kohn, A (1986). *No Contest: The Case against Competition.* New York: Houghton Mifflin Company.

Kohn, A. (1993). *Punished by Rewards: The Trouble with Gold Stars, Incentive Plans, A's, Praise and Other Bribes.* New York: Houghton Mifflin Company.

Kotler, P. (1991). *Marketing Management: Analysis, Planning, Implementation, and Control*, 7th edition. Englewood Cliffs, NJ: Prentice-Hall.

Lafley, A.G., Martin, R.L., Rivkin, J.W., and Siggelkow, N. (2012). "Bringing Science to the Art of Strategy," *Harvard Business Review* (September) pp. 57–66.

Leach, L.P. (2000). *Critical Chain Project Management*. Boston, MA: Artech House, Inc.

Levitt, T. (1983). "After the sale is over," *Harvard Business Review*, September/October, pp. 87–93.

Lewis, C.I. (1929). *Mind and the World Order*. New York: Scribners.

Maslow, A.H. (1943). "A theory of human motivation," *Psychological Review*, 50, pp. 370–396.

McGregor, D. (1960). *The Human Side of Enterprise*. New York: McGraw-Hill.

Mendelowitz, A.I. (1991). "Management Practices: U.S. Companies Improve Performance Through Quality Effort," GAO/NSIAD-91-190, Washington, D.C.: Superintendent of Documents.

Mintzberg, H. (1989). "The structuring of organizations," in Mintzberg, H. and Quinn, J.B. (1991). *The Strategy Process—Concepts, Contexts, Cases*, 2nd edition. Englewood Cliffs, NJ: Prentice-Hall, Inc.

Mintzberg, H. (1994). *The Rise and Fall of Strategic Planning*. New York: The Free Press.

Mintzberg, H. and Quinn, J.B. (1991). *The Strategy Process—Concepts, Contexts, Cases*, 2nd edition. Englewood Cliffs, NJ: Prentice-Hall, Inc.

Nelson, Lloyd S. (1984). "The Shewhart control chart-tests for special causes," *Journal of Quality Technology* 16(4), pp. 237–239.

Nelson, Lloyd S. (1988). "Control charts: rational subgroups and effective applications," *Journal of Quality Technology* 20(1), pp. 73–75.

Newbold, R.C. (1998). *Project Management in the Fast Lane: Applying the Theory of Constraints*. Boca Raton, FL: St. Lucie Press.

Nolan, M. (1996). "Job training," *The ASTD Training & Development Handbook: A Guide to Human Resources Development* in Craig, R.L. editor-in-chief, New York: McGraw-Hill, pp. 747–775.

Noreen, E., Smith, D., and Mackey, J.T. (1995). *The Theory of Constraints and Its Implications for Management Accounting*. Great Barrington, MA: North River Press.

Paton, S.M. (1995). "An interview with A. Blanton Godfrey," *Quality Digest*, November, pp. 56–58.

Peters, T. (1990). "The new building blocks," *Industry Week*, Jan 8, pp. 101–102.

Peters, T. and Austin, N. (1985). *A Passion for Excellence: The Leadership Difference*. New York: Random House.

Prince, J.B. (1994). "Performance appraisal and reward practices for total quality organizations," *Quality Management Journal*, January, pp. 36–46.

Provost, L.P. (1988). "Interpretation of Results of Analytic Studies," paper presented at 1989 NYU Deming Seminar for Statisticians, March 13, New York: New York University.

Pyzdek, T. (1985). "A ten-step plan for statistical process control studies," *Quality Progress*, April, pp. 77–81, and July, pp. 18–20.

Pyzdek, T. (1990). *Pyzdek's Guide to SPC Volume One: Fundamentals*. Tucson, AZ: Quality Publishing, Inc.

Pyzdek, T. and Keller, P. (2010). *The Six Sigma Handbook*. New York: McGraw Hill.

Reeves, M., Love, C., Tillmanns, P. (2012). "Your strategy needs a strategy," *Harvard Business Review*, September, pp. 76–83.

Reimann, C.W. and Hertz, H.S. (1993). "The Malcolm Baldrige National Quality Award and ISO 9000 Registration: Understanding Their Many Important Differences," *National Institute of Standards and Technology*.

ReVelle, J. (2000). *What Your Quality Guru Never Told You*. Tucson, AZ: QA Publishing.

Rose, K.H. (1995). "A performance measurement model," *Quality Progress*, February, pp. 63–66.

Saaty, T.L. (1988). *Decision Making for Leaders: The Analytic Hierarch Process for Decisions in a Complex World*. Pittsburgh, PA: RWS Publications.

Scholtes, P.R. (1988). *The Team Handbook*. Madison, WI: Joiner Associates.

Schragenheim, E. and Dettmer, H.W. (2000). *Manufacturing at Warp Speed*. Boca Raton, FL: St. Lucie Press.

Senge, P.M. (1990). *The Fifth Discipline: The Art and Practice of the Learning Organization*. New York: Doubleday.

Sheridan, B.M. (1993). *Policy Deployment: The TQM Approach to Long-Range Planning*. Milwaukee, WI: ASQ Quality Press.

Shewhart, W. (1931, 1980). *Economic Control of Quality of Manufacturing*. Milwaukee, WI: ASQ Quality Press.

Shewhart, W. (1939, 1986). *Statistical Method from the Viewpoint of Quality Control*. New York: Dover Publications.

Simon, J. and Bruce, P. (1992). "Using computer simulation in quality control," *Quality Progress*, Statistics Corner Column.

Slater, R. (1999). *Jack Welch and the GW Way*. New York: McGraw-Hill.

Smith, D. (2000). *The Measurement Nightmare: How the Theory of Constraints Can Resolve Conflicting Strategies, Policies, and Measures*. Boca Raton, FL: The St. Lucie Press.

Stapp, Eric H. (2001). *ISO 9001:2000 An Essential Guide to the New Standard*. Tucson: QA Publishing, LLC.

Taguchi, G. (1986). *Introduction to Quality Engineering: Designing Quality into Products and Processes*. White Plains, NY: Quality Resources.

Taylor, F.W. (1947). *Scientific Management*. New York: Harper & Brothers.

The Economist (1992). "The cracks in quality," April 18, p. 67.

Thiagarajan, S. (1996). "Instructional games, simulations, and role-plays," *The ASTD Training and Development Handbook: A Guide to Human Resources Development*, in Craig, R.L. editor-in-chief, New York: McGraw-Hill, pp. 517–533.

Thorpe, J.F. and Middendorf, W.H. (1979). *What Every Engineer Should Know About Product Liability*. New York: Marcel-Dekker.

Tuckman, B.W. (1965). "Developmental sequence in small groups," *Psychological Bulletin* 63(6), pp. 384–399.

Tufte, E.R. (1983). *The Visual Display of Quantitative Information*. Cheshire, CT: Graphics Press.

Tukey, J. (1977). *Exploratory Data Analysis*, New York: Addison-Wesley.

Vaziri, H.K. (1992). "Using competitive benchmarking to set goals," *Quality Progress*, November.

Wearring, C. and Karl, D.P. (1995). "The importance of following GD&T specifications," *Quality Progress*, February.

Weaver, C.N. (1995). *Managing the Four Stages of TQM*. Milwaukee, WI: ASQ Quality Press.

Webster's New Universal Unabridged Dictionary. (1989). New York: Barnes and Noble Books.

Wendel, P. (1996). "The European quality award: and how Texas Instruments Europe took the trophy home," *The Quality Observer*, January.

Willborn, W. (1993). *Audit Standards: A Comparative Analysis,* 2nd edition. Milwaukee, WI: ASQ Quality Press.

Wilson, P.F., Dell, L.D., and Anderson, G.F. (1993). *Root Cause Analysis: A Tool for Total Quality Management*. Milwaukee, WI: ASQ Quality Press.

Womack, J.P. and Jones, D.T. (1996). *Lean Thinking: Banish Waste and Create Wealth in Your Corporation*. New York: Simon and Schuster.

Zaremba, A.J. (1989). *Management in a New Key: Communication in the Modern Organization*. Norcross, GA: The Institute of Industrial Engineers.

Index

Note: Indexed terms within figures and tables are denoted by *f* and *t* after the page number.

E

M

O

P

R

www.ingramcontent.com/pod-product-compliance
Lightning Source LLC
LaVergne TN
LVHW082025200926
848481LV00010B/2924
9781265829230